AF581905

Recent Progress in Coordination Chemistry

Recent Progress in Coordination Chemistry

Editors

Peter Segl'a
Ján Pavlik

MDPI • Basel • Beijing • Wuhan • Barcelona • Belgrade • Manchester • Tokyo • Cluj • Tianjin

Editors

Peter Segľa
Department of Inorganic Chemistry
Slovak University of Technology
Bratislava
Slovakia

Ján Pavlik
Department of Inorganic Chemsitry
Slovak University of Technology
Bratislava
Slovakia

Editorial Office
MDPI
St. Alban-Anlage 66
4052 Basel, Switzerland

This is a reprint of articles from the Special Issue published online in the open access journal *Inorganics* (ISSN 2304-6740) (available at: www.mdpi.com/journal/inorganics/special_issues/progress_coordination_chemistry).

For citation purposes, cite each article independently as indicated on the article page online and as indicated below:

LastName, A.A.; LastName, B.B.; LastName, C.C. Article Title. *Journal Name* **Year**, *Volume Number*, Page Range.

ISBN 978-3-0365-8015-9 (Hbk)
ISBN 978-3-0365-8014-2 (PDF)

Contents

Peter Segľa and Ján Pavlik
Recent Progress in Coordination Chemistry
Reprinted from: *Inorganics* **2023**, *11*, 250, doi:10.3390/inorganics11060250 1

Milan Piroš, Martin Schoeller, Katarína Koňariková, Jindra Valentová, Ľubomír Švorc and Ján Moncoľ et al.
Structural and Biological Properties of Heteroligand Copper Complexes with Diethylnicotinamide and Various Fenamates: Preparation, Structure, Spectral Properties and Hirshfeld Surface Analysis
Reprinted from: *Inorganics* **2023**, *11*, 108, doi:10.3390/inorganics11030108 5

Patrick Nimax, Yannick Kunzelmann and Karlheinz Sünkel
Coordination Chemistry of Polynitriles, Part XII—Serendipitous Synthesis of the Octacyanofulvalenediide Dianion and Study of Its Coordination Chemistry with K^+ and Ag^+
Reprinted from: *Inorganics* **2023**, *11*, 71, doi:10.3390/inorganics11020071 35

Bálint Hajdu, Éva Hunyadi-Gulyás and Béla Gyurcsik
Interactions of an Artificial Zinc Finger Protein with Cd(II) and Hg(II): Competition and Metal and DNA Binding
Reprinted from: *Inorganics* **2023**, *11*, 64, doi:10.3390/inorganics11020064 55

Martina Kepeňová, Martin Kello, Romana Smolková, Michal Goga, Richard Frenák and Ľudmila Tkáčiková et al.
Low-Dimensional Compounds Containing Bioactive Ligands. Part XIX: Crystal Structures and Biological Properties of Copper Complexes with Halogen and Nitro Derivatives of 8-Hydroxyquinoline
Reprinted from: *Inorganics* **2022**, *10*, 223, doi:10.3390/inorganics10120223 73

Michaela Harmošová, Martin Kello, Michal Goga, Ľudmila Tkáčiková, Mária Vilková and Danica Sabolová et. al
Low-Dimensional Compounds Containing Bioactive Ligands. Part XX: Crystal Structures, Cytotoxic, Antimicrobial Activities and DNA/BSA Binding of Oligonuclear Zinc Complexes with Halogen Derivatives of 8-Hydroxyquinoline
Reprinted from: *Inorganics* **2023**, *11*, 60, doi:10.3390/inorganics11020060 91

Vladimir S. Mironov
Reaching the Maximal Unquenched Orbital Angular Momentum $L = 3$ in Mononuclear Transition-Metal Complexes: Where, When and How?
Reprinted from: *Inorganics* **2022**, *10*, 227, doi:10.3390/inorganics10120227 115

Marek Brezovan, Jana Juráková, Ján Moncol, Ľubor Dlháň, Maria Korabik and Ivan Šalitroš et al.
The Role of the Bridge in Single-Ion Magnet Behaviour: Reinvestigation of Cobalt(II) Succinate and Fumarate Coordination Polymers with Nicotinamide
Reprinted from: *Inorganics* **2022**, *10*, 128, doi:10.3390/inorganics10090128 127

Ján Titiš, Cyril Rajnák and Roman Boča
Energy Levels in Pentacoordinate d^5 to d^9 Complexes
Reprinted from: *Inorganics* **2022**, *10*, 116, doi:10.3390/inorganics10080116 145

Zdeněk Šindelář and Pavel Kopel
Bis(benzimidazole) Complexes, Synthesis and Their Biological Properties: A Perspective
Reprinted from: *Inorganics* **2023**, *11*, 113, doi:10.3390/inorganics11030113 163

Maryam Doroudian and Jürgen Gailer
Integrative Metallomics Studies of Toxic Metal(loid) Substances at the Blood Plasma–Red Blood Cell–Organ/Tumor Nexus
Reprinted from: *Inorganics* **2022**, *10*, 200, doi:10.3390/inorganics10110200 . . . . **173**

Editorial

Recent Progress in Coordination Chemistry

Peter Segl'a and Ján Pavlik *

Faculty of Chemical and Food Technology, Slovak University of Technology in Bratislava, Radlinského 9, SK-81237 Bratislava, Slovakia; peter.segla@stuba.sk
* Correspondence: jan.pavlik@stuba.sk

The following Special Issue of *Inorganics* is based on the discussions initiated at the International Conference on Coordination and Bioinorganic Chemistry (ICCBIC), which has been organized and held biennially since 1964 [1–3]. The 28th continuation of this series was held between 5 and 10 June 2022 at Smolenice Castle, Slovakia, and was named "Progressive Trends in Coordination, Bioinorganic and Applied Inorganic Chemistry". It hosted 39 scientists from Slovakia and 46 scientists from 14 European, North American, and Asian countries (Figure 1). During the conference, 66 lectures were given (out of them, 27 were a part of the "Young Scientists Section") and 18 posters were presented. We are pleased to say that the active participation of 38 colleagues younger than 30 years old is a clear sign that there is no need to worry about the future of coordination chemistry nor ICCBIC.

Figure 1. Participants of XXVIII ICCBIC.

Citation: Segl'a, P.; Pavlik, J. Recent Progress in Coordination Chemistry. *Inorganics* **2023**, *11*, 250. https://doi.org/10.3390/inorganics11060250

Received: 2 June 2023
Revised: 6 June 2023
Accepted: 7 June 2023
Published: 8 June 2023

Traditionally, the lectures are divided into four sections. In section A, (Electron, Molecular and Crystal Structures) Prof. Roman Boča (Slovakia) gave a very inspiring plenary lecture called "Magnetic Anisotropy of Hexa-, Penta- and Tetracoordinate Ni(II) Complexes". Section B (Solution and Solid-State Reactivity) was opened by Dr. Rozina Khattak (Pakistan), who spoke about "Open problems in Green Approach to Sensitizer-Mediator Interaction in DSSC: Redox Mechanism of Dicyanobis(2,2′-bipyridyl)iron(III)-hexacyanoferrate(II) in Water". Prof. Pál Sipos (Hungary) introduced section C (Applied Inorganic and Coordination Chemistry) by presenting a lecture titled "Chemists Have Solutions!—Understanding Industrial Problems by Using the Knowledge Offered by Solution Chemistry". Finally, section D (Complexes in Human Medicine and in the Environment) was led by Prof. Daniel Ruiz-Molina (Spain), who spoke about "Coordination Polymers at the Nanoscale: Old Materials New Tricks in Therapy and Diagnosis".

In 2017 the granting of the Ján Gažo Award was established at the ICCBIC, which comprises the Ján Gažo Medal and a certificate from the Slovak Chemical Society. The 2022 prize was awarded to Prof. Roman Boča for his considerable contribution to the field of inorganic and coordination chemistry and for founding the scientific school of quantum chemistry and magnetochemistry of coordination compounds in Slovakia.

As is apparent, this year's ICCBIC established a partnership with the open access journal *Inorganics*, which is published by MDPI. In this vein, *Inorganics* financially supported "The Best Oral Presentation Award for the Young Scientists Section", which is an ultimate novelty at ICCBIC. Two first prizes were awarded; one went to Ali Kaiss (Switzerland), who delivered a talk called "From Macrocycles into Assembled Chain-mails", and the other one went to Sandra Koziel (Poland), who gave a lecture called "New Heteronuclear Iridium-Copper Complexes with Phosphine Derivatives of Fluoroquinolone Antibiotics—Together is Better".

Hereby, we invite you to form a picture of the spirit of the conference by reading this Special Issue of *Inorganics*, which is devoted to XXVIII ICCBIC. In total, ten contributions were submitted, of which eight are original papers and two are perspectives.

In the work of Švorec et al., the synthesis and thorough characterization of five novel Cu(II) complexes is reported, which perform favorably as SOD mimetics [4].

Sünkel et al. made public their serendipitous synthesis of octacyanofulvalenediide dianion and its silver and potassium coordination polymers [5].

Gyurcsik et al. studied the solution chemistry of Cd(II) and Hg(II) cations as competitors of Zn(II) in a specific artificial zinc-finger protein and found them to have surprisingly different behaviors [6].

Potočňák et al. introduced the XIX and XX parts of their series on low-dimensional compounds containing bioactive ligands. They reported novel complexes with derivatives of 8-hydroxyquinoline—namely, six Cu(II) systems in their first contribution and five Zn(II) systems in the second one. Interesting bioactivity was proved in some of these systems [7,8].

In his purely theoretical work, Vladimir S. Mironov raised the question of achieving the maximal angular momentum in complexes of 3d transition metals and showed that their record case of a Co(II) system can hardly be beaten [9].

Segl'a et al. contributed via a magnetic reinvestigation of two Co(II) coordination polymers differing only in the saturation of their bridging ligands. Both of them showed a slow relaxation of magnetization, which was correlated with stiffness of the bridge [10].

The work of Boča et al. filled a gap in textbooks of theoretical chemistry by calculating the energy levels and magnetic functions for pentacoordinated $3d^5$ to $3d^9$ complexes with trigonal bipyramid and vacant octahedron geometries [11].

Finally, a perspective work on recent bis(benzimidazole) complexes and their biological activities was offered by Kopel and Šindelář [12] and, in the second perspective of this Special Issue, Gailer and Doroudian discussed the bioinorganic consequences of exposure to toxic metals and metalloids in context of the "environment-blood-organ" pathway [13].

We would like to conclude by thanking all authors for making this Special Issue possible and expressing our hope that the audience will share our passion for coordination chemistry, both remotely by reading this collection of works and also in-person by attending our upcoming XXIX ICCBIC.

Acknowledgments: Miroslav Tatarko is acknowledged for the preparation and processing of the conference attendees' pictures.

Conflicts of Interest: The authors declare no conflict of interest.

References

1. Melník, M.; Seg'a, P.; Valko, M. XXV. International Conference on Coordination & Bioinorganic Chemistry (25th ICCBiC), June 2015, Slovakia. *Chem. Papers* **2016**, *70*, 1–3. [CrossRef]
2. Seg'a, P.; Tatarko, M.; Valko, M. XXVI. International Conference on Coordination & Bioinorganic Chemistry (26th ICCBiC), June 2017, Slovakia. *Chem. Papers* **2018**, *72*, 769–771. [CrossRef]
3. Segl'a, P.; Pavlik, J.; Tatarko, M.; Valko, M. XXVII. International Conference on Coordination & Bioinorganic Chemistry (27th ICCBiC), June 2019, Slovakia. *Chem. Papers* **2020**, *74*, 3671–3672. [CrossRef]

4. Piroš, M.; Schoeller, M.; Koňariková, K.; Valentová, J.; Švorc, Ľ.; Moncoľ, J.; Valko, M.; Švorec, J. Structural and Biological Properties of Heteroligand Copper Complexes with Diethylnicotinamide and Various Fenamates: Preparation, Structure, Spectral Properties and Hirshfeld Surface Analysis. *Inorganics* **2023**, *11*, 108. [CrossRef]
5. Nimax, P.; Kunzelmann, Y.; Sünkel, K. Coordination Chemistry of Polynitriles, Part XII—Serendipitous Synthesis of the Octacyanofulvalenediide Dianion and Study of Its Coordination Chemistry with K^+ and Ag^+. *Inorganics* **2023**, *11*, 71. [CrossRef]
6. Hajdu, B.; Hunyadi-Gulyás, É.; Gyurcsik, B. Interactions of an Artificial Zinc Finger Protein with Cd(II) and Hg(II): Competition and Metal and DNA Binding. *Inorganics* **2023**, *11*, 64. [CrossRef]
7. Kepeňová, M.; Kello, M.; Smolková, R.; Goga, M.; Frenák, R.; Tkáčiková, Ľ.; Litecká, M.; Šubrt, J.; Potočňák, I. Low-Dimensional Compounds Containing Bioactive Ligands. Part XIX: Crystal Structures and Biological Properties of Copper Complexes with Halogen and Nitro Derivatives of 8-Hydroxyquinoline. *Inorganics* **2022**, *10*, 223. [CrossRef]
8. Harmošová, M.; Kello, M.; Goga, M.; Tkáčiková, L.; Vilková, M.; Sabolová, D.; Sovová, S.; Samolová, E.; Litecká, M.; Kuchárová, V.; et al. Low-Dimensional Compounds Containing Bioactive Ligands. Part XX: Crystal Structures, Cytotoxic, Antimicrobial Activities and DNA/BSA Binding of Oligonuclear Zinc Complexes with Halogen Derivatives of 8-Hydroxyquinoline. *Inorganics* **2023**, *11*, 60. [CrossRef]
9. Mironov, V. Reaching the Maximal Unquenched Orbital Angular Momentum L = 3 in Mononuclear Transition-Metal Complexes: Where, When and How? *Inorganics* **2022**, *10*, 227. [CrossRef]
10. Brezovan, M.; Juráková, J.; Moncol, J.; Dlháň, L.; Korabik, M.; Šalitroš, I.; Pavlik, J.; Seg'a, P. The Role of the Bridge in Single-Ion Magnet Behaviour: Reinvestigation of Cobalt(II) Succinate and Fumarate Coordination Polymers with Nicotinamide. *Inorganics* **2022**, *10*, 128. [CrossRef]
11. Titiš, J.; Rajnák, C.; Boča, R. Energy Levels in Pentacoordinate d^5 to d^9 Complexes. *Inorganics* **2022**, *10*, 116. [CrossRef]
12. Šindelář, Z.; Kopel, P. Bis(benzimidazole) Complexes, Synthesis and Their Biological Properties: A Perspective. *Inorganics* **2023**, *11*, 113. [CrossRef]
13. Doroudian, M.; Gailer, J. Integrative Metallomics Studies of Toxic Metal(loid) Substances at the Blood Plasma-Red Blood Cell-Organ/Tumor Nexus. *Inorganics* **2022**, *10*, 200. [CrossRef]

Article

Structural and Biological Properties of Heteroligand Copper Complexes with Diethylnicotinamide and Various Fenamates: Preparation, Structure, Spectral Properties and Hirshfeld Surface Analysis

Milan Piroš [1], Martin Schoeller [1], Katarína Koňariková [2], Jindra Valentová [3], Ľubomír Švorc [4], Ján Moncoľ [1], Marian Valko [5] and Jozef Švorec [1,*]

1 Department of Inorganic Chemistry, Faculty of Chemical and Food Technology, Slovak University of Technology, Radlinského 9, 81237 Bratislava, Slovakia
2 Institute of Medicinal Chemistry, Biochemistry and Clinical Biochemistry, Faculty of Medicine, Comenius University, Sasinkova 2, 81372 Bratislava, Slovakia
3 Department of Chemical Theory of Drugs, Faculty of Pharmacy, Comenius University in Bratislava, Kalinčiakova 8, 83232 Bratislava, Slovakia
4 Institute of Analytical Chemistry, Faculty of Chemical and Food Technology, Slovak University of Technology, Radlinského 9, 81237 Bratislava, Slovakia
5 Department of Physical Chemistry, Faculty of Chemical and Food Technology, Slovak University of Technology, Radlinského 9, 81237 Bratislava, Slovakia
* Correspondence: jozef.svorec@stuba.sk; Tel.: +421 2 59325 625

Abstract: Herein, we discuss the synthesis, structural and spectroscopic characterization, and biological activity of five heteroligand copper(II) complexes with diethylnicotinamide and various fenamates, as follows: flufenamate (fluf), niflumate (nifl), tolfenamate (tolf), clonixinate (clon), mefenamate (mef) and *N*, *N*-diethylnicotinamide (dena). The complexes of composition: $[Cu(fluf)_2(dena)_2(H_2O)_2]$ (**1**), $[Cu(nifl)_2(dena)_2]$ (**2**), $[Cu(tolf)_2(dena)_2(H_2O)_2]$ (**3**), $[Cu(clon)_2(dena)_2]$ (**4**) and $[Cu(mef)_2(dena)_2(H_2O)_2]$ (**5**), were synthesized, structurally (single-crystal X-ray diffraction) and spectroscopically characterized (IR, EA, UV-Vis and EPR). The studied complexes are monomeric, forming a distorted tetragonal bipyramidal stereochemistry around the central copper ion. The crystal structures of all five complexes were determined and refined with an aspheric model using the Hirshfeld atom refinement method. Hirshfeld surface analysis and fingerprint plots were used to investigate the intermolecular interactions in the crystalline state. The redox properties of the complexes were studied and evaluated via cyclic voltammetry. The complexes exhibited good superoxide scavenging activity as determined by an NBT assay along with a copper-based redox-cycling mechanism, resulting in the formation of ROS, which, in turn, predisposed the studied complexes for their anticancer activity. The ability of complexes **1–4** to interact with calf thymus DNA was investigated using absorption titrations, viscosity measurements and an ethidium-bromide-displacement-fluorescence-based method, suggesting mainly the intercalative binding of the complexes to DNA. The affinity of complexes **1–4** for bovine serum albumin was determined via fluorescence emission spectroscopy and was quantitatively characterized with the corresponding binding constants. The cytotoxic properties of complexes **1–4** were studied using the cancer cell lines A549, MCF-7 and U-118MG, as well as healthy MRC-5 cells. Complex **4** exhibited moderate anticancer activity on the MCF-7 cancer cells with IC_{50} = 57 μM.

Keywords: copper(II) complexes; fenamates; Hirshfeld atom refinement; interactions with DNA; SOD mimetic activity

Citation: Piroš, M.; Schoeller, M.; Koňariková, K.; Valentová, J.; Švorc, Ľ.; Moncoľ, J.; Valko, M.; Švorec, J. Structural and Biological Properties of Heteroligand Copper Complexes with Diethylnicotinamide and Various Fenamates: Preparation, Structure, Spectral Properties and Hirshfeld Surface Analysis. *Inorganics* **2023**, *11*, 108. https://doi.org/10.3390/inorganics11030108

Academic Editor: Wolfgang Linert

Received: 14 February 2023
Revised: 27 February 2023
Accepted: 28 February 2023
Published: 6 March 2023

1. Introduction

Non-steroidal anti-inflammatory drugs (NSAIDs) are a very broad class of drugs that are widely used to treat conditions associated with acute or chronic inflammation, such as pain or rheumatoid arthritis, and are also often used extensively for fever due to their analgesic or antipyretic effects [1–3]. A mode of their pharmacologic action is mostly based on the suppression of prostanoid production (important inflammatory mediators) by the inhibition of cyclooxygenase enzymes that catalyze prostanoid biosynthesis from arachidonic acid [3,4]. In addition, an alternative mechanism independent of cyclooxygenase inhibition was proposed [4]. This mechanism involves a direct effect of NSAIDs on mitochondria, leading to cellular oxidative stress and apoptosis [4]. From a chemical point of view, NSAIDs are predominantly weak acids that contain an acidic moiety together with an aromatic functional group. According to their chemical characteristics, NSAIDs can be roughly classified as derivates of carboxylic acids (salicylic, acetic and anthranilic), oxicams, sulfonamides or furanones [5].

Fenamates form a subgroup within NSAIDs and are derived from 2-anilinobenzoic (fenamic) acid. They are known to have anti-inflammatory, analgesic and antipyretic activities in animals or humans mainly through the inhibition of cyclooxygenases [6]. Typical examples of fenamates are mefenamic acid, used to treat mild or moderate pain; flufenamic acid, used to treat rheumatic disorders; and tolfenamic acid, known as the drug Clotam, used to treat migraine headaches or as a veterinary drug [7,8] (Scheme 1). The derivates of 2-phenylaminonicotinic acid, such as niflumic acid or clonixin (Scheme 1), are also formally included as fenamates. Like other fenamates, they are mostly used as analgesic and anti-inflammatory agents in the treatment of rheumatoid arthritis or for pain relief [7].

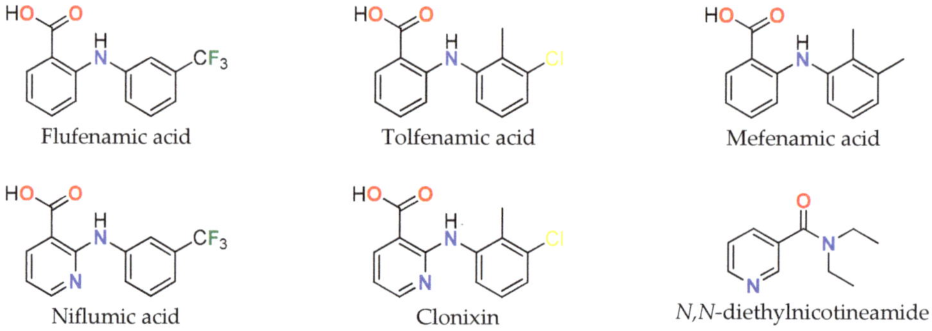

Flufenamic acid Tolfenamic acid Mefenamic acid

Niflumic acid Clonixin *N,N*-diethylnicotineamide

Scheme 1. Chemical formulae of used ligands.

Nicotinic acid derivatives, such as *N, N*-diethylnicotinamide, as well as nicotinamide or isonicotinamide, form an important class of heterocyclic pyridinecarboxamide compounds that are often used as a neutral N-donor ligand for the construction of hydrogen-bonded coordination networks and polymers [9]. Nicotinamide derivates alone exhibit interesting biological activities, including anticancer or anti-angiogenic properties, [10] and can exhibit herbicidal and antifungal activities [11].

Copper(II) complexes with NSAID ligands are now attractive objects for inorganic, pharmaceutical and medicinal chemists due to their potential to be effective anticancer, anti-inflammatory, antibacterial, antifungal or antiviral agents [12–17]. Such metal complexes often display lower toxicity and, at the same time, higher pharmaceutical efficiency than the parent NSAID drug. Ternary copper NSAID complexes containing other biologically active ancillary ligands (e.g. substituted pyridines or 1.10-phenantroline) offer a possibility how to successfully modify a coordination sphere of studied complexes toward desired activities [12,18–20]. Numerous copper complexes with NSAID, and especially with fenamates and *N*-donor ligands, such as pyridine and its derivates (2,2′-bipyridine and 1,10-phenanthroline), were studied by Psomas and coworkers [21–26]. These complexes show significant antioxidant activity, as well as an excellent ability to scavenge hydroxyl and superoxide radicals [5]. Furthermore, copper complexes with meclofenamate show potential to be a successful anti-dementia agents [25].

In order to combine the proinflammatory ROS-mediating properties of copper(II) fenamate ligands and the ability to intercalate with the DNA of phenanthroline ligands, Simunkova and coworkers [27] prepared and studied three copper fenamates (tolfenamate, mefenamate and flufenamate) with 1,10-phenanthroline as potential anticancer copper compounds, giving the best results for a complex of the composition $[Cu(fluf)_2phen]$. Furthermore, Jozefíková and coworkers synthesized and investigated copper complexes with nicotinamide [28], isonicotinamide [29] and fenamates, showing that this type of complex exhibits promising biological activity, especially in the case of niflumate and clonixinate complexes.

In this context, we decided to prepare and characterize copper(II) complexes with dena ligands and various fenamates with the general formula of either $[CuL_2(dena)_2]$ or $[CuL'_2(dena)_2(H_2O)_2]$, where L = flufenamate (**1**), tolfenamate (**3**), mefenamate (**5**) and L′= niflumate (**2**) or clonixinate (**4**) in connection with their biological activity. Although the $[CuL_2(dena)_2(H_2O)_2]$ (L = fluf [30], tolf [31] and mef [32]) and $[Cu(nifl)_2(dena)_2]$ [33] complexes were already previously prepared and crystallograhically characterized, the solution of their crystal structure showed some flaws. As an example, the crystal structures of $[Cu(tolf)_2(dena)_2(H_2O)_2]$ and $[Cu(mef)_2(dena)_2(H_2O)_2]$ were solved without the inclusion of apparently resolved disorders, and in the latter case, the coordinates of the crystal structures were missing in the CCD database [31,32]. Moreover, the crystal structure of $[Cu(nifl)_2(dena)_2]$ contains some incorrectly assigned atoms [33]. Taking these facts into account, we prepared all four complexes again. In addition, novel copper(II) complex with clonixinate anion and dena ligand with a composition of $[Cu(clon)_2(dena)_2]$ was synthesized. Subsequently, single-crystal data of all five complexes were obtained at a low temperature (100 K) and high redundancy. In turn, the crystal structures of **1–5** were refined by means of an aspheric model using the Hirshfeld atom refinement (HAR) method, thus providing more precise structural parameters. The study of intermolecular interactions in the crystal structures of all complexes was augmented by the Hirshfeld surface analyses. Moreover, the structural and spectroscopic data of the complexes are discussed in connection with biological activity to find structure–activity relationship correlations. Our choice of fenamate ligands in this study was influenced by the observation that a pyridin ring containing analogs of copper fenamates shows better biological activity than their benzene analogs (niflumic vs. flufenamic and clonixin vs. tolfenamic acid) [29]. The presence of coplanarity of the aromatic rings in copper niflumates and clonixinate, in comparison with their copper flufenamate and tolfenamate analogs, may have had a substantial positive effect on the biological activity of the complexes, e.g., they could improve the intercalation ability of the complexes into DNA. In particular, the same substituents on the benzene ring in flufenamate and niflumate (trifluoromethyl susbtituent) or tolfenamate and clonixinate (chloro and methyl susbtituent) ligands provide the additional possibility for correlating the observed biological activities of the prepared complexes (Scheme 1).

In this regard complexes **1–5** were studied via various spectroscopic methods both in a solid state and in a DMSO solution, including IR, UV-Vis and EPR spectroscopy, as well as X-ray analysis. The redox properties of the complexes were studied via cyclic voltammetry. The SOD mimetic activity of all five complexes was determined with an indirect NBT assay. In order to compare the structure and biological activity of the complexes containing 2-anilinobenzoate and 2-phenylaminonicotinate ligands, the interaction of complexes **1–4** with calf thymus DNA (ct-DNA) was studied using absorption titrations, viscosity measurements and the ethidium bromide displacement fluorescence method. The interaction of these complexes with bovine serum albumin was investigated as well. Finally, the anticancer activity of complexes **1–4** was tested against several different cancer cell lines.

2. Results and Discussion

2.1. Synthesis

The complexes under study were obtained in moderate yields (57–75%) using a complexation reaction between corresponding fenamic acid and NaOH with copper acetate dihydrate and *N*, *N*-diethylnicotinamide (in a molar ratio of 2:2:1:2) in ethanol/methanol, according to Scheme 2. All five complexes are stable in the air, and their compositions were characterized with elemental analysis and IR spectroscopy, as well as with X-ray diffraction. The elemental analysis of the complexes is in agreement with the calculated values for the corresponding formulae: $[Cu(fluf)_2(dena)_2(H_2O)_2]$ (**1**), $[Cu(nifl)_2(dena)_2]$ (**2**), $[Cu(tolf)_2(dena)_2(H_2O)_2]$ (**3**), $[Cu(clon)_2(dena)_2]$ (**4**) and $[Cu(mef)_2(dena)_2(H_2O)_2]$ (**5**). The exact crystal structures and compositions of the complexes were fully confirmed via single-crystal X-ray crystallography.

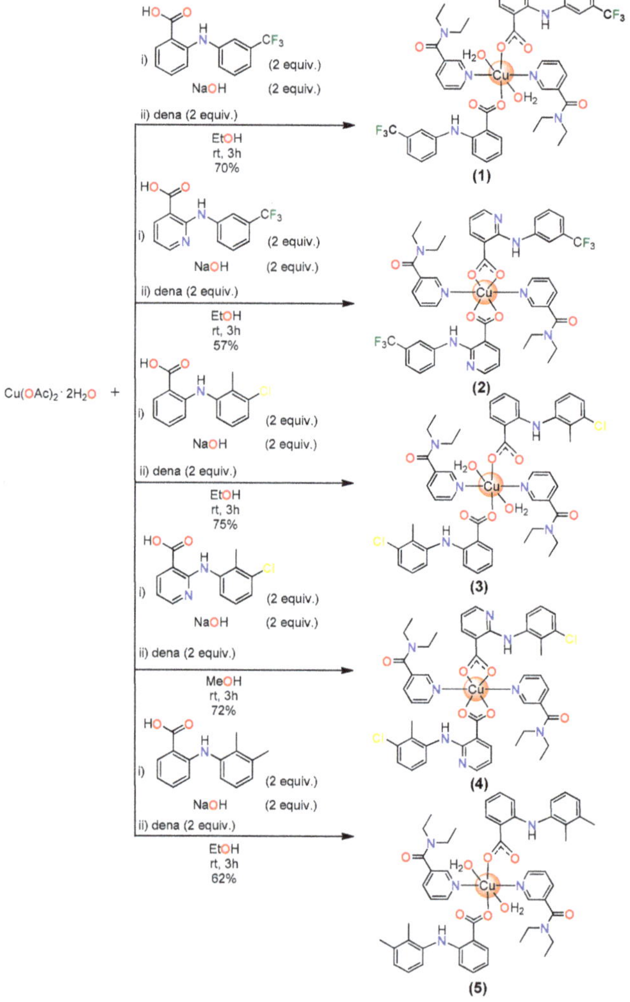

Cu(OAc)2 · 2H2O +

i) (2 equiv.) NaOH (2 equiv.) ii) dena (2 equiv.) EtOH rt, 3h 70% → (1)

i) (2 equiv.) NaOH (2 equiv.) ii) dena (2 equiv.) EtOH rt, 3h 57% → (2)

i) (2 equiv.) NaOH (2 equiv.) ii) dena (2 equiv.) EtOH rt, 3h 75% → (3)

i) (2 equiv.) NaOH (2 equiv.) ii) dena (2 equiv.) MeOH rt, 3h 72% → (4)

i) (2 equiv.) NaOH (2 equiv.) ii) dena (2 equiv.) EtOH rt, 3h 62% → (5)

Scheme 2. Schematic representation of the syntheses of **1–5**.

2.2. *IR and UV-Vis Spectroscopy*

The infrared spectra of complexes **1–5** were recorded in the region of 4000–400 cm^{-1} in a solid state with the ATR technique, and a tentative description of some important bands was performed (Table 1) on the basis of literature data [34]. The spectra of all five complexes are shown in Supplementary Figure S1. According to the composition and spectral features, we can divide the studied complexes into two distinctive groups (**1**, **3**, **5** and **2**, **4**). The IR spectra of **1**, **3** and **5** showed broad absorption bands of medium intensity (3507, 3458 and 3473 cm^{-1} and 3324, 3217 and 3206, respectively) corresponding to the OH stretching vibrations (antisymmetric and symmetric, respectively) of coordinated water molecules, which were missing in the IR spectra of **2** and **4** (Supplementary Figure S1b), in accordance with the complex compositions. In addition, the IR spectra of all complexes in the region of 3200–3100 cm^{-1} exhibited a series of weak absorptions assigned to N-H vibrations, as well as a series of weak absorption peaks corresponding to CH stretches (between 3100 and 2800 cm^{-1}). Each IR spectra also contained a strong amidic band $\tilde{\upsilon}$(C=O) at 1629 cm^{-1} for **2** and **4** and at 1619, 1613 and 1612 cm^{-1} for **1**, **3** and **5**. In the latter case, this band can be considered as a combined mixed band, caused by the vibration of the amidic C=O group and asymmetric $\tilde{\upsilon}_{as}(COO^-)$ based on its strong intensity. In parent dena ligands, the band attributed to amidic C=O stretches could be found at 1630 cm^{-1}; thus, the coordination of the ligands resulted in the lowering of the band wavenumber only in the cases of complexes **1**, **3** and **5**. C=N ring stretching vibrations in dena ligands appeared at 1592 cm^{-1}. After complex formations, this vibration moved to a lower wavenumber, with double sharp bands at 1581–1557 cm^{-1} in the cases of **1**, **3** and **5** and at 1602–1582 cm^{-1} for **2** and **4**. Symmetric and antisymmetric carboxylate stretching vibrations could serve as an indication of a type of coordination mode for the carboxylate group in the prepared complexes. According to the literature [34], if the values of parameter Δ ($\tilde{\upsilon}_{as}(COO^-)$-$\tilde{\upsilon}_{s}(COO^-)$) are higher than those in ionic complexes, such as in sodium mefenamate (Δ = 190 cm^{-1}), the coordination mode of the carboxylate group is monodentate. In the cases of complexes **1**, **3** and **5**, parameter Δ ($\tilde{\upsilon}_{as}(COO^-)$-$\tilde{\upsilon}_{s}(COO^-)$) fell in the range of 241–228 cm^{-1}, which is in a full accordance with the monodentate coordination mode of the carboxylate group. In the case of complexes **2** and **4**, parameter Δ showed values between 244 and 195 cm^{-1}, in agreement with the observed asymmetric bidentate chelating binding mode of the carboxylate group. Both bands belonging to the asymmetric and symmetric stretching vibrations of the carboxylate group were split, which could be attributed to the observed different *r*(Cu-O) bond length in the asymmetric bidentate chelating binding mode of the carboxylate group (Table 1). Moreover, bands belonging to the asymmetric stretching vibration were again found in the spectrum in the form of combined mixed bands caused by the coupled vibration of C=N and the carboxylate group based on their strong intensity.

Table 1. Infrared (in cm^{-1}) and electronic (in nm) data of complexes **1–5**.

Complex	$\tilde{\upsilon}$(O–H)	$\tilde{\upsilon}_{as}(COO^-)$	$\tilde{\upsilon}_{s}(COO^-)$	Δ	$\tilde{\upsilon}$(C=O)	$\tilde{\upsilon}$(C=N)	λ(d-d) [a]	λ(d-d) [b]
1	3057m 3324w	1619s [c] 1606s [c]	1378vs	241 228	1619s [c]	1581s 1568s	646br	795br
2	-	1595s [c] 1582s [c]	1387s 1368s	227/208 214/195	1629s	1595s [c] 1582s [c]	539br 615sh	791br
3	3458m,br 3206m,br	1613vs [c]	1377vs	236	1613s [c]	1579s 1557s	628br	789br
4	-	1602s [c] 1585vs [c]	1380s 1358vs	244/222 227/205	1629s	1602s [c] 1585s [c]	529br 612sh	794br
5	3473m,br 3217m,br	1612vs [c]	1376vs	236	1612s [c]	1575s 1561s	600br	801br

[a] nujol; [b] DMSO solution; [c] mixed bands; vs, very strong; s, strong; m, medium; w, weak; br, broad.

The electronic spectra of **1–5** were obtained in the solid state as nujol mulls, as well as in DMSO solutions. Representative examples of such spectra for complexes **4** and **5** are shown in Supplementary Figure S2. The solid-state spectra of the studied complexes showed very broad formally forbidden low-intensity d-d transitions in the visible region, with the maximum in the range of 587–648 nm, corresponding to the tetragonal bipyramidal stereochemistry around the metal center. In **2** and **4**, a shoulder at approximately 612–615 nm was observed (Table 1). In addition, the spectra also contained bands at approximately 200–400 nm, which could be considered as an intraligand transition, as well as a ligand-to-metal-charge transfer between the π electron cloud of the fenamate moiety and a central copper atom [18]. Upon dissolution in the DMSO solvent, the absorption maximum of broad *d-d* transitions shifted to higher wavelengths at a relatively constant range of 789–801 nm, which is expected for mononuclear copper complexes with distorted square planar geometry (Supplementary Figure S2 and Table 1) [35]. This shift likely indicates the potential coordination of DMSO solvent molecules in the primary coordination sphere of the complexes.

2.3. Molecular and Crystal Structures

The crystal structures of all five complexes were refined with a more accurate aspherical HAR method using data measured with high redundancy at 100 K. The crystal structures of four of the complexes, **1–3** and **5**, have been previously determined at room temperature using the standard IAM model, but the published crystal structures do not contain disordered groups and/or atomic coordinates in the CSD database [30–33]. On the other hand, complex **4** is newly synthesized, so its crystal structure is completely new. Complex **1** and the isostructural complexes **3** and **5** crystallize in a monoclinic system with the $P2_1/c$ (**1**) or $P2_1/n$ (**3**,**5**) space group, whereas complex **2** and the newly prepared complex **4** crystallize in the triclinic system with a *P*-1 space group. The molecular structures of all five complexes are shown in Figure 1, whereby the copper atoms in each case lie in a special position at the center of the symmetry. The selected bond distances of all complexes are listed in Table 2. The coordination polyhedron around the copper atom in complex **1**, as well as in isostructural complexes **3** and **5**, is in the shape of a tetragonal bipyramid. The equatorial plane is formed by a pair of oxygen atoms of monodentately bound carboxyl groups of flufenamate (**1**), tolfenamate (**3**) or mefenamate (**5**) anions (Cu1–O1 distances are in the range of 1.946–1.973 Å), and by two pyridine nitrogen atoms of *N*, *N*-diethylnicotinamide ligands (Cu1–N1 distances are in the range of 2.015–2.036 Å) in the *trans* positions. The two axial positions of the tetragonal bipyramid are complemented by two coordinated water molecules (Cu1–O1W distances are in the range of 2.435–2.488 Å). The molecular structures of complexes **1**, **3** and **5** are stabilized by intramolecular O–H···O hydrogen bonds between coordinated water molecules (O1W) and uncoordinated oxygen atoms of carboxyl groups (O2) (O1W–H1WA···O2; distances O1W···O2 are in the range of 2.728–2.738 Å; Supplementary Table S1). Fenamate (flufenamate, tolfenamate or mefenamate) anions also form intramolecular N–H···O bonds between amine nitrogen atoms (N3) and uncoordinated oxygen atoms of carboxyl groups (O2) (N3–H3···O2; distances N3···O2 are in the range of 2.631–2.658 Å). The complex molecules of **1**, **3** and **5** are connected into 1D supramolecular chains by means of intermolecular O–H···O hydrogen bonds between coordinated water molecules (O1W) and amide oxygen atoms of *N*, *N*-diethylnicotinamide ligands of neighboring complex molecules (O1W–H1WB···O3; distances O1W···O3 are in the range of 2.799–2.853 Å; Supplementary Table S1 and Supplementary Figure S8).

The crystal structures of complexes **2** and **4** are very similar and can be considered isostructural based on their similar cell parameters, same space group and similar molecular and intermolecular interactions. The coordination polyhedron around the copper atom in complexes **2** and **4** has the shape of a tetragonal bipyramid and is formed by two pairs of asymmetrically bonded oxygen atoms (O1,O2) of carboxyl groups of niflumate (**2**) or clonixinate (**4**) anions and by a pair of pyridine nitrogen atoms (N1) of *N*, *N*-diethylnicotinamide ligands in the *trans* configuration. In both cases, the equatorial plane is equally formed by a pair of oxygen atoms (O1) (Cu1–O1; distances are 1.9502(6) and 1.9296(9) Å, respectively) and a pair of nitrogen atoms (N1) (Cu1–N1; distances are 2.0086(7) and 2.0170(12) Å, respectively). However, a significant difference can be observed in the distances between the two axially bonded oxygen atoms (O2). The Cu1–O2 distances are equal to 2.6467(11) Å in the case of complex **2**, but in the case of complex **4**, they are significantly extended to a value of 2.9554(10) Å. A similar trend was reported for several copper(II) carboxylate complexes with this type of coordination, where Cu–O_{ax} varied in the range of 2.45–2.98 Å [36].

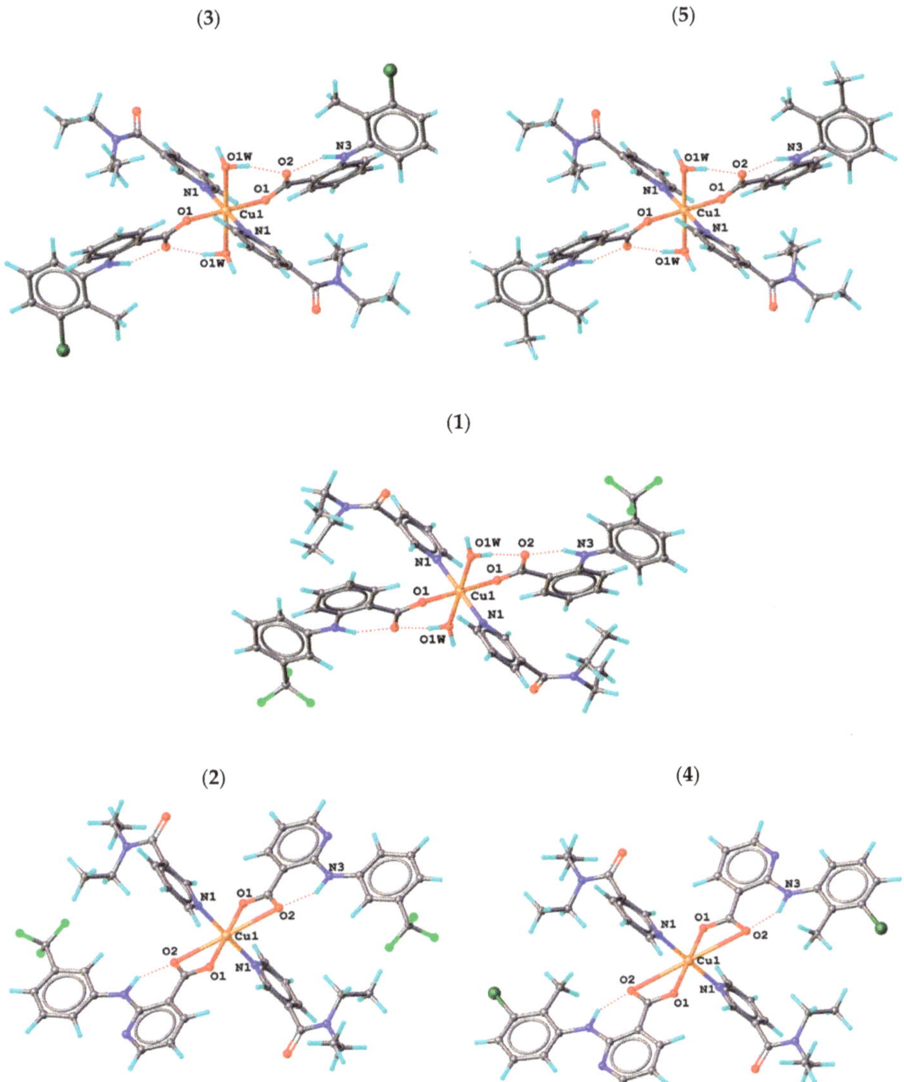

Figure 1. Molecular structures of [Cu(fluf)$_2$(dena)$_2$(H_2O)$_2$] (**1**), [Cu(nifl)$_2$(dena)$_2$] (**2**), [Cu(tolf)$_2$(dena)$_2$(H_2O)$_2$] (**3**), [Cu(clon)$_2$(dena)$_2$] (**4**) and [Cu(mef)$_2$(dena)$_2$(H_2O)$_2$] (**5**).

Table 2. Selected bond lengths (Å) for compounds (**1–5**).

	1 [i]	**3** [ii]	**5** [ii]
Cu1–O1	1.9726(9)	1.9465(5)	1.9475(7)
Cu1–N1	2.0149(11)	2.0345(6)	2.0364(9)
Cu1–O1W	2.4356(11)	2.4827(5)	2.4879(4)
	2 [iii]	**4** [iv]	
Cu1–O1	1.9502(6)	1.9296(9)	
Cu1–N1	2.0086(7)	2.0170(12)	
Cu1–O2	2.6467(11)	2.9554(10)	

Symmetry codes for a symmetrical part of a complex molecule: (i) 1−*x*, 1−*y*, 2−*z*; (ii) 1−*x*, 1−*y*, 1−*z*; (iii) 2−*x*, 1−*y*, 1−*z*; (iv)−*x*, 1−*y*, 1−*z*.

The aromatic pyridine (C12–C13–N4–C15–C16–C17) and benzene (C18–C19–C20–C21–C22–C23) rings of the niflumate (**2**) or clonixinate (**4**) anions are coplanar, which is also supported by intramolecular C–H···N hydrogen bonds between the carbon atoms of the benzene ring (C23) and the nitrogen atoms of the pyridine ring (N4) (C23–H23···N4; distances of C23···N4 are 2.913(1) and 2.899(2) Å, respectively; Supplementary Table S1). In addition, stabilization of the molecular structure can also be observed due to the intramolecular N–H···O bonds between amine nitrogen atoms (N3) and carboxylate oxygen atoms (O2) (N3–H3···O2; distances of N3···O2 are 2.669(1) and 2.673(2) Å, respectively). Coplanar pyridine and benzene rings of niflumate (**2**) or clonixinate (**4**) ligands are stacked with the neighboring complex molecules, resulting in the formation of π–π stacking interactions (Supplementary Figure S9) [37]. The angle between the plane of the pyridine ring and the plane of the benzene ring is 8.60° and 2.42°, respectively. The centroid–centroid distances are 3.72 and 3.60 Å, respectively, and the shift distances are 1.12 and 1.28 Å, respectively. Additionally, stacked complex molecules are also linked by means of C–H···O hydrogen bonds between the aromatic carbon atom (C21) and the carboxamide oxygen atom (O3) of *N, N*-diethylnicotinamide ligands of neighboring complex molecules (C21–H21···O3; distances of C21···O3 for both cases are identically equal to 3.4356(18) Å) in the 1D supramolecular chains. In the crystal structures of both complexes, other C–H···O hydrogen bonds can also be observed between the carbon atoms (C3, C4) of pyridine rings of *N, N*-diethylnicotinamide ligands and the oxygen atoms (O3) of *N, N*-diethylnicotinamide ligands of neighboring complex molecules (C3–H3A···O3 and C4–H4···O3; distances of C···O are in the range of 3.200–3.479 Å).

2.4. Hirshfeld Surface Analyses

Hirshfeld surface analysis was used to further study the intermolecular interactions of the crystal structures of all five compounds. For the illustrations, Figures 2 and 3 show the 3D Hirshfeld surfaces of **1** and **4**. The 3D Hirshfeld surfaces of other complexes are illustrated in the Supplementary Materials (Supplementary Figures S10–S12). The 3D Hirshfeld surfaces were mapped over the d_{norm} shape index (Figures 2 and 3, Supplementary Figures S10–S12). The surfaces are shown as transparent to allow for the visualization of the molecular moiety around which they were calculated. As shown in Supplementary Figures S10–S12, the deep red spots on the d_{norm} Hirshfeld surfaces indicate close-contact interactions, which were mainly responsible for the significant intermolecular hydrogen bonding interactions. The 3D Hirshfeld surface illustration of **1** (Figure 2), as well as of **3** (Supplementary Figure S10) and **5** (Supplementary Figure S11), shows deep red spots representing O–H···O hydrogen bonds. The 3D Hirshfeld surface illustration of **4** (Figure 3), as well as of **2** (Supplementary Figure S12), shows only the deep red spots that represent weak C–H···O hydrogen bonds. The Hirshfeld surfaces plotted over the shape index of **4** and **2** visualize the π–π stacking interactions by the presence of adjacent red and blue triangles (Figure 3, Supplementary Figure S12). The Hirshfeld 2D fingerprint of all complexes are illustrated in the Supplementary Materials (Supplementary Figures S13–S19). In the cases of **3** and **5**, which were strongly disordered, the structures are shown separately for the main and minor part of the disorders. Hirshfeld 2D fingerprint plots allow for the quick and easy identification of significant intermolecular interactions mapped on the molecular surface [38,39]. As shown in Supplementary Figures S13–S19, strong and medium H···O/O···H hydrogen bonding interactions covered 10.3–12.7% of the total Hirshfeld surfaces with two distinct spikes in the 2D fingerprint plots, indicating the fact that hydrogen bonding interactions were the most significant interactions in the crystal structures. In the middle of the scattered points in the 2D fingerplots, H···H interactions covered 35.0–64.0% of the total Hirshfeld surfaces; however, H···H interactions were not very strong in the crystal structures. In particular, H···C/C···H interactions covered 16.5–23.7% of the total Hirshfeld surfaces in the scattered points in the 2D fingerplots. In the scattered points in the 2D fingerplots of **1–2**, H···F/F···H interactions covered 9.8–19.6% of the total Hirshfeld surfaces, and H···Cl/Cl···H interactions covered 9.4–21.1% of the total Hirshfeld surfaces for **3–4**. Furthermore, in the 2D fingerplots of **2** and **4**, significant C···C interactions are visible, covering 4.2–4.9% of the total Hirshfeld surfaces as a result of the presence of significant π–π stacking interactions in the crystal structures.

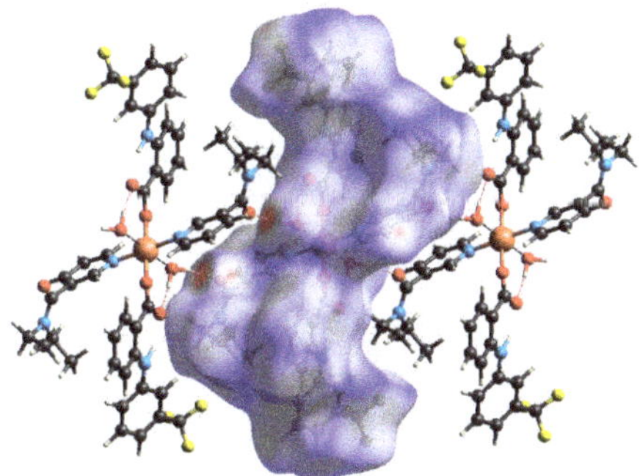

Figure 2. View of 3D Hirshfeld surface of **1** plotted over d_{norm} in the range of −0.5608 to 1.3607 a.u.

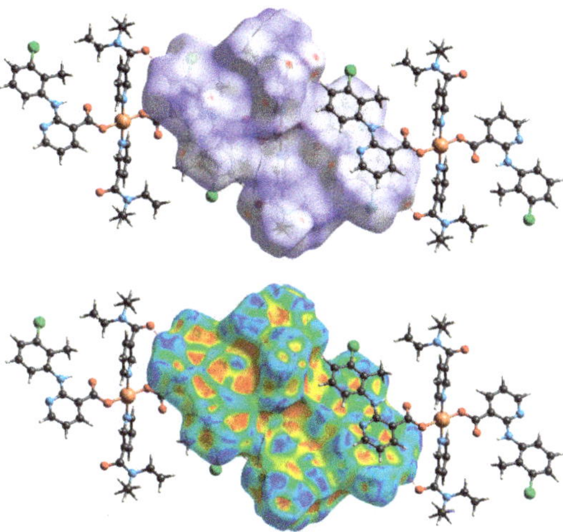

Figure 3. View of the 3D Hirshfeld surface of **4** plotted over d_{norm} in the range of −0.2000 to 1.4950 a.u. (top) and shape index (bottom).

2.5. EPR Spectroscopy

The complexes under study were investigated either as polycrystalline solids at room temperature or as frozen DMSO solutions at a low temperature (100 K). Experimental spectra were simulated using computer software in order to obtain parameters with a higher precision. The X-band EPR solid spectra of the selected complexes, **1**, **3** and **4**, are shown in Figure 4. The obtained spin Hamiltonian parameters are collected in Table 3. The EPR spectra of solid samples showed features characteristic for copper(II) monomeric complexes with S = $\frac{1}{2}$. The EPR signal was of an axial symmetry with either resolved (**3** and **4** only weakly visible) or unresolved hyperfine structures (**1**, **2** and **5**) in a parallel part of the signal as a result of the interaction of the unpaired electron with copper nuclear spin (I = 3/2) (Figure 4). The relative ordering of the axial g factors followed the usual trend ($g_{\|} > g_{\perp} \sim g_e$), indicating a $d_{x^2-y^2}$ ground state, which is characteristic for copper(II) atoms in tetragonally elongated octahedral arrangements around the central ion when the Jahn–Teller effect is operating. Similarly, values of the obtained g factors ($g_{\perp}$ = 2.055–2.082 and $g_{\|}$ = 2.29–2.36) showed only minor differences among the complexes and are in agreement with the information extracted from the crystal structures as well as from the literature for structurally similar complexes [18,27,29].

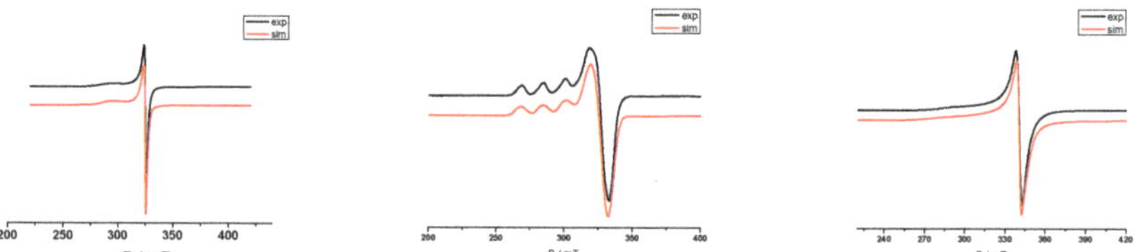

Figure 4. Room temperature solid-state spectra of **1**, **3** and **4**. Simulated spin Hamiltonian parameters are $g_{\perp}$ = 2.082, $g_{\|}$ = 2.290 and ΔB = (1.2, 8.5) mT (**1**); $g_{\perp}$ = 2.063, $g_{\|}$ = 2.305, $A_{\|}$ = 16.2 mT and ΔB = 4 mT (**3**); and $g_{\perp}$ = 2.055, $g_{\|}$ = 2.308, $A_{\|}$ = 16 mT and ΔB = (1.3, 8) mT (**4**).

Table 3. EPR spectral parameters of powders measured at RT and frozen DMSO solutions measured at 100 K.

Complex	Temperature	$g_{\perp}$	$g_{\|}$	g_{ave}	$A^{Cu}_{\|}$ /mT	G	f/cm
1	Solid RT	2.082	2.290	2.151	-	3.54	
	Sol. 100K	2.078	2.312	2.156	14.5	4.00	147
2	Solid RT	2.061	2.360	2.161	-	5.90	
	Sol. 100 K	2.077	2.303	2.152	17.5	3.94	123
3	Solid RT	2.063	2.305	2.144	16.2	4.84	132
	Sol. 100 K	2.072	2.307	2.150	16.7	4.26	128
4	Solid RT	2.057	2.308	2.140	16	5.40	134
	Sol. 100 K	2.073	2.315	2.154	15.5	4.32	138
5	Solid RT	2.055	2.320	2.143	-	5.81	
	Sol. 100 K	2.075	2.300	2.150	16.5	4.00	130

Defined as $g_{av} = (2g_{\perp} + g_{\|})/3$, $G = (g_{\|} - 2)/(g_{\perp} - 2)$ and $f = g_{\|}/A_{\|}$.

The EPR spectra of frozen solutions are usually more informative than their analogs in a solid state due to the dilution and separation of paramagnetic ions between the neighboring molecules as a result of solvation. Because the biological measurements on the complexes were performed in the liquid phase, it was reasonable to have the solution EPR spectra to obtain more appropriate structural information for correlating the experimental results. The EPR spectra of DMSO solutions measured at 100 K are depicted in Figure 5 (for selected complexes **1**, **4** and **5**).

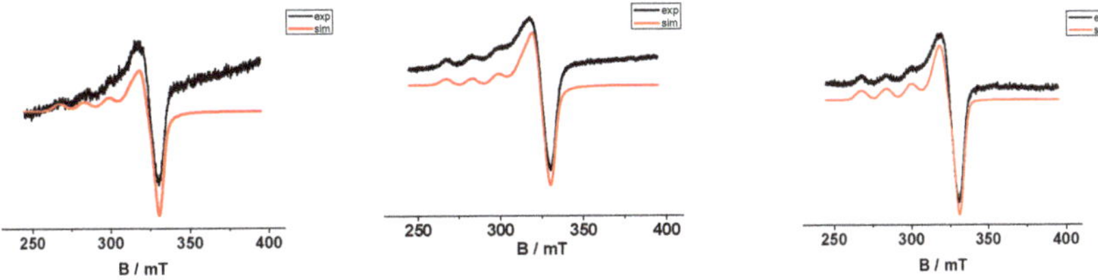

Figure 5. EPR spectra of **1**, **4** and **5** measured in DMSO solutions at 77K. Simulated spin Hamiltonian parameters are $g_{\perp}$ = 2.078, $g_{\|}$ = 2.312, $A_{\|}$ = 14.5 mT and ΔB = 3.5 mT; $g_{\perp}$ = 2.073, $g_{\|}$ = 2.315, $A_{\|}$ = 15.5 mT and ΔB = (2.8, 3.5) mT; and $g_{\perp}$ = 2.075, $g_{\|}$ = 2.300, $A_{\|}$ = 16.5 mT and ΔB = 3 mT.

For all five complexes, the EPR spectra of frozen DMSO solutions showed axial symmetry with resolved parallel hyperfine splitting showing a tree of four peaks. The spin Hamiltonian parameters obtained from the EPR simulations are collected in Table 3. The obtained EPR data show a close resemblance ($g_{\perp}$ = 2.072–2.078, $g_{\|}$ = 2.300–2.315, $A_{\|}$ = 14.5–17.5 mT) as a result of the similar structures of complexes in DMSO solutions. This fact can be clearly seen when looking at the alike values of the derived EPR parameters, such as g_{ave}, G or $g_{\|}/A_{\|}$, are also summarized in Table 3. The values of the geometric parameters G were very close to 4.00, indicating that the complexes exhibit minimal exchange interactions between copper(II) centers [35]. Similarly, values of the parameter of the tetrahedral distortion $f - g_{\|}/A_{\|}$ ranged from 123 cm for **2** to 147 cm for **1** between the studied complexes. Such values indicate that only minor tetrahedral distortion around the central copper ion existed in the primary coordination sphere, in accordance with the crystallographic information [40]. We can conclude that the observed similarity among the solid and solution EPR data suggest a close similarity in the geometries of the studied complexes.

2.6. SOD Mimetic Activity

The SOD mimetic activity of the studied complexes was characterized via an NBT assay. The superoxide radical was generated with xanthine and a xanthine oxidase biochemical system, and the ability of the complexes to scavenge the superoxide was tested indirectly via the reduction of NBT dye. Colour changes in NBT were detected at 560 nm. When the potential scavenging complex was added to the system, a competing reaction between the complex and superoxide would occur, which led to the inhibition of the reaction between NBT and the superoxide. The scavenging activity of the studied complexes was evaluated and characterized with the IC_{50} value (Figure 6).

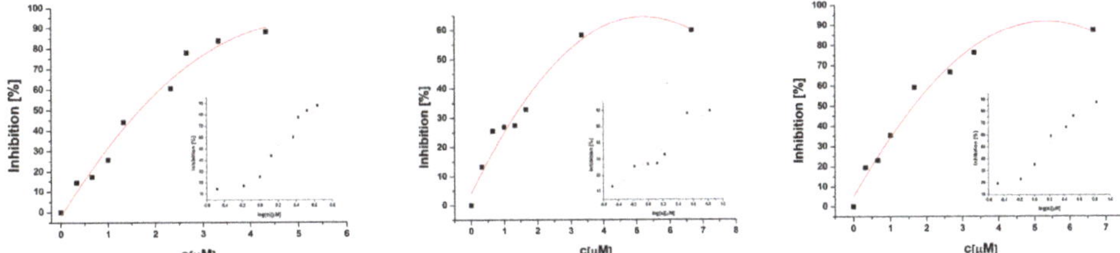

Figure 6. SOD mimetic activity of **2**, **3** and **4**.

The inhibition concentration IC_{50} corresponds to the concentration of the complex that caused 50% inhibition of the NBT reduction. The results are collected in Table 4. The complexes showed significant SOD-like activity, comparable to the SOD mimetic activity of other copper fenamates [3,27,41,42]. Complexes **2** and **4** exhibited the greatest radical scavenging effect with micromolar concentrations. On the basis of these results, the complexes could be considered as the good SOD mimetics [3].

Table 4. SOD mimetic activity of selected copper NSAIDs.

Complex	IC_{50}/μM	Reference
[Cu(fluf)$_2$(dena)$_2$(H$_2$O)$_2$] (**1**)	3.33	This work
[Cu(nifl)$_2$(dena)$_2$] (**2**)	1.69	This work
[Cu(tolf)$_2$(dena)$_2$(H$_2$O)$_2$] (**3**)	3.46	This work
[Cu(clon)$_2$(dena)$_2$] (**4**)	1.41	This work
[Cu(mef)$_2$(dena)$_2$(H$_2$O)$_2$] (**5**)	3.15	This work
[Cu(tolf)$_2$(phen)]	0.98	[27]
[Cu(mef)$_2$(phen)]	1.23	[27]
[Cu(fluf)$_2$(phen)]	0.94	[27]
[Cu(3-mesal)$_2$(dena)$_2$(H$_2$O)$_2$]	3.88	[41]
[Cu(tolf)$_2$(H$_2$O)]$_2$	1.97	[42]
Native SOD	0.04	[3]

2.7. Interactions of Complexes with KO_2 in the Presence of Neocuproine

The first step in the mechanism of action of native CuZn-SOD enzymes is the binding of the superoxide radical anion to the copper center, where Cu(II) is subsequently reduced to Cu(I) and the oxygen molecule is released [43]. Therefore, the ability of the prepared complexes to undergo reduction with the superoxide radical anion is an important step in the investigation of their potential SOD activity. To verify this assumption, we used a specific Cu(I) chelator, neocuproine, which forms a stable Cu(I)-neocuproine complex that absorbs at 458 nm [44]. The addition of potassium superoxide (KO_2) to the solutions of complexes **1–4** in the presence of neocuproine resulted in a dramatic increase in absorbance at 458 nm for complexes **1–4** and a visible reduction in the d-d band intensity of the studied complexes, as can be clearly seen in Figure 7. Thus, from these results we can assume that the prepared complexes **1–4** undergo reduction to Cu(I) under the influence of KO_2 and thus fulfill an important prerequisite to be good SOD mimetics.

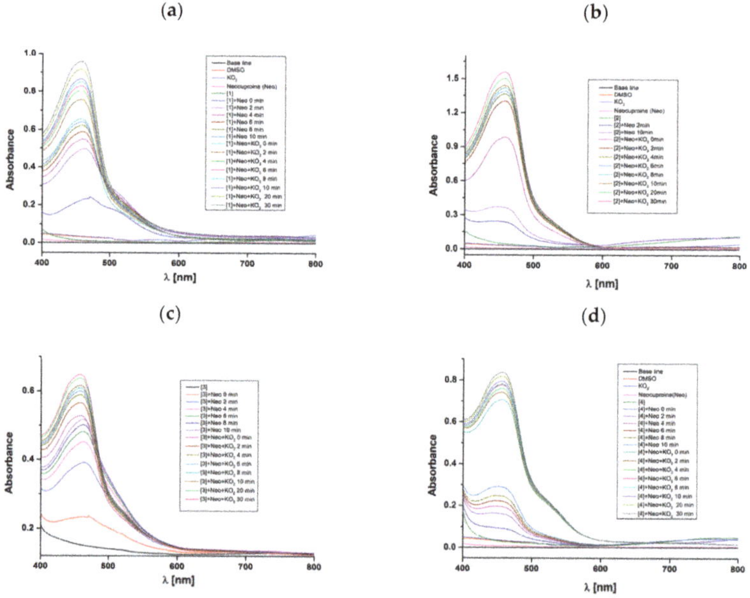

Figure 7. Time-dependent UV-Vis spectra of the interaction of studied complexes and KO_2 in DMSO in the presence of neocuproine: (**a**) **1**; (**b**) **2**; (**c**) **3**; (**d**) **4**.

2.8. Cyclic Voltammetry

The redox behavior of the studied complexes (prepared in DMSO) was investigated by means of cyclic voltammetry using a scan rate of 100 mV/s. Figure 8 displays the cyclic voltammograms of corresponding complexes **1–5** at a concentration level of 10^{-4} M, which were registered in the presence of 0.1 M NaCl as a supporting electrolyte at the boron-doped diamond (BDD) electrode. A summary of the basic redox parameters for the studied complexes are listed in the Table 5. In the cyclic voltammetric records, two potential regions were differentiated. In the first potential region between −0.136 V and –0.085 V, Cu(II)/Cu(I)-based redox transitions were observed, which showed better resolved peak currents in the anodic scan, with $E_{p,ox}$ ranging from −0.136 V (**4**) to −0.124 V (**3**). In the cathodic scan, the corresponding reduction waves could be identified at a peak potential ranging from −0.097 V (**3**) to −0.083 V (**1**). The observation of both potentials for all studied complexes indicated the quasi-reversible redox process undertaken at the BDD electrode. The same conclusion could be read from values of the $I_{p,ox}/I_{p,red}$ ratio, and the value of this ratio ranged from the lowest value of 1.3 for **3** to the highest of 2.1 for **4**. In the second potential region, quite distinctive voltammetric curves with oxidation peak potentials ranging from 0.633 V (**5**) to 0.941 V (**2**) could be noticed, which may be attributed to the redox activity within the bis(fenamate) ligand. Finally, to be good SOD mimetics, the redox potential ($E°$ vs. Ag/AgCl) should fall between −0.363 V and +0.687 V, as in the case of a native SOD enzyme [27]. This criterion was succesfully fulfilled for all studied complexes according their $E_{1/2}$ values (Table 5).

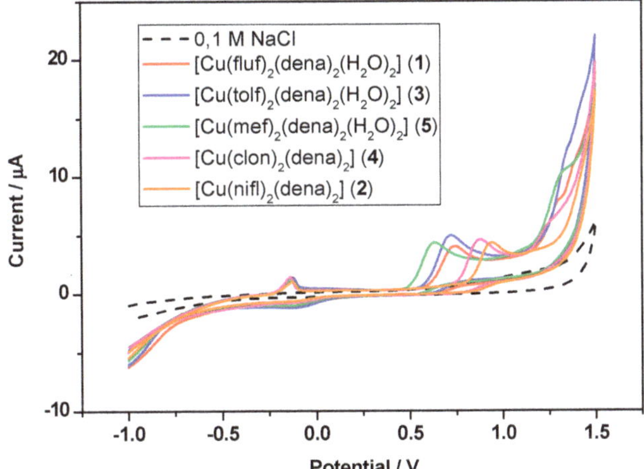

Figure 8. Cyclic voltammograms of **1–5** (c(complex) = 10^{-4} M) in 0.1 M NaCl measured at BDD electrode using scan rate of 100 mV/s.

Table 5. Redox parameters of **1–5** extracted from experimental CV data.

Comp	$E_{p,ox}$/V	$E_{p,red}$/V	$E_{1/2}$/V	ΔE/V	$I_{p,ox}$/µA	$I_{p,red}$/µA	$E_{p,ox}$/V	$I_{p,ox}$/µA
1	−0.133	−0.085	−0.109	−0.048	1.386	−0.917	0.741	4.05
2	−0.126	−0.096	−0.111	−0.030	1.064	−0.542	0.941	4.36
3	−0.124	−0.097	−0.111	−0.027	1.427	−1.104	0.734	5.00
4	−0.136	−0.096	−0.116	−0.040	1.449	−0.681	0.877	4.62
5	−0.133	−0.087	−0.110	−0.046	1.279	−0.876	0.633	4.36

2.9. *ct-DNA Interaction Studies*

Transition metal complexes, such as copper complexes, can bind to DNA and thus induce DNA cleavage, which can be exploited in the preparation of DNA structural probes, cleavage or anticacer agents [45]. Moreover, if complexes contain ligands with suitable functional groups that can be involved in hydrogen bonding or in electrostatic, hydrophilic/hydrophobic or π–π stacking interactions, then they can be rationally utilized to support the binding abilities of complexes toward the DNA [46]. Copper complexes can interact with DNA either covalently through the formation of covalent adducts (e.g., cis-platin) or non-covalently [47]. The non-covalent mode of binding between complexes and DNA includes intercalation (using mostly π–π stacking interactions), groove binding (van der Waals interactions or hydrogen bonding) and external binding (electrostatic interactions) [48]. Interactions of the prepared complexes **1–4** with calf thymus DNA were evaluated using UV-Vis absorption titrations and viscosity measurements, as well as with fluorescence emission with an ethidium bromide (EB) displacement method.

2.9.1. Absorption Titrations

The absorption spectra of the DMSO/buffer solutions of the studied complexes **1–4** exhibited a very similar pattern of absorption bands in the UV region from 250 nm to 400 nm (Figure 9). Two types of signals existed in the UV spectrum, which has different behavior upon the addition of DNA. First, for tree absorption bands with maxima located at 255, 262 and 269 nm, a sudden decrease in absorption was observed after the addition of the first amount of DNA. Then, the further addition of DNA led to band hyperchromism. Because the positions of the bands did not move after the addition of DNA, we assumed that the increase in absorbance was due to the further addition of DNA with an absorption band in this part of spectra (at 260 nm). Two other absorption peaks, coming from intraligand π to π^* transitions of aromatic NSAID moieties at around 280 nm (high-intensity discrete peak) and 320 nm (lower-intensity shoulder), showed a considerable decrease in the absorption of complexes (11.8% for **2** to 21.4% for **1**) together with a slight blue shift (1–5 nm) (Table 6). A hypochromic shift is usually associated with the stabilization of DNA secondary structures via electrostatic interactions or the intercalation of metal complexes [46,48]. The observed hypochromism and blue shift thus suggest an electrostatic or intercalative binding mode of complex–DNA interactions or their combination. However, as reported, absorption titrations give only preliminary information about complex–DNA interactions, and therefore, further measurements are necessary to clarify the binding mode [26]. The internal DNA binding K_b constants of **1–4** were determined using the most intensive and best resolved band at 288 nm with the Wolfe–Shimmer equation (inset in Figure 9).

The values of the K_b binding constants are collected in the Table 6. The obtained values of the K_b constants ranged from 1.04×10^5 (**2**) to 5.46×10^5 M^{-1} (**4**) and are in good agreement with other Cu–fenamate complexes [23,27–30]. Such values indicate relatively strong binding of the studied complexes to DNA, likely due to their ability to form hydrogen bonds with DNA in combination with partial intercalation. Based on their binding strength with DNA, the complexes can be arranged as follows: **4** > **1** > **3** > **2**. The highest values of the binding constant were obtained for complexes with clonixinate and flufenamate ligands (5.46×10^5 M^{-1} and 4.83×10^5 M^{-1}, respectively). Based on these values, there seemed to be no apparent structural trend between structurally similar complexes **2**, **4** vs. **1**, **3** with respect to the planarity of complexes **2** and **4**. Instead, the combined effect of electrostatic interactions, which is stronger for **1**, and intercalation through coplanar pyridine and benzene rings, which prefer complex **4**, could be operative. However, as was noticed, the exact mode of binding of the complexes into DNA cannot be determined using only absorption titration studies, so further measurements are necessary to confirm the obtained results [24–27].

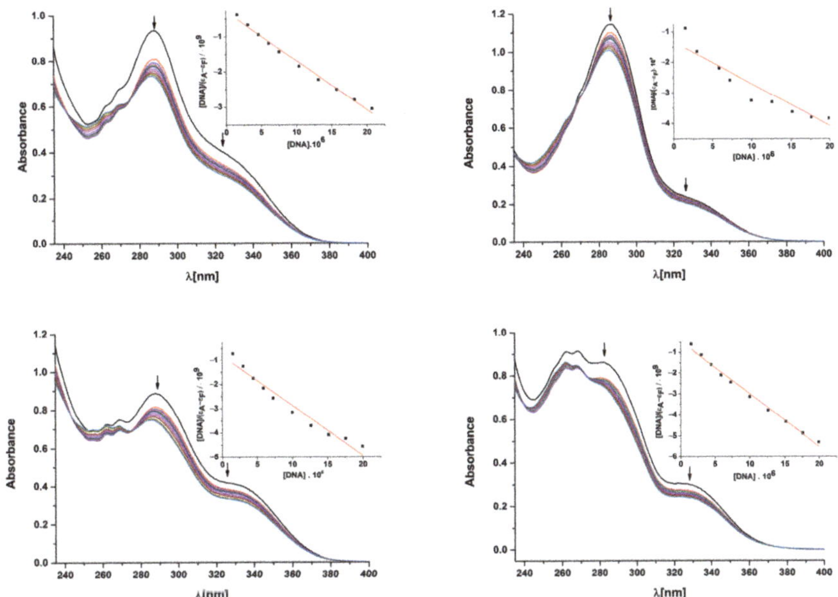

Figure 9. UV-Vis spectra of DMSO/buffer solution of **1–4** in the absence and presence of increasing amounts of DNA (r = [DNA]/[complex] = 0–2.1). Arrows show changes in intensity upon the addition of increasing amounts of DNA. Inset: Plot of [DNA]/(ε_A-ε_f) versus [DNA] for complex.

Table 6. DNA binding constant and UV spectral features of **1–4** in the presence of DNA.

Complex	K_b [M^{-1}]	R^2	λ(nm)(ΔA/A$_0$ (%), Δλ * (nm)
[Cu(fluf)$_2$(dena)$_2$(H$_2$O)$_2$] (**1**)	4.83 (±0.84) · 10^5	0.9897	287(21.4, −1)
[Cu(nifl)$_2$(dena)$_2$] (**2**)	1.04 (±0.75) · 10^5	0.8699	287(11.8, −2)
[Cu(tolf)$_2$(dena)$_2$(H$_2$O)$_2$] (**3**)	2.52 (±0.87) · 10^5	0.9551	288(15.7, −3)
[Cu(clon)$_2$(dena)$_2$] (**4**)	5.46 (±0.87) · 10^5	0.9906	281(12,3, −5)

* denotes blue shift.

2.9.2. Viscosity Measurements

Because DNA viscosity manifests sensitivity to DNA length changes in the presence of a DNA binder, it was desirable to carry out the DNA viscosity measurements in the presence of complexes with potential binding activity [48]. The DNA viscosity measurements were performed on DNA solutions in the presence of increasing concentrations of complexes **1–4**. In addition, the planar molecule of ethidium bromide (EB), which is known as a perfect nonspecific DNA intercalating agent, was used as an indicator of intercalation. As is clearly seen in Figure 10, in the presence of growing concentrations of complexes, a continual increase in the relative DNA viscosity for all four complexes was observed. This behavior supports the hypothesis about the intercalative interaction between the complex molecules and DNA. The results reveal that the best intercalating ability in this series had a complex with the flufenamate ligand (**1**). On the other hand, the relative DNA viscosity of the complex with clonixinate (**4**) gave, in this case, the lowest increase in the studied series (**1** > **2** ≈ **3** > **4**). A comparison with an EB molecule suggests that the studied complexes were weaker intercalating agents than EB, but it can be noted that all four complexes could bind to DNA via partial intercalation. Finally, no apparent trend between structurally similar complexes **2**, **4** vs. **1**, **3** was visible.

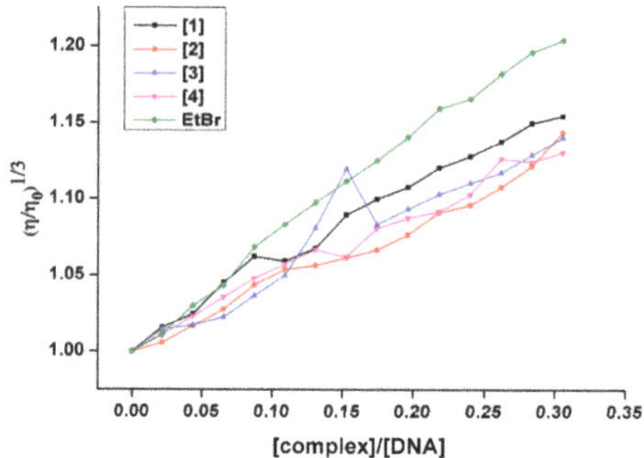

Figure 10. Relative viscosity of DNA in the buffer solution in the presence of studied complexes (**1–4**) under the condition of increasing the concentration ratio [complex]/DNA.

2.9.3. Competitive Studies with EB-DNA

Another method that was used to investigate the intercalating ability of complexes to DNA was a competitive study of the DNA interactions of the complexes with the ethidium bromide (EB) displacement method. EB represents a typical DNA intercalator that intercalates into DNA, and at the same time, it is a very effective fluorophore in the presence of DNA. When EB interacts with DNA, it creates an EB-DNA adduct that emits an intense fluorescence emission band at 615 nm when excited at 540 nm. The addition of a complex with an affinity to DNA lowers the emission intensity of the EB adduct due to competition with EB at the same binding sites in DNA. The representative fluorescence emission spectra of **4** are shown in Figure 11. The addition of increased concentrations of complexes led to a decrease in the intensity of the emission band of the EB-DNA adduct at 615 nm. The final quenching of the fluorescence reached 30–35% of the initial EB-DNA fluorescence intensity (Figure 11). Very similar results of quenching with copper fenamates were also observed by other authors [21]. The observed moderate decrease in EB-DNA fluorescence emission in the presence of complexes indicates their competitive binding ability when compared with EB. The quenching of EB bound to DNA is in good accordance with the linear Stern–Volmer equation, thus providing further proof of the observed DNA-binding ability of the studied complexes. The calculated values of the K_{sv} constant ($3.30–3.79 \times 10^3$ M^{-1}) confirmed the moderate intercalative ability of the complexes towards DNA (Table 7) when comparing these values with other copper fenamates, which show values of K_{sv} 10^5–10^6 [22,26,29]. In addition, the calculated values of the EB-DNA quenching rate constant k_q of order 10^{11} M^{-1} s^{-1} (Table 7) suggest the presence of a static quenching mechanism ($k_q > 10^{11}$ M^{-1} s^{-1}) [5].

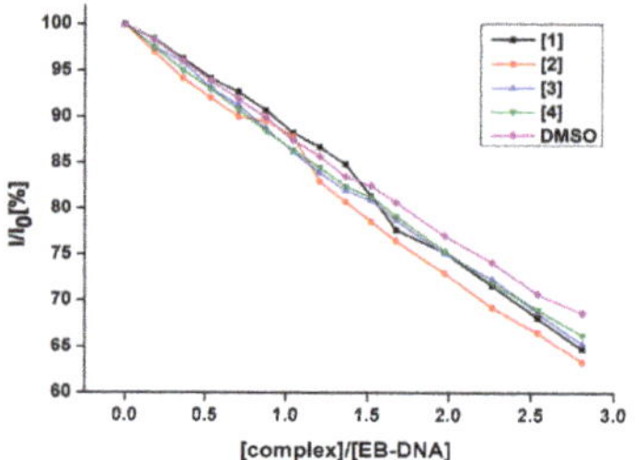

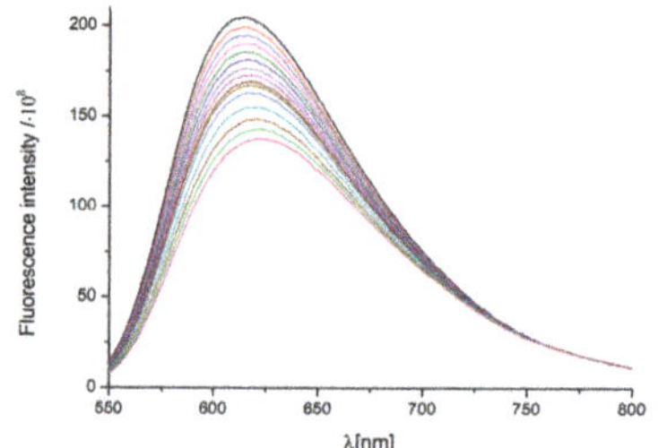

Figure 11. Graphical dependence of relative EB-DNA fluorescence emission intensity (I/I_0) at 615 nm vs. concentration ratio [complex]/DNA for **1–4**. Fluorescence emission spectra for EB-DNA in the buffer solution in the presence of increasing amounts of **4**.

Table 7. EB-DNA fluorescence (%), calculated Stern–Volmer constant K_{SV} and the quenching rate constant k_q of DMSO and complexes **1–4**.

Complex	$\Delta I/I_0$ (%)	K_{sv} (M^{-1})/10^3	k_q (M^{-1} s^{-1})/10^{11}
[Cu(fluf)$_2$(dena)$_2$(H_2O)$_2$] (**1**)	35.2	3.59 (±0.16)	1.56 (±0.07)
[Cu(nifl)$_2$(dena)$_2$] (**2**)	36.7	3.79 (±0.13)	1.65 (±0.06)
[Cu(tolf)$_2$(dena)$_2$(H_2O)$_2$] (**3**)	31.2	3.40 (±0.09)	1.48 (±0.04)
[Cu(clon)$_2$(dena)$_2$] (**4**)	33.6	3.30 (±0.08)	1.44 (±0.04)
DMSO	30.9	3.03 (±0.07)	1.32 (±0.03)

2.10. Interaction of Studied Complexes with BSA

The fluorescence emission spectra of bovine serum albumin showed intense fluorescence emission at 336 nm due to the existence of two tryptophan moieties at positions 134 and 212. The interaction of complexes **1–4** with bovine serum albumin was studied by monitoring spectral changes in tryptophan fluorescence emission after the addition of complexes [49]. The graphical dependence of the relative BSA fluorescence intensity on increasing amounts of complexes **1–4** showed significant quenching of the fluorescence up to 89–91% for all studied complexes (Figure 12).

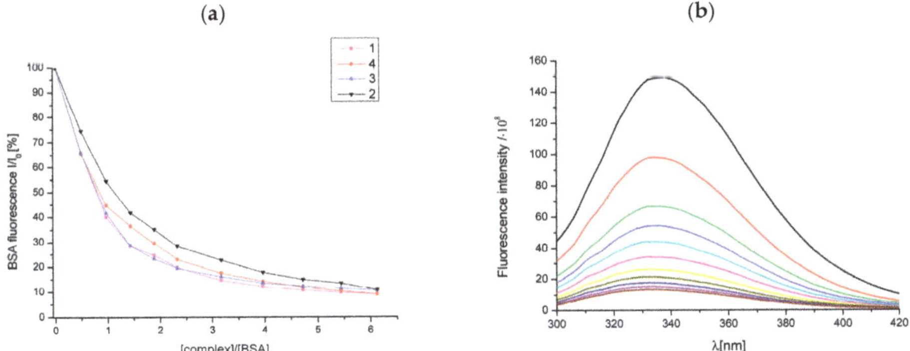

Figure 12. (**a**) Graphical dependence of relative BSA fluorescence intensity in % at λ = 336 nm vs. concentration ratio [complex]/BSA. (**b**) Fluorescence emission spectra for EB-DNA in the buffer solution in the presence of increasing amounts of **4**.

These results confirm the fact that the complexes were able to bind to serum albumin in significant amounts, likely through tryptophan residue. In addition, the interaction of complexes with serum albumin was characterized with the Stern–Volmer constant K_{sv} and the quenching constant k_q, which was calculated using the Stern–Volmer equation. Furthermore, the association binding constant K_{BSA} and the number of binding sites per albumin n, were determined using the Scatchard equation as well (Supplementary Figure S20). The obtained values are summarized in Table 8. The values of the K_{sv} constants of order 4–5 × 10^5 M^{-1} indicate an intermediate binding strength between the complexes and albumin. The presented values of the quenching constant k_q of order 10^{13} are much larger than 10^{10} M^{-1} s^{-1}, which represents a typical value for a quencher used in biopolymer quenchers. High values also indicate that quenching is performed through the static quenching mechanism [21,28]. The highest quenching ability was observed for complexes **1** and **4** according to their k_q values. As reported, the optimal range assumed for a serum albumin drug delivery system (capable of providing adequate transport and distribution in the bloodstream and reversible release of the drug to the target) should have K_{sv} values in the range of 10^2–10^8 M^{-1} and K_{BSA} values in the range of $10^4 - 10^6$ M^{-1} [49]. The calculated values of K_{sv} and K_{BSA} for the studied complexes **1–4** are within the optimal range.

Table 8. Values of the Stern–Volmer quenching constant (K_{sv}), quenching constant (k_q), association binding constant (K_{BSA}) and n (number of binding sites for albumin) obtained for the interaction of **1–4** with bovine serum albumin.

Complex	K_{sv} (M^{-1})/10^5	k_q (M^{-1} s^{-1})/10^{13}	K_{BSA}(M^{-1})/10^5	n
[Cu(fluf)$_2$(dena)$_2$(H$_2$O)$_2$] (**1**)	5.777	2.51 (±0.061)	3.39 (±0.413)	1.11
[Cu(nifl)$_2$(dena)$_2$] (**2**)	4.411	1.92 (±0.073)	2.18 (±0.164)	1.16
[Cu(tolf)$_2$(dena)$_2$(H$_2$O)$_2$] (**3**)	4.740	2.06 (±0.092)	3.36 (±0.417)	1.10
[Cu(clon)$_2$(dena)$_2$] (**4**)	5.558	2.42 (±0.054)	3.32 (±0.147)	1.08

2.11. Anticancer Activity

Copper complexes **1–4** were tested for their in vitro cytotoxicity against three cancer cell lines, including human lung cancer cells (A549), human breast cancer cells (MCF-7), human glioblastoma cells (U-118MG) and a healthy human lung fibroblast cell line (MRC-5), respectively. The cells (8×10^3 cells/200 μL well) were treated with several concentrations (20–100 μmol/L) of **1–4** for 24, 48 and 72 h, and the cytotoxicity was evaluated using an MTT assay. The measurement was repeated twice using three parallels for each concentration. The inhibitory concentration values (IC_{50}) of the studied complexes for the A549 and U-118MG cancer cell lines were higher than the highest concentration used of 100 μM at each incubation time (data not shown). In the case of the MCF-7 tumor cell line, the same results were obtained for complexes **1–3** (IC_{50} > 100 μM). On the other hand, complex **4** showed cytotoxicity against MCF-7 cells after 72 h of exposure with an IC_{50} value of 57 ± 3 μM. As can be seen from the graph (Figure 13a), the viability of the MCF-7 tumor cells decreased with increasing concentrations of **4** for 72 h of exposure. In addition, no cytotoxic effect was observed on healthy MRC-5 cells for 72 h of incubation under the same conditions (Figure 13b).

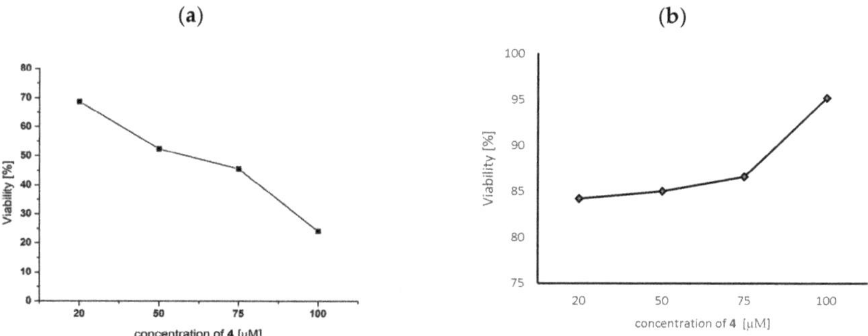

Figure 13. Cell proliferation of (**a**) MCF-7 cancer cells and (**b**) MRC-5 healthy cells in response to **4** after 72 h of exposure. Cell lines were treated (20–100 μmol/L) with complex **4**, and viable cells were evaluated using a colorimetric assay.

Complex **4** contains the ligand clonixin in its structure, which has antipyretic, antianalgesic and antirheumatic effects, and especially anti-inflammatory effects [50]. The antitumor effects of clonixin alone or of copper complexes with clonixin have not been described so far. However, some authors have focused on the effect of clonixin with platinum, as the anti-inflammatory strategy is key in the treatment of aggressive cancer diseases [51]. They investigated a platinum (IV) prodrug complex with NSAIDs (non-steroidal anti-inflammatory drugs) as ligands designed to effectively enter tumor cells due to their high lipophilicity, where they can release a cytotoxic metabolite [51]. This mechanism reduces side effects and increases the therapeutic efficacy of the drug used in chemotherapy. Copper(II) complexes containing coordinated clonixin seem to be a potential metallo-drug for closer follow-up of its biological effects, as we did not observe a cytotoxic effect on healthy MRC-5 cells for 72 h of incubation under the same conditions (Figure 13b). In addition, copper complexes with tolfenamic, mefenamic and flufenamic acids and phenanthroline can have anti-tumor effects [27]. The authors confirmed the effect of these substances according to their ability to generate intracellular reactive oxygen species (ROS) and inhibit cyclooxygenase-2 (COX-2), an enzyme that is overexpressed in breast tumors. They detected DNA damage, JNK and p38 pathway activation and an apoptosis pathway [27].

In the second step of our biological research, we were interested in the genotoxic effect of the selected copper complexes. Considering that complexes **1–3** did not show any cytotoxic activity for 72 h of incubation with A549, U-118MG and MCF-7 tumor cells in the concentration range (20–100 μmol/L), we chose complex **4** with an IC_{50} of 57 × μM. We affected MCF-7 cells with the IC_{50} value and monitored DNA damage after 72 h of incubation. Unfortunately, we did not observe any significant DNA damage compared to control cells, which were not affected by **4** (Figure 14). The DNA damage did not exceed a threshold of 10%, which is considered relevant DNA damage.

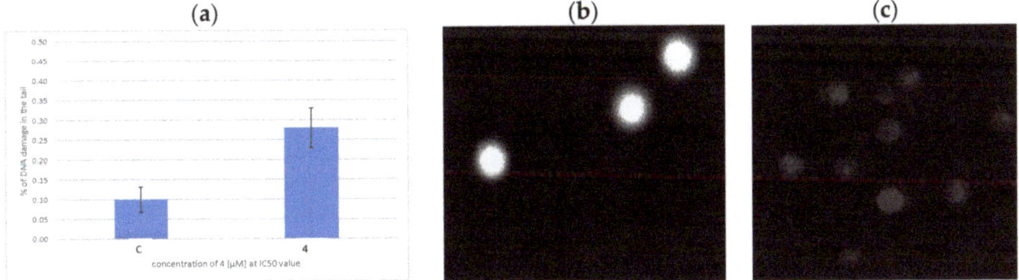

Figure 14. (**a**) Analysis of DNA damage of MCF-7 cells treated with **4**. Morphology of (**b**) control (cells were not treated) and (**c**) damaged MCF-7 cells treated with **4** at an IC_{50} of 57 ± 3 μM.

3. Materials and Methods

General procedures. All reagents and solvents were obtained from commercial sources and used as received unless noted otherwise.

3.1. Synthesis

[Cu(fluf)$_2$(dena)$_2$(H$_2$O)$_2$] (1)

Complex **1** was prepared with the following procedure. Copper acetate dihydrate (1 mmol, 0.170 g) was dissolved in 30 mL of ethanol. Then, sodium flufenamate, formed in situ by mixing an equimolar amount of flufenamic acid (2 mmol, 0.564 g) with sodium hydroxide (2 mmol, 0.080 g), was slowly poured into the solution. The solution immediately changed color from blue-green to green. After 10 min, *N*, *N*-diethylnicotinamide (2 mmol, 0.356 g, 0.4 mL) was added dropwise to form a clear dark green solution. After a while, a dark green precipitate was formed. Afterward, the mixture was stirred for 3 h at room temperature, before the crude product was filtered through smooth filtration paper and dried in the air. The pale green needle-like crystals of **1** suitable for crystallographic analyses were isolated from the mother liquor after a week.

Yield: 0.71 g (70%). Anal. calc. for $C_{48}H_{50}CuF_6N_6O_8$ (M_r = 1016.504): C 57.29, H 4.93, N 8.70 %. Found: C 56.72, H 4.96, N 8.27 %. IR (ATR, cm^{-1}): 3507 (m), 3324 (w), 3218 (w), 3091 (m), 2979 (m), 2934 (w), 1619 (s), 1606 (s), 1581 (s), 1568 (s), 1499 (s), 1456 (s), 1421 (ms), 1378 (vs), 1331 (vs), 1284 (s), 1183 (ms), 1159 (ms), 1109 (vs), 1069 (s), 1046 (ms), 997 (m), 930 (m), 872 (m), 826 (m), 753 (s), 697 (s), 650 (m). UV-Vis: λ / nm (ε/M^{-1}cm^{-1}) as nujol mulls (nm): 210, 234 (sh), 287 (sh), 322 (sh), 412 (sh), 646; in DMSO / H_2O: 255 (35440), 262 (40030), 269 (44900), 287 (62300), 320 (28300, sh), 795 (65).

[Cu(nifl)$_2$(dena)$_2$] (2)

Complex **2** was prepared with the same procedure used for complex **1**. Violet prismatic crystals suitable for X-ray analysis formed after a week.
Yield: 0.56 g (57%). Anal. calc. for $C_{48}H_{44}CuF_6N_8O_6$ (M_r = 982.448): C 57.05, H 4.70, N 11.83 %. Found: C 56.24, H 4.51, N 11.41 %. IR (ATR, cm^{-1}): 3269 (w), 3163 (w), 3117 (w), 2979 (m), 2936 (w), 2874 (w), 1629 (s), 1595 (vs), 1582 (vs), 1518 (s), 1494 (s), 1456 (s), 1420 (ms), 1387 (s), 1368 (s), 1324 (vs), 1293 (sh), 1160 (s), 1115 (s), 1096 (s), 1067 (s), 997 (w), 943 (m), 869 (m), 826 (m), 783 (s), 699 (s), 670 (m),471 (m), 413 (m). UV-Vis: λ/nm ($\varepsilon/M^{-1}cm^{-1}$) as nujol mulls (nm): 203 (sh), 286, 343 (sh), 412 (sh), 539 (br); in DMSO/H_2O: 262 (36270), 269 (46000), 287 (68950), 320 (14830, sh), 791 (95).

[Cu(tolf)$_2$(dena)$_2$(H$_2$O)$_2$] (3)

Dark green crystals of **3** suitable for X-ray analysis were isolated after two weeks.
Yield: 0.69 g (75%). Anal. calc. for $C_{48}H_{54}Cl_2CuN_6O_8$ (M_r = 977.450): C 59.70, H 5.86, N 9.82 %. Found: C 58.98, H 5.57, N 8.60 %. IR (ATR, cm^{-1}): 3458 (m), 3206 (m, br), 3094 (m), 2992 (m), 2938 (w), 1613 (s), 1579 (s), 1557 (s), 1493 (s), 1442 (s), 1377 (vs), 1281 (s), 1186 (m), 1107 (m), 1008 (m), 947 (m), 879 (m), 831 (m), 758 (s), 736 (s), 699 (s), 636 (m), 528 (m), 416 (m). UV-Vis: λ/nm ($\varepsilon/M^{-1}cm^{-1}$) as nujol mulls (nm): 221 (sh), 246 (sh), 294 (sh), 338 (sh), 412 (sh), 628 (br); in DMSO/H_2O: 255 (20820), 261 (21550), 269 (22550), 288 (26660), 325 (12520, sh), 789 (230).

[Cu(clon)$_2$(dena)$_2$] (4)

Complex **4** was prepared with the same procedure used for complex **1,** with the exception of the used solvent, which was, in this case, methanol. Brown prismatic crystals suitable for X-ray analysis formed after a week.
Yield: 0.68 g (72%). Anal. calc. for $C_{46}H_{48}Cl_2CuN_8O_6$ (M_r = 943.395): C 59.56, H 5.07, N 11.93 %. Found: C 58.56, H 5.13, N 11.88 %. IR (ATR, cm^{-1}): 3275 (w), 3186 (w), 3114 (w), 3066 (w), 2980 (w), 2937 (w), 1629 (s), 1602 (s), 1585 (s), 1519 (s), 1456 (s), 1434 (ms), 1380 (m), 1358 (s), 1315 (vs), 1256 (ms), 1189 (m), 1101 (m), 1015 (m), 924 (m), 880 (w), 823 (m), 766 (vs), 702 (s), 653 (m), 536 (m), 414 (m). UV-Vis: λ/nm ($\varepsilon/M^{-1}cm^{-1}$) as nujol mulls (nm): 224 (sh), 298, 340 (sh), 419 (sh), 529 (br), 629 (sh); in DMSO/H_2O: 262 (54960), 268 (55100), 281 (51900), 320 (18170, sh), 794 (116).

[Cu(mef)$_2$(dena)$_2$(H$_2$O)$_2$] (5).

Complex **5** was prepared as described for **1**. Dark green crystals of **5** suitable for X-ray analysis were obtained after two weeks.
Yield: 0.58 g (62%). Anal. calc. for $C_{50}H_{60}CuN_6O_8$ (M_r = 936.615): C 65.89, H 6.17, N 9.16 %. Found: C 65.38, H 6.36, N 9.15 %. IR (ATR, cm^{-1}): 3473 (w), 3217 (w, br), 2991 (w), 2934 (w), 1612 (s), 1575 (s), 1561 (s), 1496 (s), 1449 (s), 1376 (s), 1281 (s), 1214 (m), 1186 (m), 1105 (m), 947 (w), 920 (w), 878 (w), 831 (m), 783 (m), 760 (s),739 (m), 699 (m), 636 (m), 530 (m), 415 (m). UV-Vis: λ/nm ($\varepsilon/M^{-1}cm^{-1}$) as nujol mulls (nm): 216 (sh), 268 (sh), 346, 407 (sh), 600 (br); in DMSO/H_2O: 288 (22420), 330 (sh, 11400), 801 (56).

3.2. Physical Measurements

Carbon, hydrogen and nitrogen analyses were carried out on a CHNSO FlashEA^{TM} 1112 Automatic Elemental Analyzer. The electronic spectra (190–1100 nm) of the complexes were measured in a Nujol suspension with a SPECORD 250 Plus (Carl Zeiss Jena) spectrophotometer at room temperature. The infrared spectra (ATR technique, 4000–400 cm^{-1}) were recorded on a Nicolet 5700 FT-IR spectrophotometer at room temperature. Room-temperature EPR spectra of the powdered samples were recorded with an EPR spectrometer EMX Plus series (Bruker, Germany) operating at X-band (≈9.4 GHz) and simulated using Spin.exe software developed by Dr. Ozarowski [52].

3.3. X-ray Crystallography

The data collection and cell refinement of **1–5** were carried out using the four-circle diffractometer Stoe StadiVari using the Pilatus3R 300K HPD detector and the microfocused X-ray source Xenocs Genix3D Cu HF (Cu Kα radiation). The diffraction intensities were corrected for Lorentz and polarization factors. The absorption corrections were made with the LANA program [53]. The structures were solved using the ShelXT [54], Superflip [55] or Sir14 [56] program and refined using the full-matrix least-squares procedure of the Independent Atom Model (IAM) with ShelXL (version 2018/3) [57]. Hirshfeld Atom Refinement (HAR) was carried out using the IAM model as a starting point. The wave function was calculated using ORCA 4.2.0 software [58] with the basis set jorge-TZP [59] and hybrid exchange–correlation functional PBE0 [60]. The least-squares refinements of the HAR model were then carried out with olex2.refine (version 1.5) [61], while keeping the same constraints and restraints as those used for the ShelXL refinement. The NoSpherA2 implementation [62] of HAR is used for tailor-made aspherical atomic factors calculated on-the-fly from a Hirshfeld-partitioned electron density. For the HAR approach, all H atoms were refined isotropically and independently. All calculations and structure drawings were performed in the OLEX2 package [63]. Ortep-like representations of the independent part of the crystal structures of **1–5** are shown in Supplementary Figures S3–S7. The crystal data and parameters of structure refinement are listed in Table 9.

Table 9. Crystallographic data for compounds **1–5**.

	1	2	3	4	5
Chemical formula	$C_{48}H_{50}CuF_6N_6O_8$	$C_{48}H_{44}CuF_6N_8O_6$	$C_{48}H_{54}Cl_2CuN_6O_8$	$C_{46}H_{48}Cl_2CuN_8O_6$	$C_{50}H_{60}CuN_6O_8$
M_r	1016.504	982.448	977.450	943.395	936.615
Crystal system	Monoclinic	Triclinic	Monoclinic	Triclinic	Monoclinic
Space group	$P2_1/c$	P–1	$P2_1/n$	P–1	$P2_1/n$
T/K	100(1)	100(1)	100(1)	100(1)	100(1)
a/Å	7.2566(2)	7.9291(3)	7.9338(1)	7.7476(4)	7.8787(1)
b/Å	37.7315(9)	12.155(5)	10.2793(2)	11.6292(6)	10.3607(3)
c/Å	8.4952(3)	12.7150(5)	28.5066(5)	12.4652(7)	28.6114(5)
α/°	90	76.947(3)	90	82.283(4)	90
β/°	103.354(4)	72.828(3)	94.659(1)	74.404(4)	94.888(1)
γ/°	90	70.520(3)	90	76.633(4)	90
V/Å^3	2263.11(12)	1092.7(5)	2317.14(7)	1049.32(10)	2327.02(8)
Z	2	1	2	1	2
λ/Å	1.54186	1.54186	1.54186	1.54186	1.54186
Abs. correction	Multi-scan, LANA	Multi-scan, LANA	Multi-scan, LANA	Multi-scan, LANA	Multi-scan, LANA
μ/mm^{-1}	1.438	1.448	2.224	2.414	1.158
Crystal size/mm	0.35 × 0.15 × 0.12	0.36 × 0.09 × 0.09	0.35 × 0.32 × 0.25	0.25 × 0.21 × 0.05	0.32 × 0.25 × 0.18
ρ_{calc}/g.cm^{-3}	1.492	1.492	1.401	1.493	1.337
S	1.015	1.096	1.064	1.047	1.073
R_1 [$I > 2\sigma(I)$]	0.0283	0.0178	0.0191	0.0237	0.0240
wR_2 [all data]	0.0739	0.0372	0.0486	0.0581	0.0556
$\Delta\rangle_{max}$, $\Delta\rangle_{min}$/e Å^{-3}	0.74, −0.30	0.40, −0.19	0.18, −0.38	0.25, −0.38	0.34, −0.30
CCDC	2202673	2202674	2202675	2202676	2202677

The trifluoromethyl group of the niflumate ligand in the crystal structure of **2** was disordered in three parts (Supplementary Figure S4), represented by atoms with occupation factors of 0.51(2) (green lines), 0.29(2) (orange lines) and 0.20(2) (violet lines). The occupation factors were specified using the SUMP instruction. HAR refinement was carried out using restraints C–F and F···F distances using SADI instructions. All fluorine atoms were refined anisotropically with RIGU instruction restraints. Isostructural complexes **3** and **5** contained similarly disordered tolfenamate (**3**) (Supplementary Figure S5) or mefenamate (**5**) (Supplementary Figure S7) ligands in two positions (green lines for main parts, and violet lines for minor parts) with a ratio of occupancy factors of 0.850(1)/0.150(1) for **3** and 0.710(1)/0.290(1) for **5**. The disordered parts of the tolfenamate or mefenamate ligands of both compounds were modeled and refined with constraints and restraints using the SAME instructions, supplemented with SADI/DFIX instructions for C–H distances and RIGU and EADP instructions for non-hydrogen atoms.

3.4. Hirshfeld Surface Analysis

Crystal Explorer [64] was used to calculate the Hirshfeld surfaces [65] and associated fingerprint plots [39,66]. The Hirshfeld surface for complex **2** was calculated including all three orientations of the disordered –CF_3 group with their partial occupancies. The Hirshfeld surfaces of strongly disordered complexes **3** and **5** were calculated separately for the main and minor disordered components.

3.5. Electrochemical Study

The redox behavior of the studied complexes (10^{-4} M for all substances) was determined via a cyclic voltammetry study using an argentochloride reference electrode, platinum counter electrode and boron-doped diamond (BDD) working electrode (diameter of 3 mm, boron doping level of 1000 ppm, Windsor Scientific Ltd., UK). All voltammetric curves were registered at the potential range from –1.0 to +1.5 V using a scan rate of 100 mV/s.

3.6. Interactions with ct-DNA

3.6.1. Absorption Titrations

Interactions of the prepared complexes **1–4** with DNA were studied with UV-Vis monitored absorption titrations. In this experiment, a DNA stock solution was prepared by dissolving 6 mg of DNA in 5 mL of citrate buffer (containing 15 mM of sodium citrate and 150 mM of NaCl at pH = 7.0). The concentration of DNA was then determined with UV-Vis spectroscopy, where 0.150 mL of stock solution of DNA was added to 2.85 mL of citrate buffer, using a molar absorption coefficient of DNA at 260 nm (6 600 $M^{-1}cm^{-1}$) [67]. The ratio of absorbance at 260 nm and at 280 nm indicated that the DNA was sufficiently pure from proteins [68]. Increasing concentrations of DNA were then added to a buffer/DMSO solution (<1% DMSO) of the corresponding complex. The magnitudes of the binding strength of the complex–DNA interactions expressed as the intrinsic binding constant K_b were calculated with the ratio of slope to the intercept in the plots [DNA]/(ε_A-ε_f) versus [DNA], according to the Wolfe–Shimmer equation (Equation (1)):

$$\frac{[DNA]}{\left(\varepsilon_a - \varepsilon_f\right)} = \frac{[DNA]}{\left(\varepsilon_b - \varepsilon_f\right)} + \frac{1}{K_b\left(\varepsilon_b - \varepsilon_f\right)} \tag{1}$$

where the meaning of all symbols can be found elsewhere [48]. Control measurements with DMSO were performed, and no changes in the spectra of DNA were observed.

3.6.2. Fluorescence Quenching of EB-DNA Adduct

The ability of the prepared complexes to displace the standard intercalator ethidium bromide (EB) from the EB-DNA adduct was investigated with fluorescence spectroscopy. The EB-DNA adduct was prepared by adding 20 μM EB and 54 μM DNA in a buffer (15 mM of sodium citrate and 150 mM of NaCl at pH = 7.0). The possible intercalating effect of the compounds was studied by adding increasing concentrations of a corresponding complex into a solution of the DNA-EB complex. The fluorescence emission spectra were recorded in the range of 550–800 nm with an excitation wavelength of 515 nm. A decrease in the intensity of the EB-DNA emission band at 615 nm was monitored for complexes **1–4** and were correlated with the same measurement with DMSO. The quenching of the EB-DNA emission band by compounds **1–4** and DMSO was calculated via the Stern–Volmer equation (Equation (2))

$$I_0/I = 1 + k_q\tau_0\,[Q] = 1 + K_{SV}\,[Q] \tag{2}$$

where k_q ($M^{-1}\,s^{-1}$) is the quenching constant of complexes **1–4**, K_{SV} (M^{-1}) is the value of the dynamic quenching constant, τ_0 is the average lifetime of the EB-DNA adduct in the absence of the quencher (23×10^{-9} s), and [Q] is the concentration of the quencher. I_0 is the initial fluorescence intensity of the EB-DNA adduct, and I is the fluorescence intensity of the EB-DNA adduct after the addition of the complexes. K_{SV} can be obtained from the slope of the I_0/I versus [Q] plot [21].

3.6.3. Viscosimetric Studies

Changes in the viscosity of the DNA solution (0.1 mM) were measured in the presence of increasing concentrations of the compounds in a buffer solution (15 mM of sodium citrate and 150 mM of NaCl at pH = 7.0) at a constant temperature 25 °C. The measurements were carried out using an ALPHA L Fungilab rotational viscometer equipped with an 18 mL ELVAS spindle, and the measurements were performed at 60 rpm. The relation between the relative solution viscosity (η/η_0) and DNA length (I/I_0) is given by Equation (3), where η and η_0 are the viscosities of DNA in the presence and absence of the studied complex.

$$\left(\frac{\eta}{\eta_0}\right)^{1/3} = \frac{I}{I_0} \tag{3}$$

3.7. SOD Mimetic Activity

The ability of the copper complexes to scavenge superoxide radical anions was determined using the NBT (Nitro-Blue Tetrazolium) indirect colorimetric test, in which the xanthine/xanthine oxidase (X/XO) system was used as a superoxide-generating system [27]. The extent of NBT reduction was monitored spectrophotometrically by measuring the absorbance at 560 nm for 5 min. The reaction mixture contained 0.2 mM of xanthine and 0.6 mM of NBT in 0.1 mM of a sodium phosphate buffer at a pH of 7.8 and at 25 °C with a volume of 3 mL. The tested compounds were dissolved in DMSO. The concentration of xanthine-oxidase (XO) was experimentally designed to give an absorbance change (ΔA/min) between 0.035 and 0.045. Inhibitory concentrations were calculated from the slopes of individual curves with Equation (4):

$$IC = \frac{b_0 - b}{b_0} \tag{4}$$

where b_0 is the slope of the non-inhibited system, and b is the slope of the inhibited system with a corresponding complex concentration. IC_{50} values were obtained from the graphical dependence of the inhibitory concentration and the concentration of the complex [42]. An investigation of the formation of Cu(I) ions after reduction with the superoxide radical anion was conducted via UV-Vis spectroscopy using a specific Cu(I) chelating agent, neocuproine (2,9-dimethyl-1,10-phenanthroline). Potassium superoxide (KO_2) was used as a source of the superoxide anion together with 18-crown-6-ether, which acted as a stabilizing agent. The absorption spectra were measured in the range of 400–800 nm after the addition of 500 µL of 1 mM of neocuproine and 500 µL of 1 mM of potassium superoxide DMSO solution to 500 µL of 1 mM of complex **1–4** DMSO solution using the Specord 250 plus UV/Vis spectrometer [27].

3.8. Bovine Serum Albumin (BSA) Binding Studies

The albumin binding studies for complexes **1–4** were performed with tryptophan fluorescence emission quenching experiments using BSA (30 µM) in a buffer solution (containing 15 mM of trisodium citrate and 150 mM of NaCl at a pH of 7.0). The quenching of the emission intensity of the tryptophan residues of BSA at 336 nm was monitored using increasing concentrations of complexes **1–4** as quenchers [21]. The fluorescence emission spectra were recorded in the range of 300–420 nm with an excitation wavelength of 280 nm. The values of the Stern–Volmer constant K_{SV} (in M^{-1}), the BSA quenching constant k_q (in M^{-1} s^{-1}) and the BSA binding constant K_{BSA} (in M^{-1}) for the interaction of the compounds with BSA were derived with the Stern–Volmer (Equation (2)) and Scatchard equations (Equation (5)).

$$\frac{\Delta I/I_0}{[Q]} = nK - K\frac{\Delta I}{I_0} \tag{5}$$

where K (in M^{-1}) is the value of the bovine serum albumin constant K_{BSA}, n is the number of binding sites per albumin, and [Q] is the concentration of the quencher. I_0 is the initial fluorescence intensity of the tryptophan residues of albumin, and I is the fluorescence intensity of albumin after the addition of the complexes. K_{BSA} can be obtained from the slope of the $\Delta I/I_0/[Q]$ versus $\Delta I/I_0$, and n can be calculated from the intercept [29].

3.9. Anticancer Studies

3.9.1. Cell Culture

Human lung cancer cells (A549), human breast cancer cells (MCF-7) and human glioblastoma cells (U-118MG) were purchased from the American Type Culture Collection (Manassas, VA, USA) and were maintained in Dulbecco's Modified Eagle Medium (DMEM, Life Technologies, Inc., Rockville, MD, USA) containing 10% fetal bovine serum, 100 μg/mL of streptomycin and 100 U/mL of penicillin G at 37 °C in a humidified atmosphere of 5% CO_2/95% air. Human lung fibroblasts (MRC-5) (ECACC, Salisbury, UK) were cultured in MEM containing 10% fetal bovine serum, 1% non-essential amino acids and 1% L-glutamine and penicillin–streptomycin mixture at 37 °C in a humidified atmosphere of 5% CO_2/95% air. For our experiments, cells were seeded on culture dishes or plates in the amounts described below. Cells at passage numbers 10–13 were used.

3.9.2. Cytotoxic Analysis

We determined the cytotoxic effects of four copper complexes **1**–**4** on carcinoma cells and healthy cells by using the MTT [3-(4,5-dimethylthiazol-2-yl)-2,5-diphenyltetrazolium bromide] colorimetric technique [69]. Cells were seeded (8×10^3 cells/200 μL well) in individual wells of 96-multiwell plates. We added different concentrations of copper complexes (20–1000 μmol/L) to the cells and incubated them for 24, 48 and 72 h at 37 °C (humidified atmosphere of 5% CO_2/95% air). After 72 h, the cells were treated with the MTT solution (5 mg/mL) in PBS (phosphate-buffered saline) (20 μL) for 4 h. The dark crystals of formazan, formed in intact cells, were dissolved in DMSO (dimethyl sulfoxide) (200 μL). The plates were shaken for 15 min, and the optical density was determined at 490 nm using a MicroPlate Reader (Biotek, Winooski, VT, USA). All dye exclusion tests were performed three times.

3.9.3. Genotoxic Analysis

We determined DNA strand breaks in MCF-7 cells after 72 h of incubation with complex **4** at an IC_{50} value. DNA strand breaks were measured using the alkaline comet assay [70]. Cells were resuspended in 400 μL of 0.8% low-melting-point agarose in PBS at 37 °C and pipetted onto a frosted microscope slide precoated with 100 μL of 1% normal-melting-point agarose. Slides with layers of cells in agarose were incubated in a refrigerator for 10 min (4 °C) and then immersed in a lysis solution (2.5 mol/L NaCl, 100 mmol/L, Na_2EDTA, 10 mmol/l Tris, 1% Triton, pH of 10) for 1 h to remove cell membranes. After lysis, slides were placed in a horizontal electrophoresis tank containing an electrophoresis solution (1 mmol/L Na_2EDTA, 300 mmol/L NaOH, pH of 13) at 4 °C for 40 min (DNA uncoiling). Electrophoresis measurements were performed in the same solution at 25 V, 300 mA and 4 °C for 30 min. The slides were washed three times for 5 min at 4 °C with a neutralizing buffer (0.4 mmol/L Tris, pH of 7.5) before staining with 20 μL of 4′,6-diamidine-2-phenylindole dihydrochloride (DAPI, 1 μg/mL solution in distilled water). Comets were viewed with fluorescence microscopy after staining with DAPI.

3.9.4. Statistical Analysis

The results obtained from the comet assay are shown as the arithmetic means ± the standard deviation (SD). The significance of differences between values acquired with the comet assay was evaluated with Student's t-test to determine if the values were statistically different from those of the control: * $p < 0.05$.

4. Conclusions

In this report, we discuss the synthesis, structural and spectroscopic characterization and biological activity of five copper (II) complexes with *N*, *N*-dietlylnicotinamide and fenamate ligands. The following complexes were prepared: $[Cu(fluf)_2(dena)_2(H_2O)_2]$ (**1**), $[Cu(nifl)_2(dena)_2]$ (**2**), $[Cu(tolf)_2(dena)_2(H_2O)_2]$ (**3**), $[Cu(clon)_2(dena)_2]$ (**4**) and $[Cu(mef)_2(dena)_2(H_2O)_2]$ (**5**). The complexes were characterized in terms of their elemental composition, structure, physico-chemical and biological properties. The crystal structures of the studied compounds were refined using a more accurate aspherical HAR method using data measured with a high redundancy at 100 K. The crystal structures revealed the different influences of benzene versus the pyridine ring on the possibility of the coplanarity of fenamate anions and thus also the possibility of forming hydrogen bonds and/or π–π stacking interactions. The studied complexes are monomeric, forming a distorted tetragonal bipyramidal stereochemistry around a central copper ion. Complex **1** and the isostructural complexes **3** and **5** crystallize in a monoclinic system with a $P2_1/c$ (**1**) or $P2_1/n$ (**3**,**5**) space group, while the nearly isostructural complexes **2** and **4** crystallize in a triclinic system with a *P*-1 space group. The fenamate ligands are coordinated to a copper atom either monodentately (**1**, **3**, **5**) or asymmetrical chelating bidentately (**2**, **4**). The complex molecules of **1–5** are connected in *1D* supramolecular chains by means of intermolecular hydrogen bonds and π–π stacking interactions. In addition, Hirshfeld surface analysis was used to quantitatively identify the intermolecular interactions in the crystal structures of all five compounds.

The EPR spectra of solid-state and frozen DMSO solutions of **1–5** were monomeric with axial symmetry and with a relative ordering of axial *g* factors of $g_{\|} > g_{\perp} \sim g_e$, showing either resolved or unresolved copper parallel hyperfine interactions. The similarity between the solid and solution EPR data suggests a resemblance in the geometries of the studied complexes in accordance with the observed crystal structures.

The SOD mimetic activity of the complexes was studied indirectly using an NBT assay, and the complexes were characterized by means of IC_{50} (1.41–3.46 µM). The obtained inhibition concentrations showed that the complexes are good SOD mimetics, with the best results obtained for **2** and **4**. The cyclic voltammetry results confirmed the quasi-reversible nature of the redox processes on the studied complexes, with values of $E_{1/2}$ that are in agreement with the SOD mimetic activity of the complexes. The interactions of complexes **1–4** with neocuproine and KO_2 were, again, in agreement with the SOD data and support the hypothesis of the redox cycling mechanism between the studied copper complexes and superoxide.

The potential of complexes **1–4** to interact with DNA was also investigated. Absorption titration studies pointed to the intercalative binding of our complexes to DNA with a relatively strong binding constant of K_b (10^5), especially in cases of **4** and **1**. In the viscosity measurements, we observed a continual increase in the relative DNA viscosity for all four complexes, indicating a possible intercalation mechanism of interaction between the complexes and DNA. The intercalating ability of the complexes toward DNA was also studied with an ethidium-bromide-displacement-fluorescence-based method. The results revealed moderate intercalative ability toward DNA. In addition, the affinity of the complexes to interact with bovine serum albumin was studied, showing tight and reversible mutual interactions, as revealed by relatively high binding constants K_{BSA} (of order 10^5) and quenching constants k_q (of order 10^{13}) for all four complexes. In the case of comparing the structures and biological activity of structurally similar complexes **2**, **4** vs. **1**, **3**, no distinct trend was observed.

The cytotoxic activity of the studied complexes revealed that only complex **4** exhibited cytotoxic activity on the MCF-7 tumor cell line after 72 h of exposure with an IC_{50} value of 0.57×10^{-4} M.

Supplementary Materials: The following supporting information can be downloaded at https://www.mdpi.com/article/10.3390/inorganics11030108/s1: Figures S1–S20: IR spectra; UV-Vis spectra; crystal structures, 3D Hirshfeld surfaces, 2D fingerprints plots; Scatchard plots; Table S1: Hydrogen bond parameters. checkCIF and crystallographic data (excluding structure factors) for the structures reported in this paper were deposited into the Cambridge Crystallographic Data Centre as supplementary publications, nos. CCDC-2202673–2202677. Copies of the data can be obtained free of charge on application to the CCDC, 12 Union Road, Cambridge CB2 1EZ, UK [fax: (internat.) +44 1223/336033; e-mail: deposit@ccdc.cam.ac.uk].

Author Contributions: Conceptualization, J.Š. and M.P.; methodology, M.P., J.V. and J.Š.; investigation, M.P., M.S., K.K., Ľ.Š., J.M., M.V. and J.Š.; writing—original draft preparation, J.Š.; writing—review and editing, J.V., J.M. and M.V.; visualization, J.Š. and M.P.; supervision, J.Š. All authors have read and agreed to the published version of the manuscript.

Funding: Slovak grant agencies (VEGA 1/0482/20, VEGA 1/0159/20, VEGA 1/0686/23, APVV-19-0087, APVV-18-0016 and VEGA 1/0145/20) are acknowledged for their financial support.

Conflicts of Interest: The authors declare no conflict of interest. The funders had no role in the design of the study; in the collection, analyses, or interpretation of data; in the writing of the manuscript; or in the decision to publish the results.

References

1. Bindu, S.; Mazumder, S.; Bandyopadhyay, U. Non-steroidal anti-inflammatory drugs (NSAIDs) and organ damage: A current perspective. *Biochem. Pharmacol.* **2020**, *180*, 114147. [CrossRef] [PubMed]
2. Ramos-Inza, S.; Carolina, R.A.; Sanmartin, C.; Sharma, A.K.; Plano, D. NSAIDs: Old Acquaintance in the Pipeline for Cancer Treatment and Prevention-Structural Modulation, Mechanisms of Action, and Bright Future. *J. Med. Chem.* **2021**, *64*, 16380–16421. [CrossRef]
3. Weder, J.E.; Dillon, C.T.; Hambley, T.W.; Kennedy, B.J.; Lay, P.A.; Biffin, J.R.; Regtop, H.L.; Davies, N.M. Copper complexes of non-steroidal anti-inflammatory drugs: An opportunity yet to be realized. *Coord. Chem. Rev.* **2002**, *232*, 95–126. [CrossRef]
4. Matsui, H.; Shimokawa, O.; Kaneko, T.; Nagano, Y.; Rai, K.; Hyodo, L. The pathophysiology of non-steroidal anti-inflammatory drug (NSAID)-induced mucosal injuries in stomach and small intestine. *J. Clin. Biochem. Nutr.* **2011**, *48*, 107–111. [CrossRef] [PubMed]
5. Psomas, G. Copper(II) and zinc(II) coordination compounds of non-steroidal anti-inflammatory drugs: Structural features and antioxidant activity. *Coord. Chem. Rev.* **2020**, *412*, 213259. [CrossRef]
6. Sun, J.-F.; Xu, Y.-J.; Kong, X.-H.; Su, Y.; Wang, Z.-Y. Fenamates inhibit human sodium channel Nav1.7 and Nav1.8. *Neurosci. Lett.* **2019**, *696*, 67–73. [CrossRef] [PubMed]
7. Hill, J.; Zawia, N.H. Fenamates as Potential Therapeutics for Neurodegenerative Disorders. *Cells* **2021**, *10*, 702. [CrossRef] [PubMed]
8. Prasher, P.; Sharma, M. Medicinal chemistry of anthranilic acid derivatives: A mini review. *Drug Dev. Res.* **2021**, *82*, 945–958. [CrossRef]
9. Moncol, J.; Múdra, M.; Lönnecke, P.; Hewitt, M.; Valko, M.; Morris, H.; Švorec, J.; Melnik, M.; Mazúr, M.; Koman, M. Crystal structures and spectroscopic behavior of monomeric, dimeric and polymeric copper(II) chloroacetate adducts with isonicotinamide, *N*-methylnicotinamide and *N,N*-diethylnicotinamide. *Inorg. Chim. Acta* **2007**, *360*, 3213–3225. [CrossRef]
10. Choi, H.-E.; Choi, J.-H.; Lee, J.Y.; Kim, J.H.; Kim, J.H.; Lee, J.K.; Kim, G.I.; Park, Y.; Chi, Y.H.; Paik, S.H.; et al. Synthesis and evaluation of nicotinamide derivative as anti-angiogenic agents. *Bioorganic Med. Chem. Lett.* **2013**, *23*, 2083–2088. [CrossRef]
11. Peng, M.; Shi, L.; Ke, S. Nicotinamide-based diamides derivatives as potential cytotoxic agents: Synthesis and biological evaluation. *Chem. Cent. J.* **2017**, *11*, 109. [CrossRef]
12. Banti, C.N.; Hadjikakou, S.K. Non-steroidal anti-inflammatory drugs (NSAIDs) in metal complexes and their effect at the cellular level. *Eur. J. Inorg. Chem.* **2016**, *2016*, 3048–3071. [CrossRef]
13. Khan, H.Y.; Parveen, S.; Yousuf, I.; Tabassum, S.; Arjmand, F. Metal complexes of NSAIDs as potent anti-tumor chemotherapeutics: Mechanistic insights into cytotoxic activity via multiple pathways primarily by inhibition of COX-1 and COX-2 enzymes. *Coord. Chem. Rev.* **2022**, *453*, 214316. [CrossRef]
14. Boodram, J.N.; Mcgregor, I.J.; Bruno, P.M.; Cressey, P.B.; Hemann, M.T.; Suntharalingam, K. Breast Cancer Stem Cell Potent Copper(II)-Non-Steroidal Anti-Inflammatory Drug Complexes. *Angew. Chem. Int. Ed.* **2016**, *55*, 2845–2850. [CrossRef]

15. Eskandari, A.; Boodram, J.N.; Cressey, P.B.; Lu, C.; Bruno, P.M.; Hemann, M.T.; Suntharalingam, K. The breast cancer stem cell potency of copper(II) complexes bearing nonsteroidal anti-inflammatory drugs and their encapsulation using polymeric nanoparticles. *Dalton Trans.* **2016**, *45*, 17867–17873. [CrossRef] [PubMed]
16. Johnson, A.; Iffland-Muhlhaus, L.; Northcote-Smith, J.; Singh, K.; Ortu, F.; Apfel, U.P.; Suntharalingam, K. A bioinspired redox-modulating copper(II)-macrocyclic complex bearing non-steroidal anti-inflammatory drugs with anti-cancer stem cell activity. *Dalton Trans.* **2022**, *51*, 5904–5912. [CrossRef] [PubMed]
17. Zehra, S.; Tabassum, S.; Arjmand, F. Biochemical pathways of copper complexes: Progress over the past 5 years. *Drug Discov. Today* **2021**, *26*, 1086–1096. [CrossRef]
18. Kumar, S.; Sharma, R.P.; Venugopalan, P.; Ferretti, V.; Tarpin, M.; Sayen, S.; Guillon, E. New copper(II) niflumate complexes with N-donor ligands: Synthesis, characterization and evaluation of anticancer potentional against human cell lines. *Inorg. Chim. Acta* **2019**, *488*, 260–268. [CrossRef]
19. Khan, H.Y.; Zehra, S.; Parveen, S.; Yousuf, I.; Tabassum, S.; Arjmand, F. New ionic Cu(II) and Co(II) DACH-flufenamate conjugate complexes: Spectroscopic characterization, single X-ray studies and cytotoxic activity on human cancer cell lines. *ChemistrySelect* **2018**, *3*, 12764–12772. [CrossRef]
20. Nnabuike, G.G.; Salunke-Gawali, S.; Patil, A.S.; Butcher, R.J.; Obaleye, J.A.; Ashtekar, H.; Prakash, B. Copper(II) complexes containing derivative of aminobenzoic acid and nitrogen-rich ligans: Synthesis, characterization and cytotoxic potential. *J. Mol. Struct.* **2023**, *1279*, 135002. [CrossRef]
21. Malis, G.; Geromichalou, E.; Geromichalos, G.D.; Hatzidimitriou, A.G.; Psomas, G. Copper(II) complexes with non-steroidal anti-inflammatory drugs: Structural characterization, in vitro and in silico biological profile. *J. Inorg. Biochem.* **2021**, *224*, 111563. [CrossRef] [PubMed]
22. Sharma, R.P.; Kumar, S.; Venugopalan, P.; Ferretti, V.; Tarushi, A.; Psomas, G.; Witwicki, M. New copper(II) complexes of anti-inflammatory drug mefenamic acid: A concerted study including synthesis, physicochemical characterization and their biological evaluation. *RSC Adv.* **2016**, *6*, 88546–88558. [CrossRef]
23. Tolia, C.; Papadopoulos, A.N.; Raptopoulou, C.P.; Psycharis, V.; Garino, C.; Salassa, L.; Psomas, G. Copper(II) interacting with non-steroidal antiinflammatory drug flufenamic acid: Structure, antioxidant activity and binding to DNA and albumins. *J. Inorg. Biochem.* **2013**, *123*, 53–65. [CrossRef]
24. Dimiza, F.; Fountoulaki, S.; Papadopoulos, A.N.; Kontogiorgis, C.A.; Tangoulis, V.; Raptopoulou, C.P.; Psycharis, V.; Terzis, A.; Kessissoglou, D.P.; Psomas, G. Non-steroidal antiinflammatory drug-copper(II) complexes: Structure and biological perpectives. *Dalton Trans.* **2011**, *40*, 8555–8568. [CrossRef]
25. Barmpa, A.; Perontsis, S.; Hatzidimitriou, A.G.; Psomas, G. Copper(II) complexes with meclofenamate ligands: Structure, interaction with DNA and albumins, antioxidant and anticholinergic activity. *J. Inorg. Biochem.* **2011**, *217*, 111357. [CrossRef]
26. Tarushi, A.; Perontsis, S.; Hatzidimitriou, A.G.; Papadopoulos, A.N.; Kessissoglou, D.P.; Psomas, G. Copper(II) complexes with the non-steroidal anti-inflammatory drug tolfenamic acid: Structure and biological features. *J. Inorg. Biochem.* **2015**, *149*, 68–79. [CrossRef] [PubMed]
27. Simunkova, M.; Lauro, P.; Jomova, K.; Hudecova, L.; Danko, M.; Alwasel, S.; Alhazza, I.M.; Rajcaniova, S.; Kozovska, Z.; Kucerova, L.; et al. Redox-cycling and intercalating properties of novel mixed copper(II) complexes with non-steroidal anti-inflammatory drugs tolfenamic, mefenamic and flufenamic acids and phenanthroline functionality: Structure, SOD-mimetic activity, interaction with albumin, DNA damage study and anticancer activity. *J. Inorg. Biochem.* **2019**, *194*, 97–113. [CrossRef]
28. Jozefíková, F.; Perontsis, S.; Koňáriková, K.; Švorc, L'.; Mazúr, M.; Psomas, G.; Moncol, J. In vitro biological activity of copper(II) complexes with NSAIDs and nicotinamide: Characterization, DNA- and BSA-interaction study and anticancer activity. *J. Inorg. Biochem.* **2022**, *228*, 111696. [CrossRef]
29. Jozefíková, F.; Perontsis, S.; Šimunková, M.; Barbieriková, Z.; Švorc, L'.; Valko, M.; Psomas, G.; Moncol, J. Novel copper(II) complexes with fenamates and isonicotinamide: Structure and properties, and interactions with DNA and serum albumin. *New J. Chem.* **2020**, *44*, 12827–12842. [CrossRef]
30. Melník, M.; Potočňak, I.; Macášková, L'.; Mikloš, D.; Holloway, C.E. Spectral study of copper(II) flufenamates: Crystal and molecular structure of bis(flufenamato)di(*N*,*N*-diethylnicotinamide)di(aqua)copper(II). *Polyhedron* **1996**, *15*, 2159–2164. [CrossRef]
31. Švorec, J.; Lörinc, Š.; Moncol, J.; Melník, M.; Koman, M. Structural and spectroscopic characterization of copper(II) tolfenamate complexes. *Transit. Met. Chem.* **2009**, *34*, 703–710. [CrossRef]
32. Melník, M.; Koman, M.; Macášková, L'.; Glowiak, T. Spectral and magnetic properties of copper(II) mefenamates: Crystal and molecular structure of bis(mefenamato)di(N,N-diethylnicotinamide)di(aqua)copper(II). *J. Coord. Chem.* **1998**, *44*, 163–172. [CrossRef]
33. Koman, M.; Melník, M.; Glowiak, T. Structure, spectral and magnetic behaviours of copper(niflumato)$_2$(N,N-diethylnicotinamide)$_2$. *J. Coord. Chem.* **1998**, *44*, 133–139. [CrossRef]
34. Nakamoto, K. *Infrared and Raman Spectra of Inorganic and Coordination Compounds, Part B*, 6th ed.; John Wiley and Sons, Inc.: Hoboken, NJ, USA, 2009; pp. 1–199.
35. Hathaway, B.J.; Billing, D.E. The electronic properties and stereochemistry of mono-nuclear complexes of the copper(II) ion. *Coord. Chem. Rev.* **1970**, *5*, 143. [CrossRef]

36. Moncol, J.; Mudra, M.; Lönnecke, P.; Koman, M.; Melník, M. Copper(II) halogenopropionates: Low-temperature crystal and molecular structure of bis(2,2-dichloropropionato)-di(methyl-3-pyridylcarbamate)copper(II) and bis(2-bromopropionato)- di(2-pyridylmethanol)copper(II). *J. Coord. Chem.* **2004**, *57*, 1065–1078. [CrossRef]
37. Korabik, M.; Repická, Z.; Martiška, L.; Moncol, J.; Švorec, J.; Padelkova, Z.; Lis, T.; Mazur, M.; Valigura, D. Hydrogen-Bond-Based Magnetic Exchange Between mu-Diethylnicotinamide(aqua)bis(X-salicylato)copper(II) Polymeric Chains. *Z. Anorg. Allg. Chem.* **2011**, *637*, 224–231. [CrossRef]
38. Spackman, M.A.; Jayalitaka, D. Hirshfeld surface analysis. *CrystEngComm* **2009**, *11*, 19–32. [CrossRef]
39. Spackman, M.A.; McKinnon, J.J. Fingerprinting intermolecular interactions in molecular crystals. *CrystEngComm* **2002**, *4*, 378–392. [CrossRef]
40. Sakaguchi, U.; Addison, A.W. Spectroscopic and redocx studies of some copper (II) ccomplexes with biomimetic donor atoms–Implication for protein copper centers. *J. Chem. Soc. Dalton Trans.* **1979**, 600–608. [CrossRef]
41. Puchonova, M.; Švorec, J.; Švorc, L.; Pavlik, J.; Mazur, M.; Dlhan, L.; Ruzickova, Z.; Moncol', J.; Valigura, D. Synthesis, spectral, magnetic properties, electrochemical evaluation and SOD mimetic activity of four mixed-ligand Cu(II) complexes. *Inorg. Chim. Acta* **2017**, *455*, 298–306. [CrossRef]
42. Kovala-Demertzi, D.; Galani, A.; Demertzis, M.A.; Skoulika, S.; Kotoglou, C. Binuclear copper(II) complexes of tolfenamic: Synthesis, crystal structure, spectroscopy and superoxide dismutase activity. *J. Inorg. Biochem.* **2004**, *98*, 358–364. [CrossRef] [PubMed]
43. Sheng, Y.; Abreu, I.A.; Cabelli, D.E.; Maroney, M.J.; Miller, A.; Teixeira, M.; Valentine, J.S. Superoxide Dismutases and Superoxide Reductases. *Chem. Rev.* **2014**, *114*, 3854–3918. [CrossRef] [PubMed]
44. Brezova, V.; Valko, M.; Breza, M.; Morris, H.; Telser, J.; Dvoranova, D.; Kaiserova, K.; Varecka, L.; Mazur, M.; Leibfritz, D. Role of radicals and singlet oxygen in photoactivated DNA cleavage by the anticancer drug camptothecin: An electron paramagnetic resonance study. *J. Phys. Chem. B* **2003**, *107*, 2415–2425. [CrossRef]
45. Santini, C.; Pellei, M.; Gandin, V.; Porchia, M.; Tisato, F.; Marzano, C. Advances in Copper Complexes as Anticancer Agents. *Chem. Rev.* **2014**, *114*, 815–862. [CrossRef]
46. Erxleben, A. Interactions of copper complexes with nucleic acids. *Coord. Chem. Rev.* **2018**, *360*, 92–121. [CrossRef]
47. Andrezalova, L.; Orszaghova, Z. Covalent and noncovalent interactions of coordination compounds with DNA: An overview. *J. Inorg. Biochem.* **2021**, *225*, 111624. [CrossRef]
48. Sirajuddin, M.; Ali, S.; Badshah, A. Drug-DNA interactions and their study by UV-Visible, fluorescence spectroscopies and cyclic voltametry. *J. Photochem. Photobiol. B Biol.* **2013**, *124*, 1–19. [CrossRef]
49. Smolkova, R.; Smolko, L.; Samol'ova, E.; Dusek, M. Co(II) fenamato, tolfenamato and niflumato complexes with neocuproine: Synthesis, crystal structure, spectral characterization and biological activity. *J. Mol. Struct.* **2023**, *1272*, 134172. [CrossRef]
50. Arkel, Y.S.; Schrogie, J.J.; Williams, R. Effect of clonixin and aspirin on platelet-aggregation in human volunteers. *J. Clin. Pharmacol.* **1976**, *16*, 30–33. [CrossRef]
51. Spector, D.; Krasnovskaya, O.; Pavlov, K.; Erofeev, A.; Gorelkin, P.; Beloglazkina, E.; Majouga, A. Pt(IV) Prodrugs with NSAIDs as Axial Ligands. *Int. J. Mol. Sci.* **2021**, *22*, 3817. [CrossRef]
52. Ozarowski Andrzej, Spin Software (SpinP.exe). Available online: https://nationalmaglab.org/user-facilities/emr/software/ (accessed on 5 January 2016).
53. Kožišková, J.; Hahn, F.; Richter, J.; Kožišek, J. Comparison of different absorption corrections on the model structure of tetrakis(μ_2-acetato)-diaqua-di-copper(II). *Acta Chim. Slovaca* **2016**, *9*, 136–146. [CrossRef]
54. Sheldrick, G.M. *SHELXT*—Integrated space-group and crystal-structure determination. *Acta Crystallogr. A* **2015**, *71*, 3–8. [CrossRef] [PubMed]
55. Palatinus, L.; Chapuis, G. *SUPERFLIP*—A computer program for the solution of crystal structures by charge flipping in arbitrary dimensions. *J. Appl. Crystallogr.* **2007**, *40*, 786–790. [CrossRef]
56. Burla, M.C.; Caliandro, R.; Carrozzini, B.; Cascarano, G.L.; Cuocci, C.; Giacovazzo, C.; Mallamo, M.; Mazzone, A.; Polidori, G.G. Crystal structure determination and refinement *via SIR2014*. *J. Appl. Crystallogr.* **2015**, *48*, 306–309. [CrossRef]
57. Sheldrick, G.M. Crystal structure refinement with *SHELXL*. *Acta Crystallogr. C* **2015**, *71*, 3–8. [CrossRef]
58. Neese, F. Software update the ORCA program system, version 4.0. *WIREs Comput. Mol. Sci.* **2018**, *8e*, 1327. [CrossRef]
59. Balabanov, N.B.; Peterson, K.A. Systematically convergent basis sets for transition metals. I. All-electron correlation consistent basis sets for the *3d* elements Sc-Zn. *J. Chem. Phys.* **2005**, *123*, 064107. [CrossRef]
60. Adamo, C.; Barone, V. Toward reliable density functional methods without adjustable parameters: PBE0 model. *J. Chem. Phys.* **1999**, *110*, 6158–6170. [CrossRef]
61. Bourhis, L.J.; Dolomanov, O.V.; Gildea, R.J.; Howard, J.A.K.; Puschmann, H. The anatomy of a comprehensive constrained, restrained refinement program for the modern computing environment—*Olex2* dissected. *Acta Crystallogr. A* **2015**, *71*, 59–75. [CrossRef]
62. Kleemiss, F.; Dolomanov, O.V.; Bodensteiner, M.; Peyerimhoff, N.; Midgley, L.; Bourhis, L.J.; Genoni, A.; Malaspina, L.A.; Jayatilaka, D.; Spencer, J.L.; et al. Accurate crystal structures and chemical properties from NoSpherA2. *Chem. Sci.* **2021**, *12*, 1675–1692. [CrossRef]
63. Dolomanov, O.V.; Bourhis, L.J.; Gildea, R.J.; Howard, J.A.K.; Puschmann, H. *OLEX2*: A complete structure solution, refinement and analysis program. *J. Appl. Crystallogr.* **2009**, *42*, 339–341. [CrossRef]

64. Spackman, P.R.; Turner, M.J.; McKinnon, J.J.; Wolff, S.K.; Grimwood, D.J.; Jayatilaka, D.; Spackman, M.A. *CrystalExplorer*: A program for Hirshfeld surface analysis, visualization and quantitative analysis of molecular crystals. *J. Appl. Crystallogr.* **2021**, *54*, 1006–1011. [CrossRef]
65. Hirshfeld, F.L. Bonded-atom fragments for describing molecular charge densities. *Theor. Chim. Acta* **1977**, *44*, 129–138. [CrossRef]
66. Parkin, A.; Barr, G.; Dong, W.; Gilmore, C.J.; Jayalitaka, D.; McKinnon, J.J.; Spackman, M.A.; Wilson, C.C. Comparing entire crystal structures: Structural genetic fingerprinting. *CrystEngComm* **2007**, *9*, 648–652. [CrossRef]
67. Wolfe, A.; Shimer, G.H., Jr.; Meehan, T. Polycyclic aromatic hydrocarbons physically intercalate into duplex regions of denatured DNA. *Biochemistry* **1987**, *26*, 6392–6396. [CrossRef] [PubMed]
68. Pyle, A.M.; Rehmann, J.P.; Meshoyrer, R.; Kumar, C.V.; Turro, N.J.; Barton, J.K. Mixed-ligand complexes of ruthenium(II)—Factors governing binding to DNA. *J. Am. Chem. Soc.* **1989**, *111*, 3051–3058. [CrossRef]
69. Carmichael, J.; Degraff, W.G.; Gazdar, A.F.; Minna, J.D.; Mitchell, J.B. Evaluation of a tetrazolium-based semiautomated coloimetric assay—Assesament of chemosensitivity testing. *Cancer Res.* **1987**, *47*, 936–942.
70. Collins, A.R.; Dobson, V.L.; Dusinska, M.; Kennedy, G.; Stetina, R. Comet assay: What can it really tell us? *Mutat. Res.* **1997**, *375*, 183–193. [CrossRef]

Article

Coordination Chemistry of Polynitriles, Part XII—Serendipitous Synthesis of the Octacyanofulvalenediide Dianion and Study of Its Coordination Chemistry with K^+ and Ag^+

Patrick Nimax, Yannick Kunzelmann and Karlheinz Sünkel *

Department Chemistry, Ludwig-Maximilians University Munich, Butenandtstr. 5-13, 81377 Munich, Germany
* Correspondence: suenk@cup.uni-muenchen.de

Abstract: The reaction of diazotetracyanocyclopentadiene with copper powder in the presence of NEt_4Cl yields, unexpectedly, besides the known $NEt_4[C_5H(CN)_4]$ (3), the NEt_4 salt of octacyanofulvalenediide $(NEt_4)_2[C_{10}(CN)_8]$ (**5**), which can be transformed via reaction with $AgNO_3$ to the corresponding Ag^+ salt (**4**), which in turn can be reacted with KCl to yield the corresponding K^+ salt **6**. The molecular and crystal structures of **4–6** could be determined, and show a significantly twisted aromatic dianion which uses all its nitrile groups for coordination to the metals; **4** and **6** form three-dimensional coordination polymers with fourfold coordinated Ag^+ and eightfold coordinated K^+ cations.

Keywords: diazotetracyanocyclopentadiene; octacyanofulvalenediide; radical chemistry; coordination polymers

Citation: Nimax, P.; Kunzelmann, Y.; Sünkel, K. Coordination Chemistry of Polynitriles, Part XII—Serendipitous Synthesis of the Octacyanofulvalenediide Dianion and Study of Its Coordination Chemistry with K⁺ and Ag⁺. *Inorganics* **2023**, *11*, 71. https://doi.org/10.3390/inorganics11020071

Academic Editor: Peter Segľa

Received: 14 January 2023
Revised: 29 January 2023
Accepted: 30 January 2023
Published: 2 February 2023

1. Introduction

Diazotetracyanocyclopentadiene (**1**) was first reported by Webster in 1965. Reactivity studies showed that it "is a diazonium rather than a diazo compound", and that the mechanism of its reactions "is likely free radical". The presumed tetracyanocyclopentadienyl radical intermediate could be "generated polarographically at 0.23 V vs. sce in acetonitrile or by mild reducing agents" [1]. Radical reactions are in general typical of arenedediazonium ions and allow "an easy entry into the chemistry of the aryl radical". For the generation of the radical, a reducing agent is needed, and Cu(I) appeared to be the most popular one. Typical reactions of the produced aryl radicals include H atom abstractions, formal halogen atom additions (so-called "Sandmeyer reactions") and homo- and hetero-aryl coupling reactions [2]. Although this type of chemistry is very old, it is still being studied intensively, and new synthetic applications are being developed [3,4]. In the chemistry of the diazotetracyanocyclopentadiene molecule, one important difference compared to the just mentioned arenediazonium ions is that this compound is neutral, and therefore one-electron reduction generates a radical anion instead of a neutral radical. A combined theoretical/electrochemical study showed that this radical anion can be protonated to give the $[C_5H(CN)_4]$ radical, which in turn can be reduced to the $[C_5H(CN)_4]$ anion, or vice versa [5]. Calculations showed that the neutral radicals $[C_5X(CN)_4]$ (X = H, CN) are strong oxidizers and can be termed "superhalogens" [6] or "hyperhalogens" [7], which is by definition an atom or atom group that has an electron affinity higher than a halogen atom. We could show recently that the radicals $[C_5X(CN)_4]$ can be isolated and structurally characterized, when X = NH_2 [8], while for X = NO_2 only electrochemical and EPR characterization was possible [9]. Treatment of the NH_2 or NO_2- substituted anions with Fe(III) generated Fe(II), and another group showed that when trying to prepare a Cu(II) complex of the $[C_5(CN)_5]$ anion, a "yet unknown oxidation of the $[C_5(CN)_5]$ ligand" occurred [10]. The fate of the supposed intermediate $[C_5X(CN)_4]$ radicals remained unclear. For halogen radicals, dimerization to obtain the dihalogen X_2 molecules is common textbook knowledge. Thus, a dimerization of "superhalogens" might be expected as well. For radical anions such as $[TCNQ]^{-}$, a σ-dimerization to give the $[TCNQ\text{-}TCNQ]^{2-}$ dianion has been reported [11], while for $[DDQ]^{-\cdot}$ a π dimer with formation of "pancake bonds" was observed [12,13].

Although there are many hints to the formation of radicals in the chemistry of the diazotetracyanocyclopentadiene molecule, one should not forget that many diazocyclopentadienes show the typical carbene chemistry. Thus, low-temperature photolysis of $[C_5Br_4N_2]$ in an argon matrix generates the tetrabromocyclopentadienylidene carbene, which reacts with CO to give a ketene. Dimerization to octabromofulvalene is not observed at this temperature [14]. Room-temperature laser flash photolysis of $[C_5Cl_4N_2]$ in several solvents, such as alcohols, pyridine or THF, produced ylidic products derived from a triplet carbene [15], while low-temperature irradiation in an argon matrix in the presence of CF_3I produced an ylidic iodonium ion [16]. Thermolysis of $[C_5Cl_4N_2]$ in the absence of solvent produced octachloronaphthalene, while treatment of hexane solutions of $[C_5X_4N_2]$ (X = Br, Cl) with bis(μ-chloro-π-allyl palladium) at 0 °C generated the octahalofulvalenes. Both product types were regarded as products of carbene intermediates [17]. Reaction of these tetrahalodiazocyclopentadienes with some Ni(0), Pt(0) and Ru(0) coordination compounds produced complexes of the type "$L_nM(N_2\text{-}C_5X_4)$", where the diazo compound is coordinated by one or both diazo-nitrogens [18]. A high-temperature–high-pressure study of compound **1** in hexafluoro-isopropanol solution, devoted to the study of possible dinitrogen exchange, lead to the conclusion that the dediazoniation of (**1**) leads to an $^1A_2$ singlet carbene [19]. Formation of a dimerization product such as octacyanofulvalene or octacyanonaphthalene was not reported. However, thermal decomposition of $Ag[C_5H(CN)_2(OMe)_2]$ produced the corresponding fulvalene $[C_{10}(CN)_4(OMe)_4]$, which was also shown to undergo a stepwise two-electron reduction to the corresponding fulvalene dianion [20,21].

During the course of our studies on the coordination chemistry of $[C_5X(CN)_4]$ ions, we wanted to also look at the bromo and chloro derivatives (X = Br, Cl). For this purpose, we had to prepare these anions first, and we decided to employ the original synthetic pathway described by Webster in 1966. Here, we describe our experiences with this literature procedure and the unexpected results we obtained.

2. Results

2.1. Synthesis

2.1.1. Reaction of Diazotetracyanocyclopentadiene with Chloride and Bromide

The thermal reaction of diazotetracyanocyclopentadiene (1) with $NEt_4^+\ Cl^-$ in the presence of copper powder or with $NEt_4^+\ Br^-$ without additives was reported by Webster to yield the halo-tetracyanocyclopentadienides 2a/b in high yields ([1], Scheme 1).

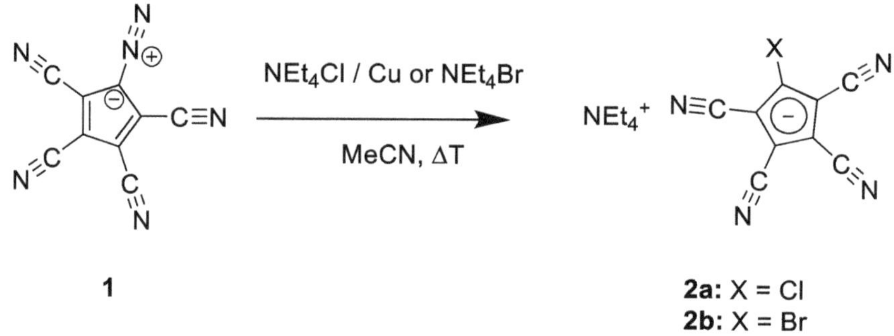

NEt₄Cl / Cu or NEt₄Br

MeCN, ΔT

1

2a: X = Cl
2b: X = Br

Scheme 1. Synthesis of the halo-tetracyanocyclopentadienides **2a/b** according to Webster.

When we tried to repeat the synthesis of **2a**, we obtained a product mixture. The ^{1}H NMR spectrum (see Figure S1 of the Supporting Information) showed, besides the two NEt_4 multiplets, one singlet at $\delta = 6.93$. The ^{13}C NMR spectrum (Figure S2) showed (besides the two NEt_4 carbon atoms) 10 signals, hinting to two different substances **A** and **B** of the type $[C_5X(CN)_4]$ (three signals for the ring carbons and two for the cyano groups each). Comparison with the NMR data of $Ag[C_5H(CN)_4]$, which we had prepared before on a different route [22], suggests that one of these compounds (**A**) is the tetraethylammonium salt of tetracyanocyclopentadienide, NEt_4^+ $[C_5H(CN)_4]^-$ (**3**). Our attempt to reproduce the synthesis of compound **2b** also yielded a product mixture. The ^{1}H- NMR spectrum (Figure S3) looked very similar to the one obtained with NEt_4Cl, suggesting that **3** had also been formed. This interpretation was supported by the ^{13}C NMR spectrum (Figure S4), which showed (besides the two NEt_4 signals) 15 signals, suggesting three different substances **A**, **B** and **C** all of the type $[C_5X(CN)_4]$. Two of the signal sets were identical to the ones obtained with NEt_4Cl, and therefore the formation of compound **3** can be assumed for this reaction as well. The identity of compound **B** remained unclear, as its formation in both reactions excluded the presence of a halide. Since recrystallization attempts did not turn out successful in both cases (but see Section 2.1.3), both products were treated with $AgNO_3$ in acetone.

From the original NEt_4Cl product, a mixture was obtained again which could be separated by column chromatography. A first fraction turned out to be the already known $Ag[C_5H(CN)_4]$ (Figures S5 and S6), while the second was obviously the silver analogue **B′** of the second product of the NEt_4 starting material because of the similarity with the ^{13}C NMR data (Figure S7). Recrystallization of this product from toluene/acetonitrile yielded X-ray quality crystals. The crystal structure determination identified the product **B′** as the complex $Ag_2[C_{10}(CN)_8]$ (**4**) which contains the so-far unknown octacyanofulvalenediide dianion. It is therefore very likely that the unidentified product **B** is the NEt_4 analogue **5** of the silver complex **4**.

2.1.2. Synthesis of $K_2[C_{10}(CN)_8]$ (6)

Salt metathesis of the silver complex **4** with KCl in MeOH yielded the corresponding potassium complex salt $K_2[C_{10}(CN)_8]$ (**6**) as a colorless solid. Its purity was confirmed by ^{1}H- and ^{13}C-NMR spectra (Figures S8 and S9). Recrystallization allowed the isolation of X-ray quality crystals. Our results are summarized in Scheme 2.

2.1.3. Reaction of Diazotetracyanocyclopentadiene with $[Ru(C_5H_5)Cl(PPh_3)_2]$

It was mentioned in the Introduction that treatment of $C_5X_4N_2$ (X = Br, Cl) with a palladium complex yielded the corresponding octahalofulvalenes. We reasoned that compound **1** might also react with catalytically active metals. Just by chance, we chose the ruthenium complex $[Ru(C_5H_5)Cl(PPh_3)_2]$ for our study. We studied the reaction of **1** both with stoichiometric or catalytic amounts of this ruthenium compound in MeCN as the solvent, using either reflux temperature (100 °C) or 0 °C. After addition of NEt_4Cl or NEt_4Br and chromatographic work-up, the main product was compound **3**. However, MS analysis (ESI or FAB) showed that octacyanofulvalenediide had also been formed, i.e., compound **5**. This was confirmed by crystal structure analysis.

2.2. *Cyclovoltammetry*

Cyclic voltammetry (CV) of crystalline **6** reveals one oxidation step at 1.25 V during oxidative scanning and a broad shoulder at 1.4 V in the reverse scan, which could be interpreted as partial reversibility similar to observations in several other tetracyanocyclopentadienides (Figure 1). Since **6** takes the form of a dianion in its initial stage, oxidation leads to a radical mono-anion. While the possibility of a two-electron transfer to a neutral compound (analogous to a fulvalene compound) cannot be excluded due to the broad shoulders of the voltagramm, it would not be supported by results from the preparative part of this work, which gave no indication of such a compound being present. The lack of a neutral fulvalene analogue in CV might indicate that the neutral form of this compound is highly unfavorable, while the radical and dianionic forms are better stabilized. This would fit the electron-deficient nature of the system, which in theory should favor the charged states.

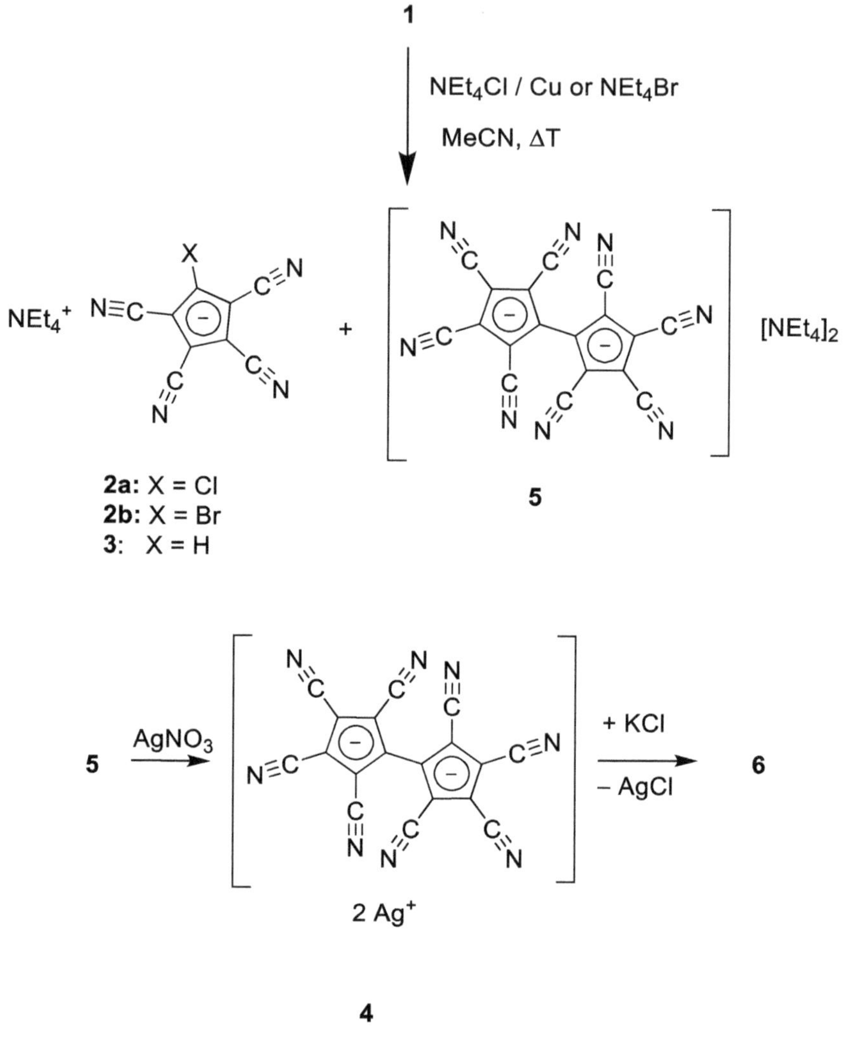

Scheme 2. Summary of the reactions described in Sections 2.1.1 and 2.1.2. The formation of **2a** was not observed in our study, but was reported in the literature.

2.3. EPR Spectroscopy

EPR measurements show signals in both the compounds **4** and **6** with g values of 2.001 and 2.002 when subjected to UV or sunlight, similar to measurements of other 1,2,3,4 tetracyanocyclopentadienide derivatives, indicating free radical formation by photoinduced electron transfer (Figure 2). While the signals of **6** are short-lived, **4** shows less degradation over time, a property also observed in other derivatives of this ligand class.

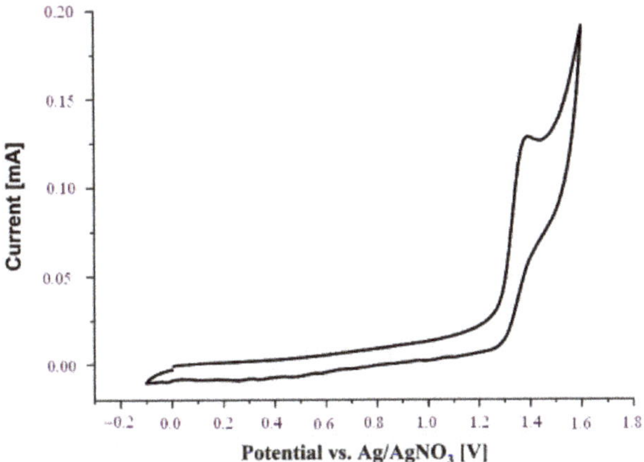

Figure 1. CV curve of crystalline **6** (0.1 mM) in MeCN at 25 °C (0.1M $[NBu_4][PF_6]$ supporting electrolyte; scan rate 100 mV s^{-1}; Pt-wire as working electrode, $Ag/AgNO_3$ as reference electrode.

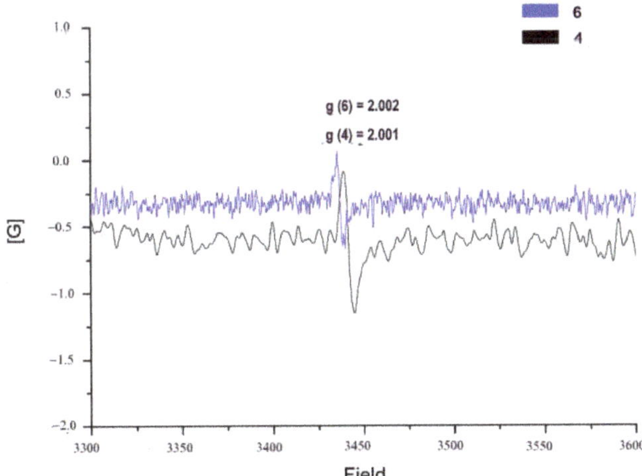

Figure 2. EPR spectra of compounds **4** and **6** after being subjected to sunlight.

2.4. Crystallography

2.4.1. Molecular and Crystal Structure of Compound 1

Compound **1** crystallizes in the orthorhombic space group *Pbca* with one molecule in the asymmetric unit. Figure 3 shows an ORTEP3 view of the molecular unit. Table 1 collects some important metric parameters of the molecule. Although there is still some variation in the C–C bond lengths of the ring (two "short" bond lengths of ca. 1.393(1) Å and three "long" ones with ca. 1.418(1) Å), they are definitely more delocalized than in the structure of diazo-tetraphenylcyclopentadiene [23]. In this compound, the "short" bonds are at 1.373(3) Å and the "long" bonds average at 1.450(3) Å. In addition, the N–N bond of this compound is with 1.121(3) Å, which is substantially longer, while the C–N_2 bond is significantly shorter (1.316(3) Å). Unfortunately, there are no more diazocyclopentadiene structures available for comparison. However, there are some DFT calculations on the parent compound, which yield C–C bond lengths of 1.369 and 1.447 Å and a C–N_2 bond length of 1.312 Å [24].

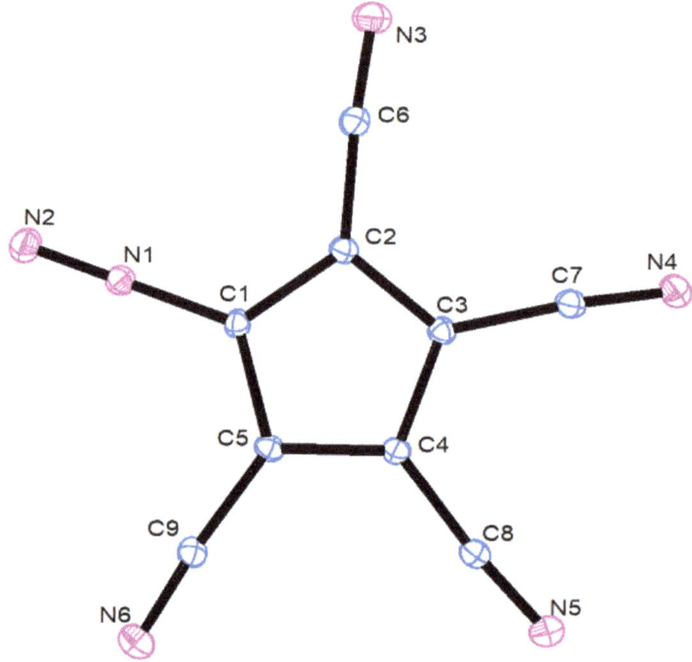

Figure 3. ORTEP3 view of the molecular unit of compound **1**.

Table 1. Important bond parameters of compound **1**.

Bond	Length [Å]/Angle [°]	Bond	Length [Å]/Angle [°]
N1–N2	1.1017(13)	C1–N1	1.3565(13)
C1–C2	1.4202(14)	C2–C3	1.3919(14)
C3–C4	1.4164(14)	C4–C5	1.3937(14)
C5–C1	1.4212(14)		
C_{cp}–C_{CN}	1.4225(15)–1.4252(15)	C_{CN}–N_{CN}	1.1475(15)–1.1502(15)
C1–N1–N2	179.7(1)	C_{cp}–C–N	174.3(1)–175.4 [1]

[1] C_{cp} are the carbon atoms of the ring, while C_{CN} are the carbon atoms of the nitrile groups.

When looking at intermolecular interactions, there are several short N. . . N contacts, particularly between atoms N1 and N5′ (1/2 +*x*, 1.5–*y*, 1–*z*) which are closer by 0.23 Å than the sum of their Van der Waals radii. Thus, an infinite 1D chain forms with the base vector [1 0 0] (Figure S10).

2.4.2. Crystal and Molecular Structure of Compound 4, Toluene Solvate

Compound **4** crystallizes as a bis-toluene solvate $Ag_2[C_{10}(CN)_8] \times 2(C_7H_8)$ in the monoclinic space group *C2/c* with half a molecule in the asymmetric unit. All crystals were twinned and were refined against a HKL5 dataset; in addition, the toluene molecules were disordered (50:50) across an inversion center. Two crystals were measured, one at 298 K and one at 100 K. Since the results of both refinements did not produce significant differences, here only the low temperature results are discussed. The results of the room-temperature determination are reported in the Supplementary Information.

Figure 4 shows an ORTEP3 view of the asymmetric unit (Figure S11 shows the corresponding plot of the r.t. determination), Figure 5 shows the coordination sphere of the unique silver atom and Figure 6 displays the anion with all coordinated silver ions.

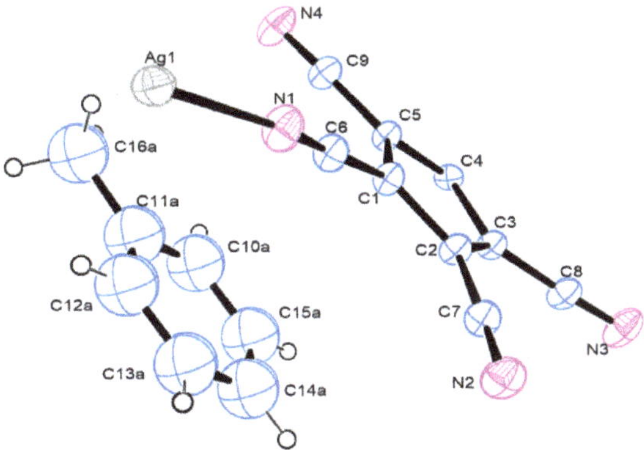

Figure 4. The asymmetric unit of compound **4** (only one orientation of the toluene solvent shown).

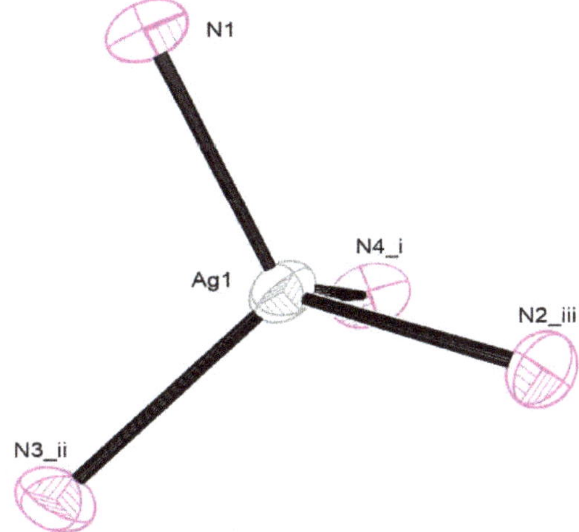

Figure 5. The coordination sphere of Ag1. Symmetry operators: i: 1/2–*x*, 1/2–*y*, 1–*z*; ii: *x*–1/2, 1.5–*y*, *z*–1/2; iii: 1/2–*x*, *y*–1/2, 1.5–*z*.

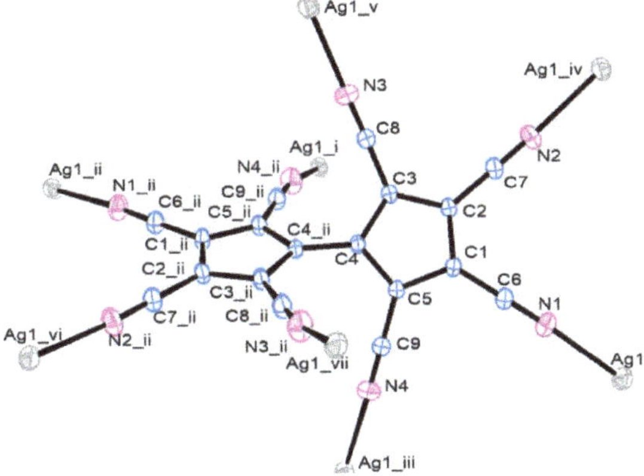

Figure 6. The octacyanofulvalenediide dianion with its eight coordinated Ag^+ ions.

The silver ion is tetrahedrally coordinated by four different octacyanofulvalenediide anions, while each dianion is coordinated to eight silver ions using all of its eight cyano nitrogen atoms. The two halves of the dianion are joined by a single bond between C4 and C4_ii (symm. op.: 1–*x*, *y*, 1.5–*z*) and are twisted by ca. 54° (which is an equivalent description of the torsion angle calculated as −126.1°). Other important bond parameters are collected in Table 2.

Table 2. Important bond parameters of compound **4** (low-temperature determination).

Bond	Length [Å]/Angle[°]	Bond	Length [Å]/Angle [°]
Ag1–N1	2.248(4)	Ag1–N2_iii	2.246(4)
Ag1–N3_ii	2.301(5)	Ag1–N4_i	2.309(5)
$(C–C)_{cp}$	1.399(6)–1.417(6)	C_{cp}–C_{CN}	1.412(6)–1.424(6)
(C–N)	1.134(7)–1.147(7)	C4–C4_ii	1.462(8)
C1–C6–N1	178.6(5)	C2–C7–N2	177.9(5)
C3–C8–N3	177.1(6)	C5–C9–N4	177.1(6)
Ag1–N1–C6	169.2(4)	Ag1–N2–C7	168.7(4)
Ag1–N3–C8	164.1(5)	Ag1–N5–C9	164.2(5)
C3–C4–C4_ii–C3_ii	−126.1(5)		

The symmetry operators given in the Ag–N distances refer to Figure 5, while the torsion angle refers to Figure 6.

Compound **4** forms a three-dimensional polymer structure, with base vectors [1 0 0], [0 1 0] and [0 0 1]. Figure 7 shows a packing plot watched along the crystallographic *b* axis, and Figure 8 a packing plot along *a*.

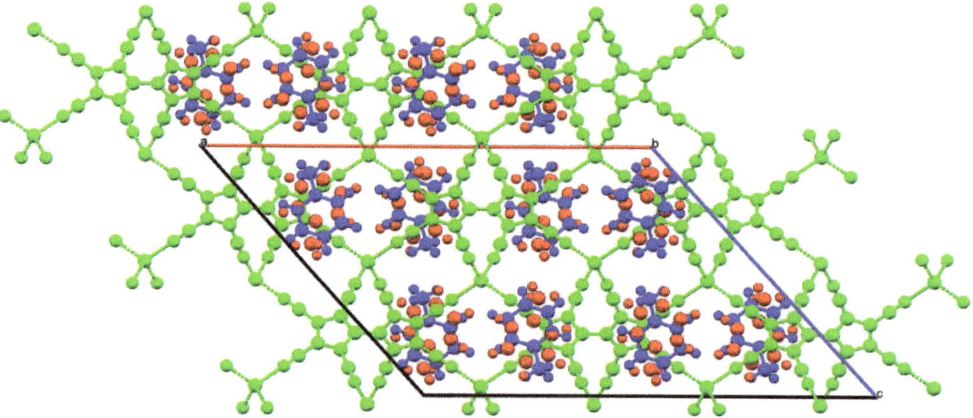

Figure 7. Packing plot (MERCURY) of compound **4**, watched along *b*. Color coding: green is the coordination polymer, while blue/red are the disordered toluene molecules.

Although we were not able to remove the toluene chemically, we just removed it virtually from the crystal structure. Its exclusion reveals a network of pores with a diameter of 5.8 Å at the bottleneck. The calculation of solvent accessible voids shows a volume of 1664 $Å^3$ per cell, corresponding to 56% of cell volume. The space takes the form of helical channels running parallel to the crystallographic *b* axis (Figure S12). At the same time, exclusion of the toluene guests from the packing diagrams shows that there exist a series of fused $[Ag_2(NC–(C_n)–CN)_2]$ ring systems, as they are quite often observed in the structures of silver polycyanocyclopentadienides. Figure S13 shows the common 14-membered rings (involving nitrile nitrogens N2 and N3) and two different 20-membered rings involving either nitrogens N1 and N3 or N2 and N4. In addition, there are helices winding down the *b* axis involving nitrogens N1 and N2. Ag... Ag distances across the rings are 7.023 and 8.799 Å, respectively, and across the helix the distance is 7.363 Å.

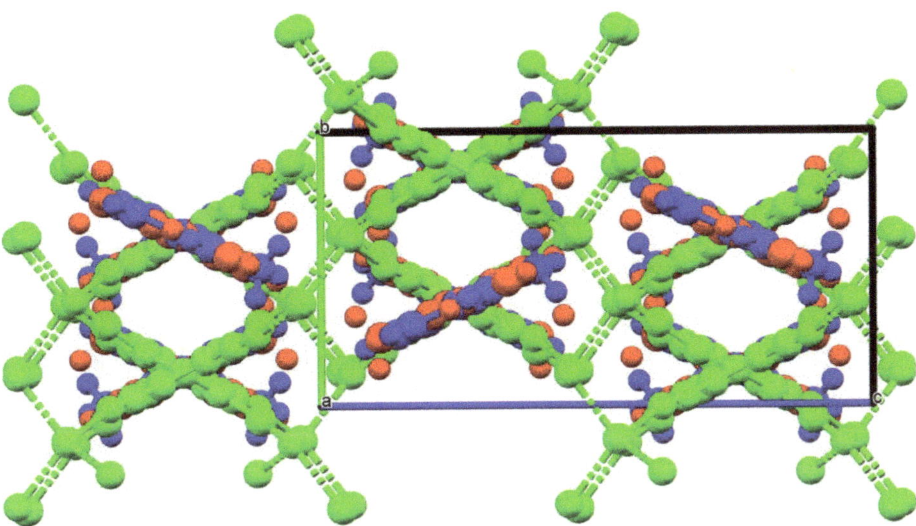

Figure 8. Packing plot (MERCURY) of compound **4**, watched along *a*. Same color coding as in Figure 7.

2.4.3. Crystal and Molecular Structure of Compound 5

Compound **5** crystallizes in the orthorhombic space group *Pbcn* with two independent half molecules in the unit cell. Figure 9 shows an ORTEP3 plot of the asymmetric unit.

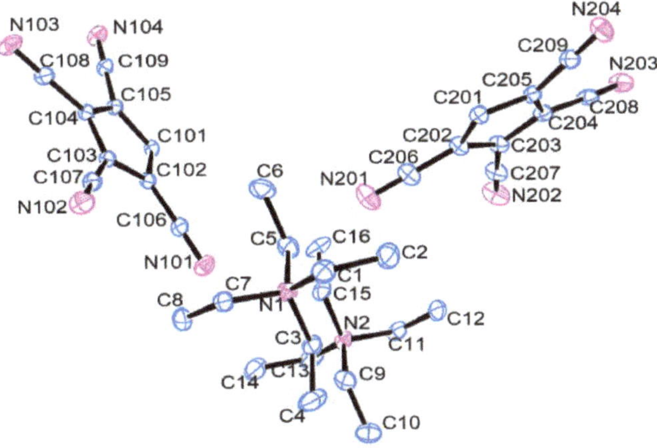

Figure 9. The asymmetric unit of compound **5**.

The half molecules are completed via single bonds between atoms C101–C101[i] and C201–C201[i], respectively, created by a twofold rotation axis through the middle of these bonds (symmetry operator: 1–*x*, *y*, 1.5–*z*). The complete octacyanofulvalenediide dianion of molecule A is shown in Figure 10. The two rings are twisted by ca. 54°. The bond parameters of the two independent molecules are very similar and are collected in Table 3.

The unit cell contains small cavities at the corners and the center, making up for approximately 1.1% of the volume. They are, however, too small to accommodate solvent molecules. As there is no metal ion in the structure, and no "classical" H-bond donor, all anions are isolated and separated from each other. Figure 11 shows a packing plot of this structure.

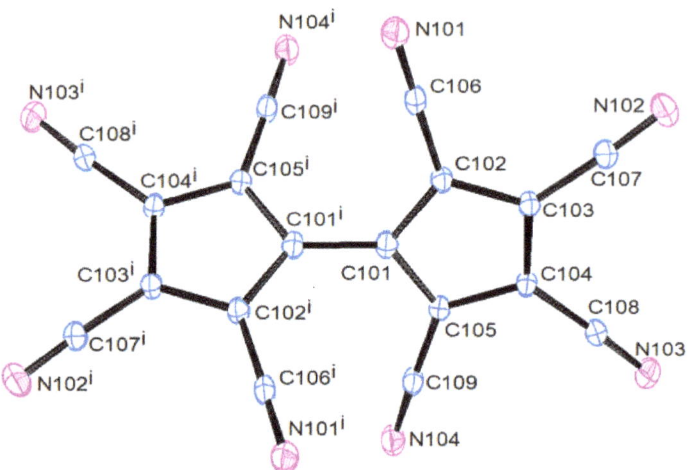

Figure 10. The complete octacyanofulvalenediide dianion of molecule A of compound **5**.

Table 3. Bond parameters of compound **5**.

Bond	Length [Å]/Angle [°]	
	Molecule A	Molecule B
$(C–C)_{cp}$	1.399(3)–1.418(3)	1.397(3)–1.428(3)
(C–N)	1.145(3)–1.152(3)	1.143(3)–1.151(3)
C_{cp}–C_{CN}	1.418(3)–1.428(3)	1.417(3)–1.429(3)
C*n*01–C*n*01^i	1.467(4)	1.472(4)
C*n*02–C*n*06–N*n*01	178.3(3)	179.5(3)
C*n*03–C*n*07–N*n*02	176.4(3)	177.7(3)
C*n*04–C*n*08–N*n*03	176.8(3)	177.5(3)
C*n*05–C*n*09–N*n*04	178.8(3)	179.2(3)
C*n*02–C*n*01–C*n*01^i–C*n*02^i	−125.9(3)	−54.9(3)

$(C–C)_{cp}$ are the C–C bond lengths within one cyclopentadienyl ring; C_{cp} and C_{CN} are the individual carbon atoms of the cyclopentadienyl ring and the attached carbon atoms of the nitrile groups; C*n*0x (n = 1 or 2, x = 1–9) are the corresponding C atoms of molecules A and B.

2.4.4. Crystal and Molecular Structure of $[K(H_2O)]_2[C_{10}(CN)_8]$, 6a

The aqua complex **6a** crystallizes in the orthorhombic space group *Ccca* with a quarter molecule in the asymmetric unit (Figure 12).

The potassium ions are coordinated by six nitrile nitrogens (from five different dianions) and two water oxygens (Figure 13). Pairs of K^+ ions related by [x, y, z] and [$-x-1/2$, $-y$, z] form zig-zag chains of face-fused (distorted) octahedra along the x direction, with metal–metal distances of 3.99 Å (Figure S14). The dianion is coordinated to ten K^+ ions, of which two are bridging two nitrile functions of the same dianion (Figure 14). Important bond parameters are collected in Table 4.

Compound **6a** forms a 3D polymer with base vectors [1 0 0], [0 1 0] and [0 0 1]. The unit cell contains ca. 3% solvent-accessible voids. The coordinated water molecules form weak hydrogen bonds to nitrile nitrogens N2. A packing plot is displayed in Figure 15.

Pairs of independent zig-zag chains of potassium ions (viewed in superposition in Figure 15, or displayed separately in Figure S15), as also shown for a single chain in Figure S14, run along the a axis at $y = 0$ and $y = 0.5$, and are bridged by octacyanofulvalenediides via nitrogens N1 in the b direction and N2 in the c direction (Figure S16). The closest approach of two K^+ ions in the b direction is 8.756 Å, while in the direction of the ab diagonal it is 7.601 Å.

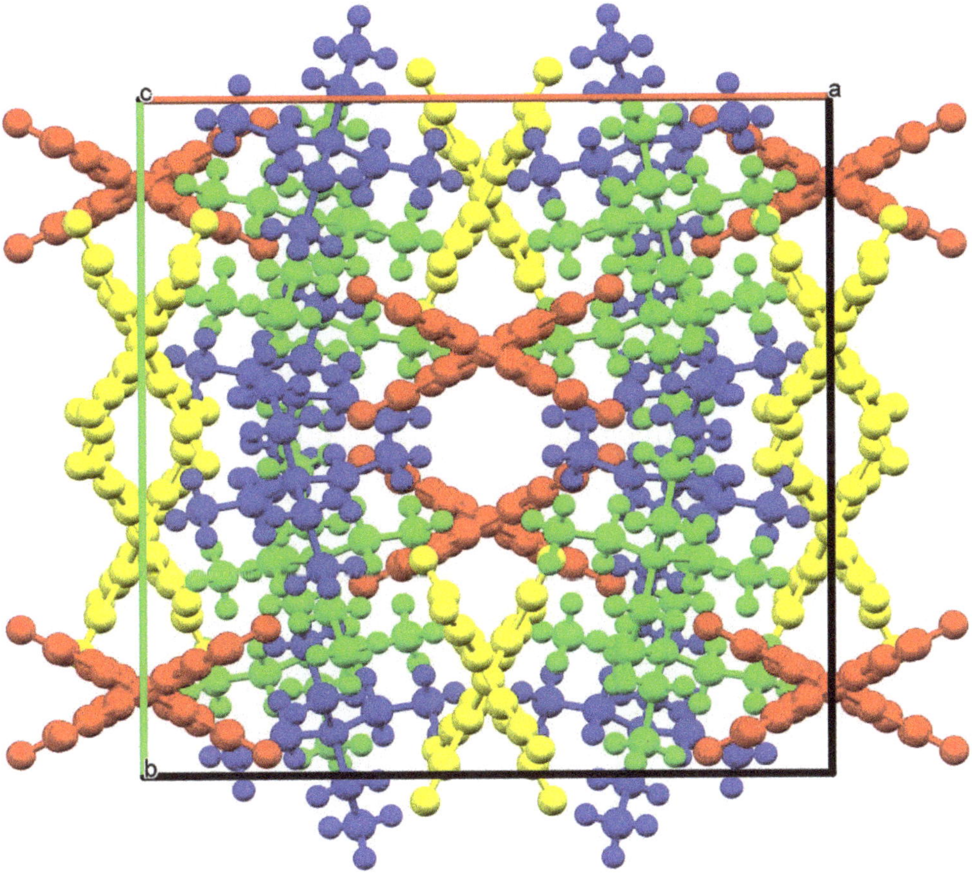

Figure 11. Packing plot (MERCURY) of compound **5**, watched along the crystallographic *c* axis. Color coding: blue and yellow are the NEt_4 cations, yellow and red the $[C_{10}(CN)_8]$ dianions.

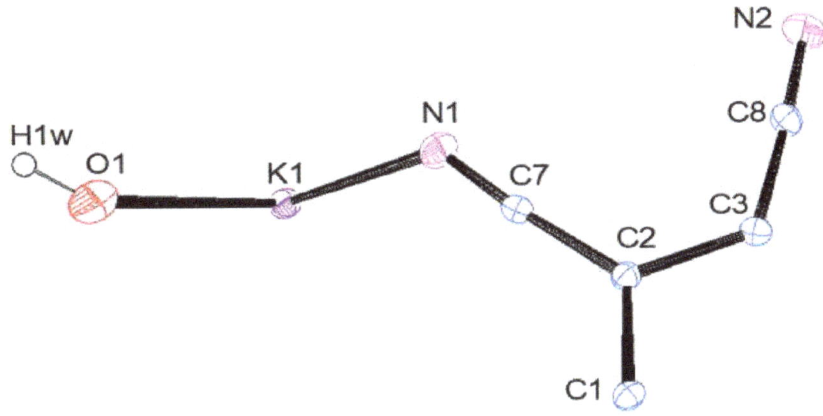

Figure 12. ORTEP3 view of the asymmetric unit of **6a**.

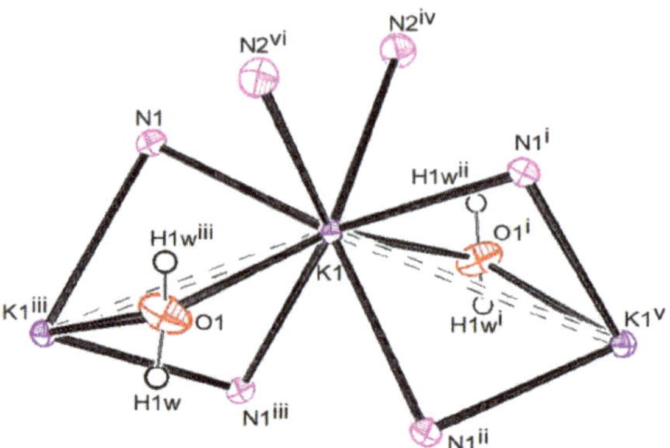

Figure 13. The coordination sphere of K^+ in compound **6a**. Symmetry operators: i: $-1-x, y, 1/2-z$; ii: $x-1/2-y, 1/2-z$; iii: $-x-1/2, -y, z$; iv: $x-1/2, y, -z$; v: $-x-1.5, -y, z$; vi $-x-1.5, y, 1/2-z$.

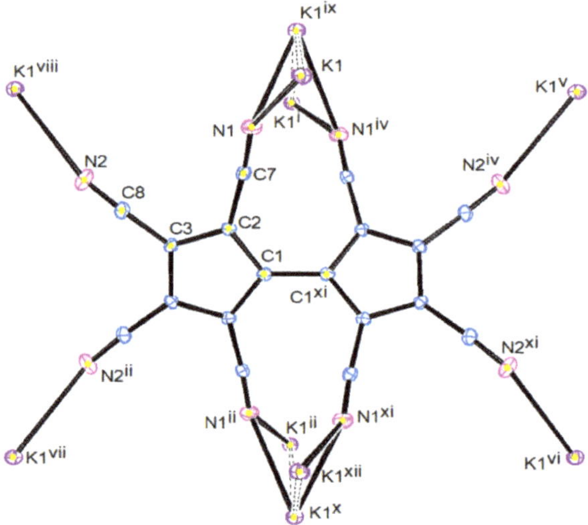

Figure 14. A complete octacyanofulvalenediide dianion with all coordinated K^+ ions. Symmetry operators: i: $1+x$, y, z; ii: $-x$, $-y-1/2$, z; iv: $-x, y, 1/2-z$; v: $1/2+x, y, 1-z$; vi: $-x-1/2, -y-1/2, 1-z$; vii: $-x-1/2, -y-1/2, -z$; viii: $1/2+x, y, -z$; ix: $-x-1/2, -y, z$; x: $1/2+x, y-1/2, z$; xi: $x, -y-1/2, z$; xii: $-1-x, -y-1/2, z$.

Table 4. Important bond parameters of compound **6a**.

Bond	Length [Å]/Angle[°]	Bond	Length [Å]/Angle [°]
$K1–N1/K1–N1^{i}$	2.839(2)	$K1–N2^{iv}/K1–N2^{vi}$	2.870(2)
K1–O1	2.874(2)	$K1–N1^{ii}/K1–N1^{iii}$	2.915(2)
$K1–K1^{iii}$	3.9929(6)		
$(C–C)_{cp}$	1.400(4)–1.418(3)	$C_{cp}–C_{CN}$	1.418(3)–1.426(3)
(C–N)	1.146(3)–1.147(3)	$C1–C1^{xi}$	1.477(6)
C2–C7–N1	175.3(2)	C3–C8–N2	176.8(3)
K1–N1–C7	132.3(2)	$K1–N1^{ii}–C7^{ii}$	131.6(2)
K1–N2–C8	150.0(2)		
$C2–C1–C1^{xi}–C2^{xi}$	−140.3(2)		

The symmetry operators in the distances and angles involving K1 refer to Figure 13, while the ones involving only the anion refer to Figure 14.

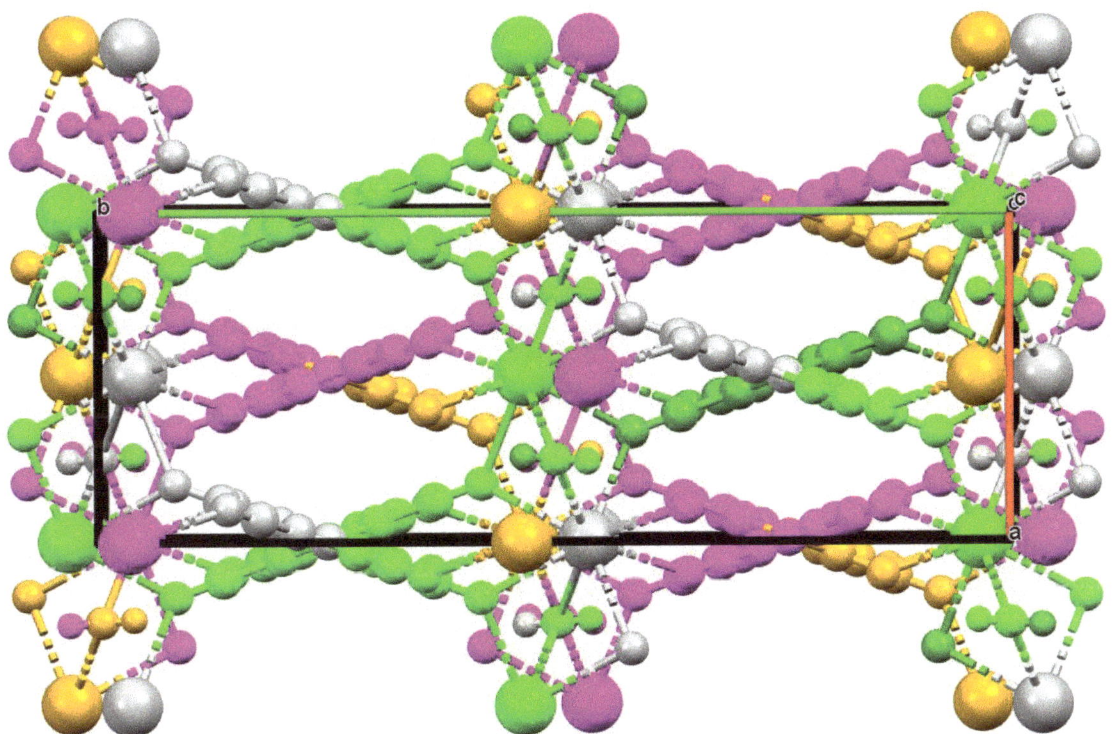

Figure 15. Packing plot of compound **6a**, watched along the crystallographic *c* axis. Color coding "by symmetry operation" (default settings of MERCURY).

3. Discussion

Diazotetracyanocyclopentadiene had been reported to behave as an aromatic diazonium ion, based on its reactivity [1] and its ^{13}C- and ^{15}N-NMR data [25]. Our crystal structure determination supports this view. Its bond length alternation parameter Δ_R is only 0.018 Å, much smaller than the 0.078 Å calculated for $C_5H_4N_2$, which was already regarded as small and as having a strong indication of aromaticity [24]. We also performed some DFT calculations (B3LYP/6-311G(dp) level) using the program *CrystalExplorer* [26]. Figures S17–S19 show HOMO/LUMO contours, an electron density visualization and two views of the electrostatic potential. In addition, these results support the high aromaticity of the cyclopentadienyl unit. With this background, the typical reactivity of arenediazonium ions towards metals and metal salts, i.e., radical reactions, should be expected for compound **1**. Unfortunately, the major reaction of these radicals is H abstraction to give the $[C_5(CN)_4H]$ anion. The usual main products of the Cu(I)-catalyzed coupling of aromatic diazonium ions, i.e., biaryls and azo compounds [27,28], were formed only in small yields or not at all. (It should be mentioned, however, that the original publication on the synthesis and reactivity of compound **1** reports the formation of an azo compound in the thermal reaction of **1** with CuCN in MeCN [1].) Although our study was not a true mechanistic one, we assume that the low yields of coupling products are the consequence of both steric and electronic repulsions: the two "ortho" cyano groups and the negative charges inhibit the mutual approach of the two radical anions, and thus the H abstractions dominate the reaction kinetics. The fact that rather high yields of coupling products can be obtained from the reaction of **1** with $[RuCp(PPh_3)_2Cl]$ remains rather mysterious. The only comparable reaction in the literature is the reaction of $[RuCp^*(P(OR)_3)_2Cl]$ with 1,1-diaryldiazomethanes, which was reported to give cationic N-coordinated Ru diazoalkane complexes [29] with no indication of any coupling products. This result is reminiscent of the N-coordinated $C_5Cl_4N_2$ complexes mentioned above [18]. Now, the only "explanation" might be that the aromaticity of compound **1** drives the reaction into another direction compared with "true" diazoalkanes and less aromatic diazocyclopentadienes.

The structure of compound **5** can be regarded as the structure of the "free" octacyanofulvalenediide anion because there are no coordinating cations or strong H-bond donors (there are, however, very weak interactions with aliphatic C–H bonds, but their influence on bond parameters can be neglected). The cyclopentadienyl rings are aromatic, as can be derived from the bond alteration parameter Δ_R, which amounts to 0.019 Å in molecule A and 0.031 Å in molecule B. In addition, the bonds to the nitrile carbon atoms have the same lengths (within 2σ) as the bonds within the ring. The central C–C bond between the two cyclopentadienyl halves is a typical single bond with ca. 1.47 Å. The two cyclopentadienyl rings are twisted by ca. 55°. In the Ag^+ complex **4** (low temperature), these bond parameters hardly change. Δ_R amounts to 0.018 Å and the bonds within the ring and between the ring and nitrile carbons are identical within 2σ. Both halves of the fulvalenediide anion are bound via a single bond and are twisted by ca. 54°. In the K^+ complex **6**, the bond alteration parameter also amounts to 0.018 Å. The exocyclic C–C bonds are slightly longer than the endocyclic ones. The single bond between the two cyclopentadienyl rings is ca. 1.48 Å long, with the two halves being twisted by ca. 40°. This significantly smaller twist angle is a consequence of the fact that the potassium ion bridges two of the "ortho" nitriles.

The different sizes of Ag^+ ("effective ionic radius" 1.00 Å for CN 4, [30]) and K^+ (1.51 Å for CN 8, [30]) lead to significant differences in the coordination spheres. While the Ag^+ ion is tetrahedrally coordinated by four nitrile nitrogens with bond distances ranging from 2.25–2.31 Å, the K^+ ion has coordination number 8, KN_6O_2 polyhedra, with bond distances between 2.84 and 2.92 Å. The Ag–N–C bonds deviate only slightly (11–16°) from linearity, but the K–N–C bonds can be regarded as "bent", particularly at atoms N1. Both structures might be compared with the structures of $[AgC_5(CN)_5]$ and $[KC_5(CN)_5]$, respectively [31,32]. These compounds show different coordination polyhedra dependent on the solvent the crystals came from. Crystals of the Ag^+ salt from EtOH show severely distorted AgN_4 tetrahedra, with Ag–N distances ranging from 2.236(5) to 2.378(6) Å, N–Ag–N angles between 88.0(2) and 129.7(2)°, and Ag–N–C angles between 149.7(6) and 169.9(5)°. The pentacyanocyclopentadienide anion uses only four of its nitrile groups for coordination. The compound forms an interwoven 3D network with very large pores, similar to the situation found in **4** [31]. $[KC_5(CN)_5]$, when grown from isopropanol/pentane, shows an octahedral metal ion with K–N bonds ranging from 2.777(1) to 2.9537(6) Å and a pentacyanocyclopentadienide using all five nitrile groups for coordination, with one bridging two K^+ ions in 4.975 Å distance, and K–N–C angles ranging from 122.1 to 138.0° [32]. Comparison of these compounds with our compounds **4** and **6** shows a wider range of metal-nitrogen distances in the $[C_5(CN)_5]$ salts and also a higher degree of bending in the M–N–C groups. Alternatively, one might use for comparison the structures of the tricyanomethanides $M[C(CN)_3]$. While the silver salt has a coordination number of three with Ag–N distances at 2.16 and 2.27 Å and Ag–N–C angles of 172.8° and 153.7° [33], the potassium salt has seven nitrile nitrogens coordinated at 2.86–2.98 Å and K–N–C angles ranging from 114 to 159° [34]. The $K[C(CN)_3]$ contains triply N-bridged K^+ ions with a K. . . K distance of 3.89 Å.

Octacyanofulvalenediide uses, in both compounds **4** and **6**, all of its eight nitrile groups for coordination; in compound **6**, four of them coordinate to two symmetry-related K^+ ions each, which is similar to the situation found in $[KC_5(CN)_5]$. Therefore, both compounds form three-dimensional coordination polymers, with the potassium structure being more "complicated".

4. Materials and Methods

4.1. Starting Materials and Instrumentation

Reagents (Cu, NEt_4Cl, NEt_4Br, $AgNO_3$, $[Ru(C_5H_5)Cl(PPh_3)_2]$) and solvents (MeCN, acetone, toluene) were obtained commercially in analytical grade and used as obtained, except for anhydrous MeCN, which was saturated with argon. Diazotetracyanocyclopentadiene (**1**) was prepared according to the literature [1].

CV measurements were performed with an Autolab potentiostat/galvanostat (PGSTAT302N) with a FRA32M module operated with Nova 1.11 software and a conventional three-electrode setup. Two platinum wires were used as the working and counter electrode, respectively. A Ag/0.01M $AgNO_3$ electrode was utilized as reference electrode.

4.2. Reaction of 1 with Cu and NEt_4Cl in MeCN

A suspension of powdered copper (0.57 g, 9.0 mmol) and NEt_4Cl (1.58 g, 9.5 mmol) in acetonitrile (30 mL) was heated to 90 °C. A solution of **1** (1.13 g, 5.9 mmol) in MeCN (19 mL) was added to the boiling mixture within one hour. After cooling down to room temperature and filtering, the solvent was removed in vacuo. A brown powder was obtained (2.18 g). ^{1}H- and ^{13}C{^{1}H}-NMR spectra (Figures S1 and S2) suggested the presence of two compounds: **A** (=**3**) and **B** (=**5**).

Compound **3**: ^{1}H NMR (400 MHz, DMSO-d^6): δ= 6.93 (s, $C_5\underline{H}$), 3.19 (q, $NC\underline{H}_2$), 1.15 (tt, $C\underline{H}_3$). ^{13}C-NMR (101 MHz, DMSO-d^6): δ = 123.9 ($\underline{C}$H), 116.0, 114.8 (2 × $\underline{C}$N), 99.4, 96.3 (2 × $\underline{C}$CN), 51.3 (t, $N\underline{C}H_2$), 7.0 (s, $\underline{C}H_3$). MS: FAB(+): m/z = 130.2 (NEt_4^+); FAB(−): m/z = 165.3 ($[C_9HN_4^-]$).

Compound **5**: ^{1}H NMR (400 MHz, DMSO-d^6): δ= 3.19 (q, $NC\underline{H}_2$), 1.15 (tt, $C\underline{H}_3$). ^{13}C-NMR (101 MHz, DMSO-d^6): δ = 127.5 ($\underline{C}$H), 115.1, 114.5 (2 × $\underline{C}$N), 99.4, 96.3 (2 × $\underline{C}$CN), 51.3 (t, $N\underline{C}H_2$), 7.0 (s, $\underline{C}H_3$). MS: FAB(+): m/z = 130.2 (NEt_4^+); FAB(−): m/z = 164.2 ($[C_{18}N_8^{2-}]$).

4.3. Reaction of 1 with NEt_4Br in MeCN

A solution of **1** (3.00 g, 15.6 mmol) and NEt_4Br (10.10 g, 48.1 mmol) in MeCN (160 mL) was heated at 50 °C for 30 min. Then, the solvent was evaporated in vacuo, and water (150 mL) was added and the resulting suspension was filtered. The residue on the filter was washed with water (50 mL) and dried in vacuo. A brown powder was obtained (3.06 g). ^{1}H- and ^{13}C{^{1}H}-NMR spectra (Figures S3 and S4) showed the formation of three compounds: **A** (=**3**), **B** (=**5**) and **C** (=**2b**).

Compound **2b**: ^{13}C-NMR (101 MHz, DMSO-d^6): δ = 114.9, 114.0 (2 × $\underline{C}$N), 107.2 ($\underline{C}$Br), 100.0, 98.4 (2 × $\underline{C}$CN), 51.3 (t, $N\underline{C}H_2$), 7.0 (s, $\underline{C}H_3$). MS: FAB(+): m/z = 130.2 (NEt_4^+); FAB(−): m/z = 243.1/ 245.1 ($[C_9BrN_4^-]$)

4.4. Reaction of the Crude Product from Section 4.2 with $AgNO_3$ in Acetone–Water

A solution of the product mixture obtained in 4.2. (1.53 g) in acetone (55 mL) was treated with a solution of $AgNO_3$ (2.69 g, 15.8 mmol) in water (15 mL). Stirring was continued at r.t. with exclusion of light for 4 days. After removal of the solvents in vacuo, the residue was taken up in the minimum amount of MeCN and placed on top of a silica gel column. MeCN: toluene 3:7 eluted a first fraction, while elution with MeCN: toluene 1:1 produced a second fraction. After evaporation to dryness, yellow-brown powders were obtained. Both fractions were examined by NMR spectroscopy. The first turned out to be $Ag[C_5H(CN)_4]$, Figures S5 and S6, while the second was identified as **B′** (=**4**) (Figure S7). Recrystallization of the second fraction from toluene/acetonitrile produced X-ray quality crystals.

Compound **4**: ^{13}C-NMR (101 MHz, DMSO-d^6): δ = 128.0 ($\underline{C}$–$\underline{C}$), 115.2, 114.6 (2 × $\underline{C}$N), 99.9, 96.9 (2 × $\underline{C}$CN), MS: MALDI(-): m/z = 329.1 ($[HC_{18}N_8^-]$). IR: [cm^{-1}]: ν = 2218 vs, 1465 m, 1437 m, 730 vs, 695 m (+ many weak and very weak absorptions between 1900 and 750); EA: calc. for $Ag_2C_{18}N_8$ × 1.68 toluene × MeCN: C 51.57, N 17.03 H 2.24; found: C 51.54, N 16.97 H 2.29.

4.5. Reaction of **1** with Catalytic Amounts of $[Ru(C_5H_5)Cl(PPh_3)_2]$ and NEt_4Br

A solution of **1** (2.0 g, 10.1 mmol) in anhydrous MeCN (50 mL) was added to a suspension of $[Ru(C_5H_5)Cl(PPh_3)_2]$ (0.030 g, 4.1 µmol) in anhydrous MeCN (5 mL) at 0 °C within two hours. Then, water (20 mL) was added first and then solid NEt_4Br (6.00 g, 28.8 mmol). The obtained suspension was filtered, and the residue taken up in the minimum amount of MeCN. Chromatography on silica gel using MeCN: toluene 1:9 as eluent produced two major fractions. NMR spectroscopy identified the first fraction as $NEt_4[C_5H(CN)_4]$. The second fraction produced after evaporation a brown powder (0.40 g) which contained, according to its mass spectra, compound **5**. Recrystallization from MeCN/toluene produced a small number of crystals of low quality (pseudomerohedral twins).

4.6. Reaction of **4** with KCl in MeOH

A suspension of **4** (68 mg, 0.125 mmol) in methanol (10 mL) was treated with a solution of KCl (4.7 mg, 0.063 mmol) in methanol (5 mL). The mixture was stirred at room temperature under the exclusion of light for 8 h and the solvent was evaporated to a volume of 1 mL. The solution was purified in a syringe filter before the product was precipitated with dichloromethane as a colorless solid (23 mg). NMR spectra showed the presence of small amounts of a $[C_5H(CN)_4]$ salt besides an octacyanofulvalenediide (presumably with K^+ as cation, Figures S8 and S9). The compound was recrystallized from MeCN/toluene by vapor diffusion at 25 °C to yield X-ray quality crystals that were also subjected to further characterization.

Compound **6**: ^{13}C-NMR (101 MHz, DMSO-d^6): δ = 128.0 ($\underline{C}$–$\underline{C}$), 114.6, 113.9 (2 × $\underline{C}$N), 99.9, 96.9 (2 × $\underline{C}$CN). MS (MALDI*): m/z = 328.1 ($C_{18}N_8^-$). calcd. for $K_2C_{18}N_8$ × 0.49 toluene × 2 H_2O:: C 52.79, N 22.98, H 1.64; found C 52.26, N 23.00, H 2.21.

4.7. Crystal Structure Determinations

All crystals were measured on a BRUKER D8VENTURE system. Compound **4** was obtained as twins. Refinement was possible using a HKLF 5 file with a BASF factor of ca. 0.25. Compound **5** also showed twinning (pseudomerohedral), which could, however, not be properly resolved. Still, refinement was possible. The experimental details of the structure determinations are collected in Table 5. The software package WINGX [35] was used for structure solution (SHELXT, [36], refinement (SHELXL 2018/3, [37]), evaluation (PLATON, [38]), and graphical representation (ORTEP3 and MERCURY [35]). Carbon-bound hydrogen atoms were treated with a riding model using the AFIX command of SHELXL.

Table 5. Experimental details of the crystal structure determinations.

Compound	1	4-LT	4-RT	5	6
Empirical formula	C_9N_6	$C_{16}H_8AgN_4$	$C_{16}H_8AgN_4$	$C_{34}H_{40}N_{10}$	$C_9H_2KN_4O$
Formula weight	192.15	364.13	364.13	588.76	221.25
Temperature [K]	100(2)	100(2)	297(2)	100(2)	110(2)
Crystal system	Orthorhombic	Monoclinic	Monoclinic	Orthorhombic	Orthorhombic
Space group	*P bca*	*C* 2/*c*	*C* 2/*c*	*P bcn*	*C cca*
Unit cell dimensions					
a [Å]	12.5735(4)	26.679(1)	26.715(3)	17.8082(6)	7.4600(7)
b	9.8840(3)	7.3110(4)	7.4955(7)	17.7182(6)	20.3621(19)
c	13.6977(4)	20.031(1)	19.993(2)	22.0972(7)	13.2449(12)
ß [°]		131.793(1).	131.929(3)		
Volume [Å^3]	1702.30(9)	2912.9(3)	2978.5(5)	6972.3(4)	2011.9(3)
Z	8	8	8	8	8
ρ_{calc} [g cm^{-3}]	1.499	1.661	1.624	1.122	1.461
μ [mm^{-1}]	0.104	1.381	1.350	0.070	0.503
F(000)	768	1432	1432	2512	888
Crystal size [mm^3]	0.060 × 0.050 × 0.040	0.090 × 0.060 × 0.020	0.090 × 0.060 × 0.020	0.100 × 0.080 × 0.050	0.100 × 0.030 × 0.010
Θ range	3.240–28.273°	2.046–28.303°	2.049–30.494°	2.287–26.387°	3.290–26.372°
Index ranges	$-16 \leq h \leq 15, -11 \leq k \leq 13, -18 \leq l \leq 18$	$-35 \leq h \leq 26, 0 \leq k \leq 9, 0 \leq l \leq 26$	$-38 \leq h \leq 28, 0 \leq k \leq 10, 0 \leq l \leq 28$	$-22 \leq h \leq 21, -21 \leq k \leq 22, -27 \leq l \leq 27$	$-8 \leq h \leq 9, -25 \leq k \leq 20, -16 \leq l \leq 16$
Refl. coll.	18615	3608	6594	68882	8675
Indep. Refl. [R_{int}]	2108 [0.0452]	3608 [0.0440]	6594 [0.0347]	7143 [0.0708]	1036 [0.0617]
Absorpt. correction			Semi-empirical from equivalents		
T_{max}/T_{min}	0.7457/0.6852	0.7457/0.6621	0.6478/0.5642	0.9705/0.8778	0.9281/0.7917
Data/restr./param.	2108/0/136	3608/7/148	6594/7/148	7142/0/405	1036/0/70
GOOF	1.027	1.108	1.046	1.144	1.070
R1/wR2 [I>2σ (I)]	0.0365/0.1077	0.0445/0.1173	0.0580/0.1590	0.0626/0.1401	0.0413/0.0886
R1/wR2 (all data)	0.0494/0.1215	0.0508/0.1221	0.0742/0.1709	0.0933/0.1537	0.0563/0.0942
$\Delta\rho_{el}$ [e Å^{-3}]	0.351/−0.225	1.571/−0.904	2.171/−1.112	0.243/−0.216	0.307/−0.291
CCDC-#	2236241	2236239	2236240	2236242	2236238

Supplementary Materials: The following supporting information can be downloaded at https://www.mdpi.com/article/10.3390/inorganics11020071/s1, Table S1: Important bond parameters of compound **4** (room-temperature determination); Figure S1: ^{1}H-NMR (400 MHz, DMSO-d^6) of the crude reaction product of reaction 4.2; Figure S2: ^{13}C{^{1}H}-NMR (101 MHz, DMSO-d^6) of the crude reaction product of reaction 4.2.; Figure S3: ^{1}H-NMR (400 MHz, DMSO-d^6) of the crude reaction product of reaction 4.3; Figure S4: ^{13}C{^{1}H}-NMR (101 MHz, DMSO-d^6) of the crude reaction product of reaction 4.3; Figure S5: ^{1}H-NMR (400 MHz, DMSO-d^6) of the first chromatography fraction of reaction 4.4.; Figure S6: ^{13}C{^{1}H}-NMR (101 MHz, DMSO-d^6) of the first chromatography fraction of reaction 4.4; Figure S7: ^{13}C{^{1}H}-NMR (101 MHz, DMSO-d^6) of the second chromatography fraction of reaction 4.4; Figure S8: ^{1}H-NMR (400 MHz, DMSO-d^6) of the purified product of reaction 4.6; Figure S9: ^{13}C{^{1}H}-NMR spectrum (101 MHz, DMSO-d^6) of the purified product of reaction 4.6; Figure S10: The infinite 1D chain in the structure of compound **1**; Figure S11: ORTEP3 plot of the asymmetric unit of **4**, at r.t; Figure S12: Two views of the pore structure of **4**, generated by virtual removing the disordered toluenes; Figure S13: Two packing views of compound **4**; Figure S14: ORTEP3 view of the zig-zag chain of face-fused KN_4O_2 octahedra along the x direction; Figure S15: Packing plots of compound **6**, watched along c; Figure S16: Packing plot of compound **6**, watched along b; Figure S17: HOMO and LUMO of compound **1**, as calculated with *CrystalExplorer*; Figure S18: Electron density plot of compound **1**; Figure S19: Two views of the electrostatic potential distribution in compound **1**.

Author Contributions: Conceptualization, P.N. and K.S.; methodology, P.N.; software, K.S.; validation, P.N., and K.S.; formal analysis, P.N.; investigation, P.N. and Y.K.; resources, K.S.; data curation, K.S.; writing—original draft preparation, P.N. and Y.K.; writing—review and editing, K.S.; visualization, K.S. and P.N.; supervision, K.S.; project administration, K.S.; funding acquisition, Y.K. All authors have read and agreed to the published version of the manuscript.

Funding: This research received no external funding.

Data Availability Statement: CCDC-2236238-2236242 contain the supplementary crystallographic data for this paper. These data can be obtained free of charge via www.ccdc.cam.ac.uk/data_request/cif, by emailing data_request@ccdc.cam.ac.uk, or by contacting The Cambridge Crystallographic Data Centre, 12, Union Road, Cambridge CB2 1EZ, UK; fax: +44-1223336033.

Acknowledgments: We thank T. Klapötke for providing the NMR and D. Fattakhova-Rohlfing for providing the CV instrumentation. We also acknowledge P. Mayer for performing the X-ray diffraction measurements and F. Zoller for performing the CV experiments.

Conflicts of Interest: The authors declare no conflict of interest.

References

1. Webster, O.W. Diazotetracyanocyclopentadiene. *J. Amer. Chem. Soc.* **1966**, *88*, 4055–4061. [CrossRef]
2. Galli, C. Radical Reactions of Arenediazonium Ions: An Easy Entry into the Chemistry of the Aryl Radical. *Chem. Rev.* **1988**, *88*, 765–792. [CrossRef]
3. Mo, F.; Qiu, D.; Zhang, L.; Wang, J. Recent Development of Aryl Diazonium Chemistry for the Derivatization of Aromatic Compounds. *Chem. Rev.* **2021**, *121*, 5741–5829. [CrossRef] [PubMed]
4. Ghigo, G.; Bonomo, M.; Antenucci, A.; Damin, A.; Dughera, S. Ullmann homocoupling of arenediazonium salts in a deep eutectic solvent. Synthetic and machanistic aspects. *RSC Adv.* **2022**, *12*, 26640–26647. [CrossRef] [PubMed]
5. Shkrob, I.A.; Marin, T.W.; Wishart, J.F. Ionic Liquids Based on Polynitrile Anions: Hydrophobicity, Low Proton Affinity, and High Radiolytic Resistance Combined. *J. Phys. Chem.* **2013**, *117*, 7084–7094. [CrossRef] [PubMed]
6. Child, B.Z.; Giri, S.; Gronert, S.; Jena, P. Aromatic Superhalogens. *Chem. Eur. J.* **2014**, *20*, 4736–4745. [CrossRef]
7. Zhong, M.; Zhou, J.; Jena, P. Rational Design of stable Dianions by Functionalizing Polycyclic Aromatic Hydrocarbons. *ChemPhysChem* **2017**, *18*, 1937–1942. [CrossRef]
8. Nimax, P.R.; Zoller, F.; Blockhaus, T.; Küblböck, T.; Fattakhova-Rohlfing, D.; Sünkel, K. An aminotetracyanocyclopentadienide system: Light-induced formation of a thermally stable cyclopentadienyl radical. *New J. Chem.* **2020**, *44*, 72–78. [CrossRef]
9. Nimax, P.R.; Rotthowe, N.; Zoller, F.; Blockhaus, T.; Wagner, F.; Fattakhova, D.; Sünkel, K. Coordination polymers of 5-substituted 1,2,3,4-tetracyanocyclopentadienides: Structural and electrochemical properties of complex compounds of 5-amino- and 5-nitro-tetracyanocyclopentadienide. *Dalton Trans.* **2021**, *50*, 17643–17652. [CrossRef]
10. Less, R.J.; Wilson, T.C.; McPartlin, M.; Wood, P.T.; Wright, D.S. Transition metal complexes of the pentacyanocyclopentadienide anion. *Chem. Commun.* **2011**, *47*, 10007–10009. [CrossRef]
11. Zhao, H.; Heintz, R.A.; Dunbar, K.R. Unprecedented Two-Dimensional Polymers of Mn(II) with $TCNQ^{\cdot-}$ (TCNQ = 7,7,8,8,-Tetracyanoquinodimethane. *J. Am. Chem. Soc.* **1996**, *118*, 12844–12845. [CrossRef]

12. Milasinovic, V.; Krawczuk, A.; Molcanov, K.; Kojic-Prodic, B. Two-Electron Multicenter Bonding (‚Pancake Bonding') in Dimers of 5,6-Dichloro-2,3-dicyanosemiquinone (DDQ) Radical Anions. *Cryst. Growth Des.* **2020**, *20*, 5435–5443. [CrossRef]
13. Preuss, K.E. Pancake bonds: π-Stacked dimers of organic and light-atom radicals. *Polyhedron* **2014**, *79*, 1–15. [CrossRef]
14. Dunkin, I.R.; McCluskey, A. Tetrabromocyclopentadienylidene: Generation and reaction with CO in low temperature matrices. *Spectrochim. Acta* **1993**, *49A*, 1179–1185. [CrossRef]
15. Olson, D.R.; Platz, M.S. The reaction of cyclopentadienylidene, fluorenylidene and tetrachlorocyclopentadienylidene with alcohols. A laser flash photolysis study. *J. Phys. Org. Chem.* **1996**, *9*, 759–769. [CrossRef]
16. Henkel, S.; Trosien, I.; Mieres-Perez, J.; Lohmiller, T.; Savitsky, A.; Sanchez-Garcia, E.; Sander, W. Reactions of Cyclopentadienylidenes with CF_3I: Electron Bond Donation versus Halogen Bond Donation of the Iodine Atom. *J. Org. Chem.* **2018**, *83*, 7586–7592. [CrossRef]
17. Abel, E.W.; Pring, G.M. The Preparation of Perhalogenopentafulvalenes utilising a Palladium Catalyst. *Inorg. Chim. Acta* **1980**, *44*, L161–L163. [CrossRef]
18. Schramm, K.D.; Ibers, J.A. Synthesis and Characterization of Some -Bonded Diazo Complexes of Nickel(0), Platinum(0), and Ruthenium(0): Molecular Structure of $Ru(CO)_2(N_2C_5C1_4)(P(C_6H_5)_3)_2$-$CH_2C12$. *Inorg. Chem.* **1980**, *19*, 2441–2448. [CrossRef]
19. Grieve, D.M.A.; Lewis, G.E.; Ravenscroft, M.D.; Skrabal, P.; Sonoda, T.; Szele, I.; Zollinger, H. Reactivity of Carbenes and Related Compounds towards Molecular Nitrogen. *Helv. Chim. Acta* **1985**, *68*, 1427–1439. [CrossRef]
20. Aquad, E.; Leriche, P.; Mabon, G.; Gorgues, A.; Khodorkovsky, V. Fulvalene derivatives: Strong proaromatic electron acceptors. *Tetrahedron Lett.* **2001**, *42*, 2813–2815. [CrossRef]
21. Aquad, E.; Leriche, P.; Mabon, G.; Gorgues, A.; Allain, M.; Riou, A.; Ellern, A.; Khodorkovsky, V. Base-catalyzed condensation of cyclopentadiene derivatives. Synthesis of fulvalene analogues: Strong proaromatic electron acceptors. *Tetrahedron* **2003**, *59*, 5773–5782. [CrossRef]
22. Nimax, P.; Sünkel, K. Structural diversity in the alkaline earth metal compounds of tetra and pentacyanocyclopentadienide. *Dalton Trans.* **2018**, *47*, 409–417. [CrossRef]
23. Ramm, M.; Schulz, B.; Thurner, J.-U.; Tomaschewski, G. Crystal and Molecular Structure of 1-Diazo-2,3,4,5-tetraphenylcyclopentadiene. *Cryst. Res. Technol.* **1990**, *25*, 405–410. [CrossRef]
24. Najafian, K.; von Rague Schleyer, P.; Tidwell, T.T. Aromaticity and antiaromaticity in fulvenes, ketocyclopolyenes, fulvenones, and diazocyclopolyenes. *Org. Biomol. Chem.* **2003**, *1*, 3410–3417. [CrossRef] [PubMed]
25. Duthaler, R.O.; Förster, H.G.; Roberts, J.D. ^{15}N and ^{13}C Nuclear Magnetic Resonance Spectra of Diazo and Diazonium Compounds. *J. Amer. Chem. Soc.* **1978**, *100*, 4974–4979. [CrossRef]
26. Spackman, P.R.; Turner, M.J.; McKinnon, J.J.; Wolff, S.K.; Grimwood, D.J.; Jayatilaka, D.; Spackman, M.A. *CrystalExplorer*, a program for Hirshfeld Surface Analysis, visualization and quantitative analysis of molecular crystals. *J. Appl. Cryst.* **2021**, *54*, 1006–1011. [CrossRef] [PubMed]
27. Cohen, T.; Lewarchik, R.J.; Tarino, J.Z. Role of Radical and Organocopper Intermediates in Aromatic Diazonium Decomposition induced by Cuprous Ion. *J. Am. Chem. Soc.* **1974**, *96*, 7753–7760. [CrossRef]
28. Cepanec, I.; Litvic, M.; Udikovic, J.; Pogorelic, I.; Lovric, M. Copper(I)-catalysed homo-coupling of aryldiazonium salts: Synthesis of symmetric biaryls. *Tetrahedron* **2007**, *63*, 5614–5621. [CrossRef]
29. Albertin, G.; Antoniutti, S.; Bortoluzzi, M.; Botter, A.; Castro, J. Pentamethylcyclopentadienyl Half-Sandwich Diazoalkane Complexes of Ruthenium: Preparation and Reactivity. *Inorg. Chem.* **2016**, *55*, 5592–5602. [CrossRef]
30. Shannon, R.D. Revised Effective Ionic Radii and Systematic Studies of Interatomic Distances in Halides and Chalcogenides. *Acta Cryst.* **1976**, *A32*, 751–767. [CrossRef]
31. Nimax, P.R.; Reimann, D.; Sünkel, K. Solvent effects on the crystal structure of silverpentacyanocyclopentadienide: Supramolecular isomerism and solvent coordination. *Dalton Trans.* **2018**, *47*, 8476–8482. [CrossRef] [PubMed]
32. Less, R.J.; Wilson, T.C.; Guan, B.; McPartlin, M.; Steiner, A.; Wood, P.T.; Wright, D.S. Solvent Direction of Molecular Architectures in Group 1 Metal Pentacyanocyclopentadienides. *Eur. J. Inorg. Chem.* **2013**, *2013*, 1161–1169. [CrossRef]
33. Batten, S.R.; Hoskins, B.F.; Robson, R. Structures of [Ag(tcm)], $[Ag(tcm)(phz)_{1/2}]$ and [Ag(tcm)(pyz)] (tcm = tricyanmethanide, $C(CN)_3^-$, phz = phenazine, pyz = pyrazine). *New J. Chem.* **1998**, *22*, 173–175. [CrossRef]
34. Witt, J.R.; Britton, D. The crystal structure of potassium tricyanomethanide $KC(CN)_3$. *Acta Cryst.* **1971**, *B27*, 1835–1837. [CrossRef]
35. Farrugia, L.J. WINGX and ORTEP for Windows: An update. *J. Appl. Crystallogr.* **2012**, *45*, 849–854. [CrossRef]
36. Sheldrick, G.M. Crystal Structure Refinement with SHELXL. *Acta Crystallogr. Sect. C Struct. Chem.* **2015**, *C71*, 3–8. [CrossRef]
37. Sheldrick, G.M. SHELXT-integrated space group and crystal structure determination. *Acta Crystallogr. Sect. A Found. Adv.* **2015**, *A71*, 3–8. [CrossRef]
38. Spek, A.L. PLATON squeeze: A tool for the calculation of the disordered solvent contribution to the calculated structure factors. *Acta Crystallogr. Sect. C Struct. Chem.* **2015**, *C71*, 9–18. [CrossRef]

Article

Interactions of an Artificial Zinc Finger Protein with Cd(II) and Hg(II): Competition and Metal and DNA Binding

Bálint Hajdu [1], Éva Hunyadi-Gulyás [2] and Béla Gyurcsik [1,*]

[1] Department of Inorganic and Analytical Chemistry, University of Szeged, H-6720 Szeged, Hungary
[2] Laboratory of Proteomics Research, Biological Research Centre, Eötvös Loránd Research Network (ELKH), H-6726 Szeged, Hungary
* Correspondence: gyurcsik@chem.u-szeged.hu; Tel.: +36-62-54-4335

Abstract: Cys2His2 zinc finger proteins are important for living organisms, as they—among other functions—specifically recognise DNA when Zn(II) is coordinated to the proteins, stabilising their $\beta\beta\alpha$ secondary structure. Therefore, competition with other metal ions may alter their original function. Toxic metal ions such as Cd(II) or Hg(II) might be especially dangerous because of their similar chemical properties to Zn(II). Most competition studies carried out so far have involved small zinc finger peptides. Therefore, we have investigated the interactions of toxic metal ions with a zinc finger proteins consisting of three finger units and the consequences on the DNA binding properties of the protein. Binding of one Cd(II) per finger subunit of the protein was shown by circular dichroism spectroscopy, fluorimetry and electrospray ionisation mass spectrometry. Cd(II) stabilised a similar secondary structure to that of the Zn(II)-bound protein but with a slightly lower affinity. In contrast, Hg(II) could displace Zn(II) quantitatively ($\log\beta' \geq 16.7$), demolishing the secondary structure, and further Hg(II) binding was also observed. Based on electrophoretic gel mobility shift assays, the Cd(II)-bound zinc finger protein could recognise the specific DNA target sequence similarly to the Zn(II)-loaded form but with a ~0.6 log units lower stability constant, while Hg(II) could destroy DNA binding completely.

Keywords: zinc finger proteins; cadmium; mercury; metal binding affinity; DNA binding; electrospray ionisation mass spectrometry; spectroscopy; EMSA; FluoZin-3

Citation: Hajdu, B.; Hunyadi-Gulyás, É.; Gyurcsik, B. Interactions of an Artificial Zinc Finger Protein with Cd(II) and Hg(II): Competition and Metal and DNA Binding. *Inorganics* **2023**, *11*, 64. https://doi.org/10.3390/inorganics11020064

Academic Editors: Peter Segľa and Ján Pavlik

Received: 29 December 2022
Revised: 25 January 2023
Accepted: 27 January 2023
Published: 29 January 2023

1. Introduction

Zinc finger proteins (ZFPs) are present in various living organisms, such as amphibians, reptiles and mammals [1–4]. ZFPs are involved in DNA transcription, translation, error correction, metabolism, stimulus generation, cell division and cell death by interacting with other proteins, small molecules, RNA and DNA [5,6]. Zinc finger (ZF) motifs of a ZFP are involved in molecular recognition, while the rest of the protein is most commonly responsible for its function [7–13]. The structure of a ZF motif is stabilised by the tetrahedral coordination of Zn(II) and by the formation of a hydrophobic core [14]. Similar tetrahedral coordination was found in self-assembling peptides offering a Cys2His2 binding site [15]. Cys2His2-type proteins form the most populous family of ZFPs [16]. Their biotechnological significance is highlighted by the fact that a ZF unit recognises three subsequent nucleobases in a double strand (ds) DNA, and several ZF units can be linked together to increase the specificity of the interaction. ZF arrays were first applied as the DNA recognition domains of artificial nucleases linked to the FokI nuclease domain [17]. Since then, numerous gene modification experiments have been performed with nucleases of this type [18–24]. The recognition sequence can further be extended by the chimera of ZF and other DNA binding motifs [25].

Cys2His2 ZFPs can only bind to DNA specifically in their Zn(II)-bound form. Other metal ions inside the living organism may substitute Zn(II), form mixed complexes and/or promote oxidation of the cysteines. In addition, toxic metal ions may also react with ZFPs, rendering the investigation of these interactions crucial. The coordination chemistry and biophysical properties of ZFPs [11,26–28] have been extensively studied, but still there are open questions regarding the stabilities of their complexes and the competition between Zn(II) and non-native metal ions [29].

Cd(II) has a stronger "soft" character than Zn(II). Therefore, it forms the most stable complexes with thiolate ligands, while it can also interact with nitrogen and oxygen donor atoms in biological systems. ZF peptides modelling a Cys2His2 type ZF unit bind Zn(II) about two–three orders of magnitude more strongly than Cd(II). The affinities of the two metal ions are comparable towards the Cys3His binding site, while the Cys4 ZF motifs bind Cd(II) two–three orders of magnitude stronger than Zn(II) [30–33]. Heinz et al. investigated the metal ion coordination of the consensus peptide 1 (CP1) Cys2His2 ZF model. Starting from the apo-peptide, they reported that a biscomplex forms with Cd(II) when the ligand is in excess through the cysteine thiolates, while Cys2His2 coordination is favoured in the monocomplex formed at a 1:1 initial metal to ligand ratio [34]. Cd(II) may disturb DNA recognition of a ZFP and thus the related biological processes, which is one of the possible mechanisms of its toxic effect [35]. Petering et al. and Hanas et al. found that Cd(II) could inhibit DNA binding of the Zn(II)-bound TFIIIA ZFP [35–37]. On the other hand, it was shown that both the high- and low-stability binding sites of TFIIIA bind Cd(II) weaker than Zn(II) by ~2.5 and one order of magnitude, respectively [38]. Investigations with the 3rd ZF subunit of TFIIIA ZFP revealed that once formed, the Cd(II) complex had a similar secondary structure and just a slightly lower DNA binding affinity (not specific) than the Zn(II) complex [39]. The situation is even more complicated in the Sp1 transcription factor consisting of three ZF subunits. In a few publications, Cd(II) has been shown to inhibit DNA binding of Zn(II) saturated Sp1 [40–42], while others did not observe such effect [43]. Kuwahara et al. reported that the Cd(II) complex of Sp1 was also capable of recognizing the specific target DNA, but with slightly lower affinity than the Zn(II) complex [44]. Malgieri et al. reported comparable affinity of Ros87, an eukaryotic Cys2His2-type ZFP, towards Cd(II) and Zn(II). The two complexes shared a similar secondary structure based on UV–Vis, CD and NMR measurements. Furthermore, the Cd(II) complex could recognise the same DNA target as the Zn(II)-loaded Ros87. However, it must be emphasised that this protein had only a single ZF unit linked to other protein elements, which also played a role in DNA binding [45]. A similar phenomenon was observed with the Tramtrack ZFP (consisting of two ZF subunits), although the α-helix content and DNA binding affinity of the Cd(II) complex was lower than that of Zn(II) bound protein [46] based on CD and EMSA measurements. The MTF-1 (metal response element-binding transcription factor-1) consists of six Cys2His2 ZF subunits. Cd(II) could inhibit the DNA-binding of this ZFP both when added to the apo-protein or to the Zn(II)-loaded MTF-1 [43,47]. The three unusual C-terminal fingers (4th–6th) of MTF-1 were investigated by Giedroc et al., where NMR and UV–Vis data revealed that although Cd(II) could bind these subunits, the obtained secondary structure differed significantly from the native $\beta\beta\alpha$-type, most probably due to the unusual amino acid composition of the 5th subunit between the two cysteines [48].

The high affinity of Hg(II) towards sulphur donor groups is well known. However, the fact that Hg(II) can form a very stable complex with Cl^- ions under physiological conditions ($\lg\beta_{ML_2}$ = 13.23) makes it difficult to compare the results of equilibrium studies [49]. Depending on the chloride content of the medium, some studies consider the affinity of Hg(II) and Zn(II) for Cys2His2 ZF motifs to be comparable, while in the absence of Cl^- ions, Hg(II) forms more stable complexes [50]. Depending on the applied medium and measurement conditions, it has been shown that Hg(II) could not inhibit the DNA binding of TFIIIA during a DNase I footprinting assay [37], while more studies revealed that the secondary structure of the Zn(II)-loaded ZFPs collapsed in the presence of both organic and inorganic Hg(II), and the Hg(II)-bound ZFPs were unable to recognise DNA target sequences [40,44,51].

Solution equilibria of Cys2His2 zinc finger motifs have been widely investigated using the CP1 model peptide and the metal binding properties of naturally occurring ZFP subunits are usually compared to this model. However, CP1 could only give information regarding the metal binding properties of a single ZF subunit but not the protein–DNA interactions, since one ZF subunit cannot provide significant selectivity and affinity towards DNA. Therefore, the trends predicted on the basis of CP-1 and the behaviour experienced with natural ZFPs are contradictory [35–37,40–43,47].

Recently, we quantitatively characterised the Zn(II) and DNA binding properties of 1MEY#, an artificial ZFP consisting of three CP1-like subunits [52]. The amino acid sequence of 1MEY# can be found in Figure S1. With the knowledge of Zn(II) and DNA affinities, here we investigated the interaction between the 1MEY# artificial ZFP and toxic metal ions by spectrometric and electrophoretic methods. The competition of Cd(II) and Hg(II) with Zn(II) for the ZFP is described quantitatively to better understand the possible mechanisms of toxicity.

2. Results and Discussion

2.1. Interaction of 1MEY# Zinc Finger Protein with Cd(II)

The investigated 1MEY# ZFP consists of three CP1-like ZF subunits. The ZF units in 1MEY# differ only in a few amino acids responsible for DNA recognition. Therefore, it is assumed that they have similar metal binding properties. For this reason, the metal binding affinities of a single "average" 1MEY# binding site, referred to as 1MEY# bs, are presented throughout the text, unless otherwise stated. A Zn(II)-loaded Cys2His2 ZFP displays $\beta\beta\alpha$-type secondary structure, while the apo-protein turns into an unordered structure, both resulting in characteristic circular dichroism spectra [53]. This provides a great opportunity to apply circular dichroism (CD) spectroscopy to investigate the effect of Cd(II) on 1MEY# ZFP. First, Zn(II) was removed from 1MEY# by treatment with ~25× excess of EDTA (Section 3.1). In a subsequent ultrafiltration, the apo-protein was transferred into an EDTA-free buffer. The initial Zn(II)-loaded holo-1MEY# protein displayed an ordered $\beta\beta\alpha$ structure as revealed by its CD spectrum with two negative peaks around 220 and 205 nm and a positive one at 190 nm (Figure 1a). The CD spectrum of apo-1MEY# represented an unfolded protein with a single negative peak around 200 nm. By adding three equivalents of Cd(II) (i.e., one equivalent per 1MEY# bs), the CD spectrum of the protein adopted a similar pattern to the Zn(II)-loaded protein. This indicated that Cd(II) could induce folding of the Cys2His2 ZF units into an ordered structure. Furthermore, no additional change in the CD spectra was observed upon an increase in the Cd(II) to protein ratio of up to ~120 fold (40 fold compared to the binding site) (Figure 1b). Cd(II) binding to the thiol groups of the ZFP was observed via ligand-to-metal charge transfer bands [54,55] in the 230–250 nm region of the UV–Vis absorption spectra as well (Figure S2).

It has to be mentioned that there are slight differences in the CD spectra of the Cd(II)- and Zn(II)-bound 1MEY#. The negative peak around 220 nm disappeared and the intensity of the 205 nm negative peak increased for the Cd(II)-loaded 1MEY#. For a better understanding, we have evaluated the CD spectra using the BeStSel software [56], by means of which the secondary structure compositions of the Zn(II)- and Cd(II)-loaded proteins were obtained (Figure 1c,d). According to the best fit of the data, the percentage of antiparallel β-sheet increased by ~6%, while the percentage of α-helices decreased by ~8.5% in the Cd(II)-bound protein. The cysteines favoured by Cd(II) are located in the antiparallel β-sheet region, and tight binding to these ligands might have caused extension of these structural elements, while the percentage of the helices decreased. Nevertheless, the most significant changes in the spectra occurred in the 220–240 nm region, which may also be attributed to the chiral contribution of the charge transfer transitions [54,55,57]. The programs used for evaluation of the protein CD spectra do not consider these contributions separately; thus, these are finally detected as the change in the secondary structural elements. In our case, this may result in overestimation of the β-sheet content of the protein. Based on the above discussion, we can conclude the 1MEY# ZFP could fold into an ordered secondary structure in the presence of Cd(II), which is most probably similar to that of the Zn(II)-loaded protein. This finding is consistent with the observations of Malgieri et al. for Ros87 ZFP [45], and Krepkiy et al. for the 3rd finger of TFIIIA [39]. In contrast, in case of the Tramtrack ZFP, Roesijadi et al. could only observe the 220–240 nm changes without the intense peak around 190 nm dedicated mostly to the α-helices [46]. The CD spectra of Cd(II)-1MEY# also suggested that under the measurement conditions, binding to the Cys2His2 site was favoured over the Cys3 or Cys4 coordination mode, which could have also been a possibility for 1MEY# containing altogether six cysteines in the three ZF units. The exclusive coordination to the cysteines would, however, result in the collapse of the finger structures, which did not occur.

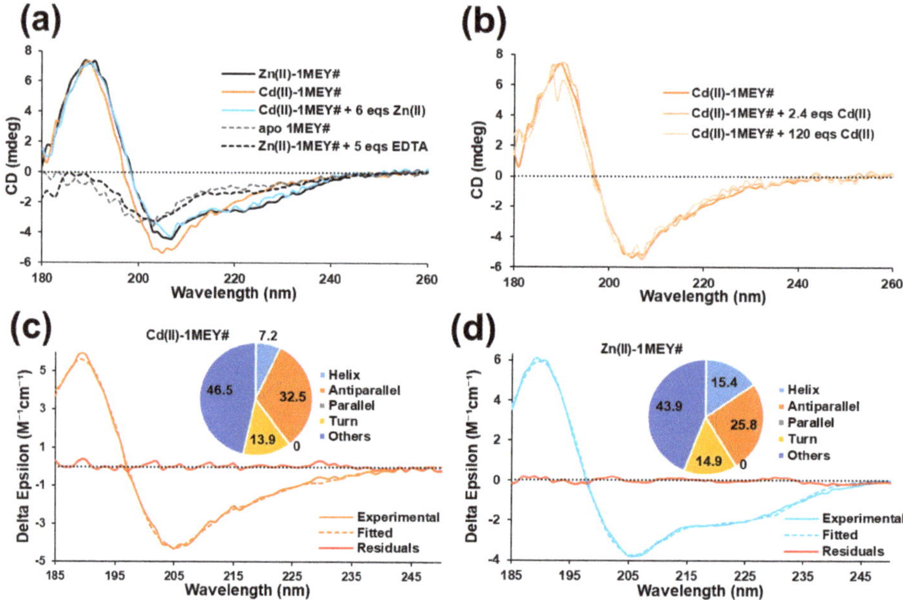

Figure 1. (**a**) Circular dichroism spectra of Zn(II)-loaded (black), Cd(II)-loaded (orange) and metal-free (dashed grey) 1MEY# ZFP. The Cd(II)-loaded form in the presence of six equivalents (eqs) Zn(II) per 1MEY# (two eqs per binding site) (light blue) and the Zn(II)-depleted form using five eqs of EDTA per 1MEY# (1.7 eqs per binding site) (dashed black) are also presented. (**b**) CD spectra of Cd(II)-loaded 1MEY# in the presence of excess Cd(II). All CD spectra were normalised to the intensity of the starting Zn(II)-loaded 1MEY# spectrum recorded at 18.8 μM protein concentration. Measured and fitted CD spectra of (**c**) Cd(II)-1MEY# and (**d**) Zn(II)-1MEY# in the 185–250 nm wavelength range. Residual curves showing the differences between the fitted and measured spectra are marked with a red colour. Insets represent the estimated secondary structure composition of the complexes. The fitting was performed by BeStSel program suite [56].

By adding six eqs of Zn(II) to the Cd(II)-saturated 1MEY# protein (two eqs per 1MEY# bs), the CD spectrum of the initial Zn(II)-loaded ZFP was recovered, indicating that Zn(II) has significantly higher affinity towards the Cys2His2 binding site than Cd(II). The similarity of the resulting spectrum with that of the initial holoprotein also demonstrated that in the series of the above described experiments, no oxidation of the cysteines of 1MEY# occurred.

FluoZin-3 is considered to be a Zn(II) selective fluorescent probe, which can be applied to detect free Zn(II) replaced by Cd(II) in a competition assay with holo-1MEY#. However, Cd(II) also binds to FluoZin-3. Therefore, the stability of this complex was determined first by titrating four samples containing Zn(II) and FluoZin-3 at various molar ratio with Cd(II). Assuming only the formation of a monocomplex and excluding formation of the ternary complex formation, a pH-independent stability constant of $\log\beta = 7.44 \pm 0.01$ was determined by evaluating the titration curves with the PSEQUAD program [58]. Recalculating this value at pH = 7.0, the obtained value of 7.18 was close to the available literature data for the Cd(II)-FluoZin-3 monocomplex ($\log\beta' = 6.9$ [59]) (Figure S3a).

While starting from the apo-1MEY# ZFP, it was demonstrated that a protein coordinated to three Cd(II) could be established. Competition experiments were also performed to monitor the Cd(II) vs. Zn(II) exchange within 1MEY#. Based on the results of the CD spectroscopic, fluorometric and ESI-MS measurements, it was not possible to completely substitute Zn(II) with Cd(II) in the applied concentration range (Figure 2). A gradual metal–ion exchange was observed in the mass spectra with the subsequent formation of Zn_2Cd_1 and Zn_1Cd_2 mixed complexes and the Cd_3 species upon an increase in the Cd(II) excess. Since previously we could not distinguish the Zn(II) binding ability of the three subunits of 1MEY# ZFP [52], the ZF units (1MEY# bs) were considered to be identical here as well. The evaluation of the fluorometric titrations revealed that the apparent stability constant of the Cd(II)-1MEY# bs complex is ~2 orders of magnitude lower than that of the Zn(II)-1MEY# bs, characterised by a $\log\beta'_{\text{Zn(II)-1MEY\#bs, pH 7.4}}$ of 12.2 [52]. This is a similar effect to that observed for the CP-1 model ZF peptide, where the stability decreased by ~2.5 orders of magnitude (Table 1) [33]. From the CD titration data, a one order of magnitude higher stability value was calculated, but here, only very small changes were observed in the spectra during the metal–ion exchange, decreasing the sensitivity of the method. By fitting the mass spectrometric data, an average $\log\beta'_{\text{Cd(II)-1MEY\# bs pH 7.4}}$ of 10.75 was obtained. Taking into account that the results of the ESI-MS measurements may not always correlate with the solution equilibria due to the different measurement conditions and the potentially different ionisation rates of different species, this value is in a very good agreement with those determined by fluorometric and CD experiments in Table 1. It is also worth mentioning that average $\log\beta'$ values and single binding site models could not be directly used in the calculation of the Cd(II) binding affinity of the protein from the ESI-MS results, since here, the whole protein is observed. Therefore, statistical considerations were applied (Figure 2d) (Supplementary Section S1). A good agreement between the fitted and experimental ESI-MS data supported the hypothesis of the identity of the ZF units within 1MEY#.

These results indicated reversible Cd(II)/Zn(II) exchange within the CP-1-based 1MEY# ZFP. No cooperativity was observed during the titrations, suggesting that the ZF subunits behaved independently. Furthermore, the secondary structures of the formed complexes were almost identical (Figure 2) (Scheme 1a). The Cd(II) binding affinity of 1MEY# was found to be the highest among the available literature data with Cys2His2 ZFPs (Table 1), although it was still ~1–2 $\log\beta'$ units lower than the Zn(II) binding under similar conditions ($\log\beta'_{\text{Zn(II)}} = 12.2$) [52]. The difference between the Zn(II) and Cd(II) binding affinity of CP1 and TFIIIA was reported to be 2.5 [33,38], while in case of the Ros87, it was only 1.2 $\log\beta'$ units [60].

In good correlation with the results of CD measurements, no signals related to Zn_1Cd_1 or Cd_2 species were observed during the analysis of the ESI-MS spectra of 1MEY#. Such species could have been characterised by Cys4 or Cys3 coordination where Cd(II) would bind to the cysteine sidechains of more than one ZF subunit, resulting the collapse of secondary structure and presumably the loss of the DNA-binding function (Scheme 1b). This effect might be responsible for the observed function loss in natural ZFPs with low Zn(II) and Cd(II) affinity towards the Cys2His2 coordination site. For example, in the case of TFIIIA ZFP, out of the nine subunits, only between two and three had higher Zn(II) affinity, and while it was possible to purify a protein with ~9 Zn(II) per ZFP, the purifications in the presence of Cd(II) yielded up to ~4 Cd(II) per ZFP products [35–38] (Table 1). This might be due to the formation of Cd(II)-Cys3 and Cd(II)-Cys4 coordination where Cd(II) bound the cysteines of multiple ZF subunits [32,36].

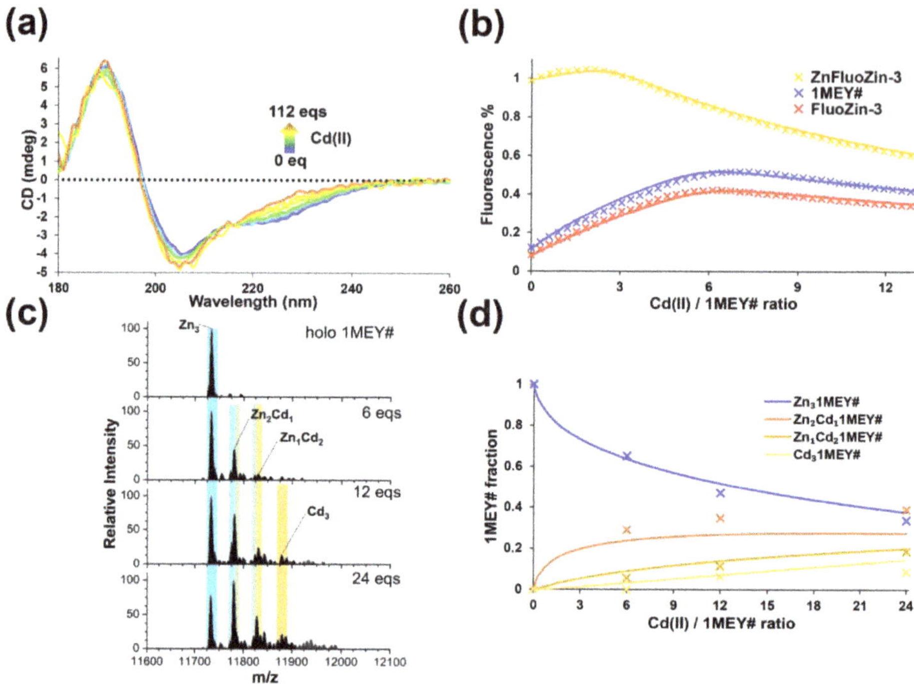

Figure 2. (**a**) Measured (full lines) and calculated (dashed lines) CD spectra of 1MEY# in the course of titration with $Cd(ClO_4)_2$; $c_{1MEY\#}$ = 16.4 μM. The endpoint of the titration (fully Cd(II)-loaded 1MEY#, orange line) was established separately starting from apo-1MEY#. (**b**) Measured (symbols) and simulated (full lines) relative fluorescence curves obtained by titrating holo-1MEY# (blue) with Cd(II). Reference titrations were simultaneously conducted to obtain data for the relative fluorescence of the system in the absence of 1MEY# (red) or in the presence of an amount of $Zn(ClO_4)_2$ equal to the Zn(II) content of 1MEY# (yellow). $c_{FluoZin\text{-}3}$ = 6 μM and $c_{1MEY\#}$ = 1 μM. (**c**) ESI-MS spectra of Zn(II)-loaded 1MEY# in the presence of increasing amounts of Cd(II). Sample amount: 20 μL $c_{1MEY\#}$ = 2 μM. (**d**) The species distribution diagram calculated from the ESI-MS data (separate points).

Table 1. Apparent average Cd(II) and Zn(II) affinities of some Cys2His2 ZFs and ZFPs in logβ' units. cITC: competition with complexones monitored by ITC; FTc: competition with fluorescent complexones monitored by fluorescence spectroscopy; rCD: circular dichroism spectroscopy followed reverse titration; ESI-MS: reverse titration followed by ESI-MS; RT: spectroscopic reverse titration; DT: spectroscopic direct titration; ED: equilibrium dialysis.

	log$\beta'_{Cd(II)}$	log$\beta'_{Zn(II)}$	Δlogβ' [1]	Method	References
1MEY#		12.2		cITC	[52]
	10.11 ± 0.03		2.11	FTc	Present work
	11.14 ± 0.03		1.06	rCD	Present work
CP1	8.7	11.2	2.5	RT	[33]
TFIIIA	5.6 [a]	8.0 [a]	2.4	ED	[38]
	3.8 [b]	4.6 [b]	0.8	ED	[38]
Ros87	8.0	9.2	1.2	RT	[51,61]
	7.7			DT	[45]

[1] Difference between Zn(II) and Cd(II) affinity, if the measurement conditions were identical. [a] High-affinity binding sites. [b] Low-affinity binding sites.

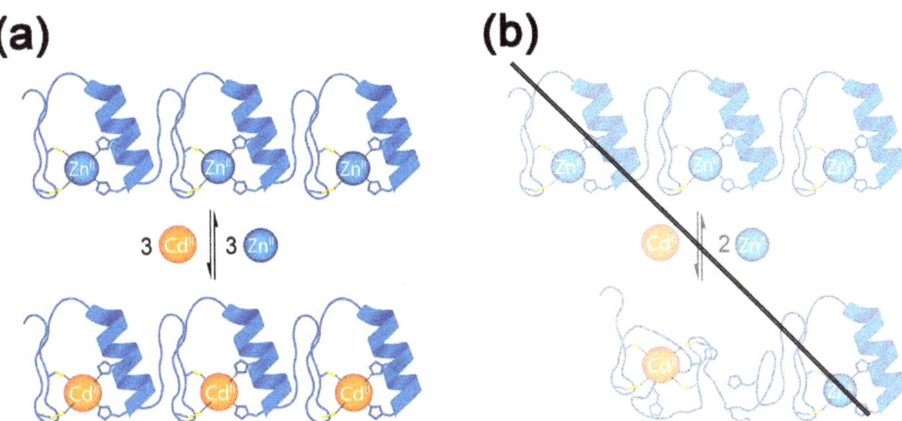

Scheme 1. (**a**) Schematic representation of the Zn(II)/Cd(II) exchange in the 1MEY# ZFP. The extended scheme including the microspecies that may form during the replacement of the first and second Zn(II) is included in Supplementary Scheme S1. (**b**) A hypothetical reaction scheme where Cd(II) coordinates to the cysteine sidechains of multiple ZF subunits inside the ZFP, resulting the collapse of the secondary structure. Such a reaction has been proven not to take place in the case of the highly stable 1MEY#, but might occur in the case of lower stability natural ZFPs such as TFIIIA.

2.2. Hg(II) Binding of the 1MEY# Zinc Finger Protein

A perchlorate salt of Hg(II) was used in experiments with 1MEY# ZFP, aiming to avoid the interference with Cl^- ions, i.e., to observe the direct interaction of the toxic metal ion with the ZFP. A significant change in the CD spectra of 1MEY# was observed upon addition of three equivalents of Hg(II) (one eq. per 1MEY# bs), referring to the collapse of the ββα secondary structure of the protein, similar to other investigations [51]. The positive peak assigned to the α-helix around 190 nm completely disappeared (Figure 3a,b). Thus, the formed species, assumed to be the $Hg_3$1MEY# complex, displayed an unordered structure. Since Hg(II) shows extreme affinity towards the cysteine thiolates, this was the expected outcome, independent of whether Zn(II) was completely displaced from the complex or a ternary species was formed.

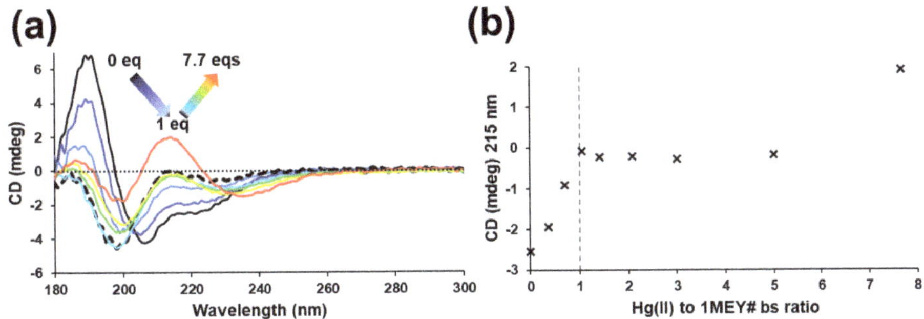

Figure 3. (**a**) CD spectra of 1MEY# ZFP in the presence of increasing equivalents of $Hg(ClO_4)_2$, starting from the Zn(II)-loaded protein (black full line). The black dashed spectrum belongs to the system containing 1 eq. Hg(II) per 1MEY# bs. (**b**) The plot of the ellipticity values (×) recorded at 215 nm vs. Hg(II) equivalents per 1MEY# bs. The breakpoint at 1 eq. is indicated by a vertical dashed grey line. $c_{1MEY\#}$ = 16.4 μM.

By the addition of further Hg(II), up to ~5 eq per 1MEY# bs, only small, but continuous, changes were observed in the CD spectra, indicating that further thermodynamic events took place. Furthermore, the shape of the CD spectrum changed completely and an intense positive band appeared around 215 nm, most probably due to a charge transfer band related to Hg(II) at ~8 eq Hg(II) per 1MEY# bs [57,62,63].

The sharp breakpoint in the plot of the CD intensity at 215 nm after adding one equivalent of Hg(II) per 1MEY# bs suggested that Hg(II) is indeed a much stronger competitor for ZF subunits than Cd(II) (Figure 3). The sharp breakpoint at one equivalent Hg(II) per ZF binding site also suggests that during the sequential exchange of Zn(II) for Hg(II), the unfolding of one subunit did not weaken the Zn(II) binding (and the secondary structure) of the remaining Zn(II)-bound subunit(s). This again indicated a high degree of independence of the subunits inside 1MEY#.

Hg(II) interactions with 1MEY# ZFP was also investigated by fluorometric titrations. There are no published data characterising the solution equilibria in the Hg(II)–FluoZin-3 system. Therefore, we carried out competitive fluorometric measurements prior to the experiments without protein, titrating the Zn(II)–FluoZin-3 system at various ratios of Hg(II). The best fit of these titration curves was achieved by assuming the formation of mono (HgA logβ = 6.8 $\pm$ 0.1), bis (HgA_2 logβ = 12.8 $\pm$ 0.3) and ternary complexes (HgZnA logβ = 14.0 $\pm$ 0.1), although the chemical composition and coordination mode of such species remained ambiguous, but as it turned out there was no need for the use of these constants in further calculations (see below) (Figure S3b).

As the next step, the Hg(II)–holo-1MEY# system was studied in the presence of FluoZin-3. As it turned out, Hg(II) bound very strongly to the 1MEY# ZFP, so that the first three equivalents (one equivalent per binding site) of Hg(II) displaced Zn(II) in the ZFP almost quantitatively, and therefore, did not interact with the fluorophore (Figure 4a). This behaviour indicated that while Cd(II) could not compete efficiently with Zn(II) for 1MEY# ZFP, inorganic Hg(II) is an extremely strong competitor. Another interesting fact is that, contrary to expectations, no increase in fluorescence could be observed even after continuing the titration; the sample practically behaved as if it had been diluted with a buffer. This suggested that the extra added Hg(II) also bound to the 1MEY# ZFP with high affinity. This phenomenon provided an opportunity to estimate the lower limits of affinities of the ZF units as binding sites not only towards the first but also towards additional Hg(II) (Table 2). Based on the obtained stability constants, we could construct a distribution diagram, where the dashed section is based on the estimated limiting constants related to the binding of further Hg(II) (Figure 4b). Although this model may not be accurate in describing metal ion cluster formation, it could be successfully applied to describe the phenomena in this complicated system. Oligomerisation may also occur in these systems, but no species related to oligomers could be identified by ESI-MS.

Table 2. Estimated stability constants for Hg(II):1MEY# bs system based on the fluorometric method. The presented average constants were calculated for a single subunit of 1MEY# that was assumed to bind 1, 2, 3 and 4 Hg(II) in the 1MEY# protein binding 3, 6, 9 and 12 Hg(II), respectively.

	p*K*′
$Hg_1$1MEY# bs	≥16.7
$Hg_2$1MEY# bs	≥9.3
$Hg_3$1MEY# bs	≥8.5
$Hg_4$1MEY# bs	≥8.2

Mass spectrometric measurements confirmed the fluorometric results. A gradual displacement of Zn(II) was detected through the Zn_2Hg1MEY#, $ZnHg_2$1MEY# and $Hg_3$1MEY# complexes. In addition, it was possible to detect the presence of 1MEY# species with even 13 coordinated Hg(II) while increasing the metal ion excess up to 24 eqs (eight eqs per binding site); however, the exact mode of coordination remained unknown (Figure 4c).

The stability constant of the monocomplex estimated from ESI-MS data was identical to the value obtained from fluorometric titration (Table 2); however, for the binding of the second Hg(II), a value three orders of magnitude lower was assigned, which is presumably due to the previously mentioned uncertainty of the mass spectrometry. It can be assumed that Hg(II) only coordinates to the cysteines [51]. Thus, in theory, the histidine residues remain available for the coordination of Zn(II). Despite this, no mixed metal species within the same ZF subunit could be seen under the conditions of the ESI-MS measurements (Scheme 2). In the $Zn_2Hg1MEY\#$ and $ZnHg_21MEY\#$ ternary complexes, the metal ions bind to different subunits and thus the binding events are independent.

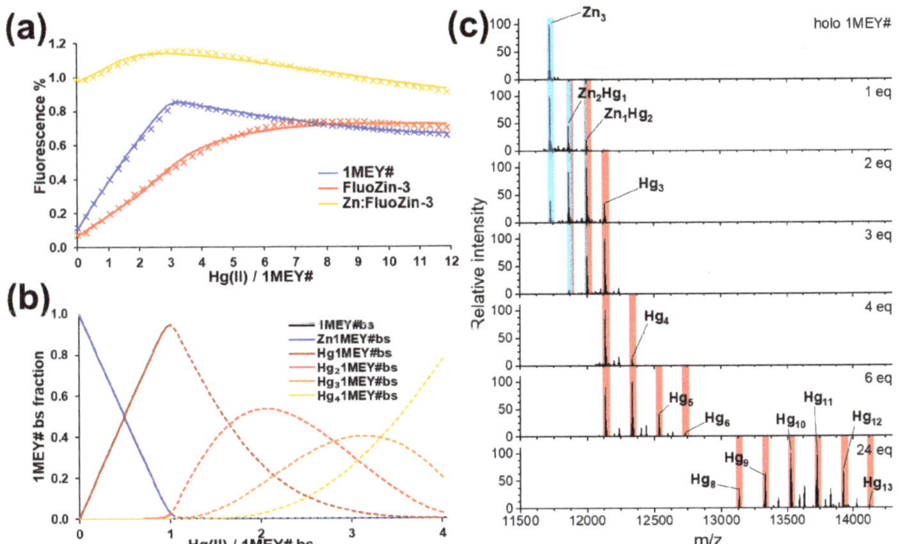

Figure 4. (**a**) Measured (symbols) and simulated (continuous line) relative fluorescence obtained during the titration of holo-1MEY# ZFP with Hg(II) (blue). Reference titrations were simultaneously performed to obtain data for the relative fluorescence of the system in the absence of 1MEY# (red) or in the presence of an amount of $Zn(ClO_4)_2$ equal to the Zn(II) content of 1MEY# (yellow). $c_{FluoZin\text{-}3}$ = 6 μM and $c_{1MEY\#}$ = 1 μM. (**b**) Distribution diagram obtained by the fitting of the fluorescence titration data. (**c**) Zn(II)-loaded 1MEY# in the presence of increasing amounts of Hg(II) followed by ESI-MS. Sample amount: 20 μL and $c_{1MEY\#}$ = 2 μM.

The obtained conditional stability constant for the binding of the first Hg(II) ($\log\beta'_{Hg(II)\text{-}1MEY\#\,bs} \geq 16.7$) is quite a high value on the scale of ZFP metal binding. Among the peptides containing the CXXC-amino acid sequence, only those had a similar or higher affinity for Hg(II), where the cysteines were in favourable position and the secondary structure was not completely unordered [64–67]. Although the CD spectra of 1MEY# in the presence of one Hg(II) eq. per binding site significantly differed from the unfolded structure (Figure S4), the contribution of the thiolate–Hg(II) charge transfer bands might be significant. Therefore, an accurate evaluation of the secondary structure composition of the Hg(II)-1MEY# complex is not possible [57,62,63]. Nevertheless, based on the NMR analysis of the similarly behaving Ros87 ZFP, it can be assumed that the structure is not completely disordered [51]. Sivo et. al. also determined the Hg(II) affinity of Ros87 using $HgCl_2$ ($\log\beta'_{Hg(II)\text{-}Ros87}$ = 6.1) [51].

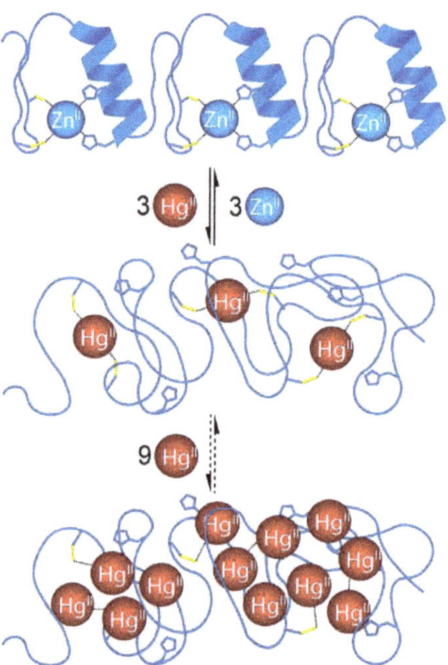

Scheme 2. Schematic representation of Zn(II)/Hg(II) exchange in 1MEY# ZFP. The extended scheme, including the microspecies that may form during the replacement of the first and second Zn(II), is included in Supplementary Scheme S2.

2.3. DNA Binding of 1MEY# ZFP Is Influenced by Toxic Metal Ions

2.3.1. Cd(II)

Previously, we have shown that Zn(II)-saturated 1MEY# could bind DNA by EMSA experiments. A clear difference in the affinity of the protein towards the 34 bp DNA probes with and without the specific 5′-GAGGCAGAA-3′ sequence was observed [52]. Here, we studied the interaction of DNA with the Cd(II)-loaded 1MEY# ZFP obtained from the apo-protein. EMSA experiments revealed a single, well defined shifted DNA band with the nonspecific DNA probe. By fitting the quantified band intensities, a stability constant could be determined as $\log\beta' = 6.04 \pm 0.02$ (Figure 5a,c). This corresponds to ~0.2 log units weaker binding compared to the interaction of the Zn(II)-loaded protein with nonspecific DNA. A more significant ~0.6 log unit decrease was found in the affinity of the Cd(II)-loaded 1MEY#, with $\log\beta' = 7.62 \pm 0.04$ (Figure 5b,d), compared to that of the Zn(II)-loaded ZFP ($\log\beta' = 8.20$). Nevertheless, the above listed stability constants are still substantially high values. It has to be mentioned that the titration curves for nonspecific DNA binding show a slight sigmoidal pattern instead of the saturation curve expected from the simple binding scheme. This, however, might be attributed to the ambiguity of the gel staining when the amount of bound DNA is too small. It was also visible that the DNA binding of the Cd(II)-loaded 1MEY# resulted in similar changes in the CD spectra of the system around 190 nm, as in case of the Zn(II)-loaded protein. However, the rate of the increase in ellipticity for the Cd(II)-1MEY# was approximately half of that of Zn(II)-1MEY#, which may indicate that the interaction of the Cd(II)-loaded ZFP with DNA induces smaller structural rearrangements, i.e., the process is weaker (Figure 5g).

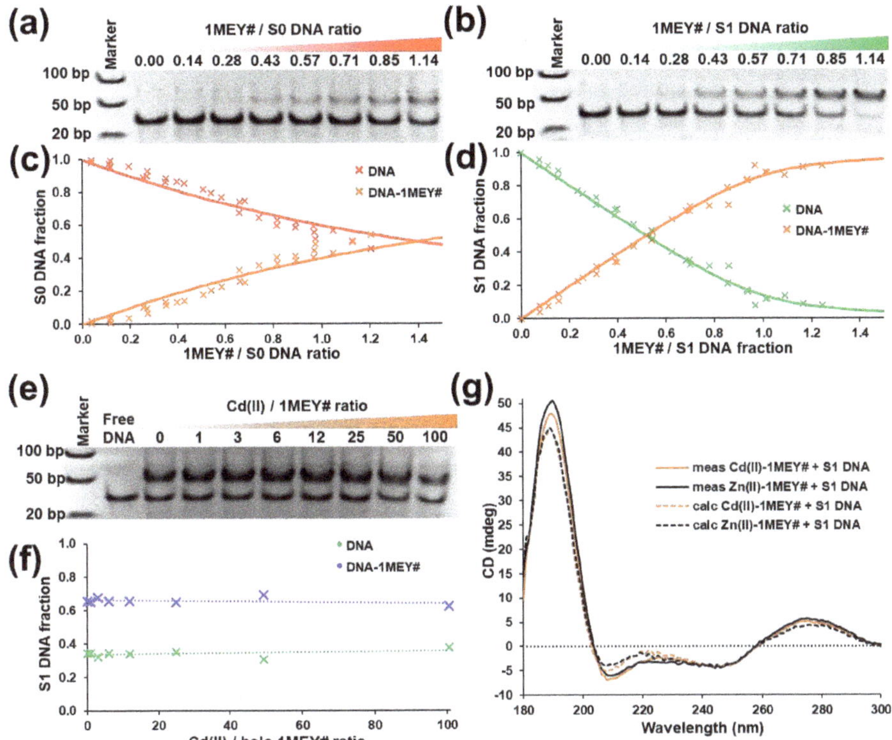

Figure 5. Representative electrophoretic gel mobility shift assays of (**a**) nonspecific S0 DNA and (**b**) specific S1 DNA in the presence of increasing equivalents of Cd(II)-loaded 1MEY# zinc finger protein. c_{DNA} = 0.88–1 μM. (**c**) Distribution of S0 DNA or (**d**) S1 DNA in the presence of increasing equivalents of Cd(II)-1MEY# ZFP. DNA fractions (separate points) were calculated based on the intensities of three independent electrophoretic gel mobility shift assays. Band intensity calculations were performed in ImageJ [68]. (**e**) Electrophoretic gel mobility shift assay of Zn(II)-1MEY# with specific S1 DNA in the presence of increasing equivalents of Cd(II). c_{DNA} = 1 μM, $c_{Zn(II)\text{-}1MEY\#}$ = 0.8 μM. (**f**) Distribution of S1-DNA among the free and protein-bound forms as calculated from the band intensities of the electrophoretic gel mobility shift assay image. (**g**) Comparison of the CD spectra of S1 DNA in the presence of 0.5 eq Zn(II)-1MEY# (black) and 0.5 eq Cd(II)-1MEY# (orange). The dashed line represents the CD spectrum calculated by summing the appropriate protein and DNA component spectra. All CD spectra were normalised to the intensity of the starting Zn(II)-loaded 1MEY# spectrum recorded at 18.8 μM protein concentration.

After it was proven that a fully Cd(II)-saturated ZFP can bind its DNA target, competition reactions were performed, where the Zn(II)-loaded 1MEY# ZFP in complex with the specific DNA probe was titrated with increasing amounts of Cd(II). During the process, no change could be seen in the DNA binding ability of the protein, which suggests that if mixed $Zn_2Cd_1$1MEY# and $Zn_1Cd_2$1MEY# complexes form, these can also bind DNA with a similar affinity to the Zn(II)-loaded protein (Figure 5e,f). On the other hand, previously we have shown that the interaction with specific DNA increases the stability of the Zn(II) complex of 1MEY# [52]. Therefore, it can be assumed that metal ion exchange is a minor process in this experiment.

Based on these findings, we could assume that the toxicity of cadmium in the living organism cannot be directly attributed to competition with high-stability Cys2His2 ZFPs. If the protein is in the Zn(II)-loaded form, a large excess of Cd(II) would be necessary to compete with Zn(II) and it is unlikely to completely substitute Zn(II). The Cd(II)-loaded 1MEY# and the mixed metal complexes can also bind DNA, the only visible differences during our measurements were in their affinity, similar to the limited literature data [43,44,46]. Although the competition with the high stability Zn(II)-loaded ZFP is not significant, if Cd(II) meets with the apo-ZFP then it can form stable complexes due to its high affinity towards to protein. Regardless, it cannot be ruled out that such a Cd(II)-loaded high stability ZFP may still be able to perform its function, since once it is coordinated in Cys2His2 mode, the structure and function may differ only slightly compared to the Zn(II)-loaded protein. Furthermore, even if the structure of the protein differs, it does not necessarily mean that the DNA binding function is also completely vanished. In the case of the Tramtrack ZFP, where based on CD, the α-helix content of the protein was reduced significantly during Cd(II) binding, the complex could still recognise its target DNA but ~1 order of magnitude more weakly [46].

2.3.2. Hg(II)

Electromobility shift assay titrations revealed that the protein binding to its DNA target had no inhibitory effect on Hg(II) competition. Three equivalents of Hg(II) (one eq. per 1MEY# bs) could completely eliminate the DNA binding of the ZFP (Figure 6a,b). Thus, the effect of Hg(II) was unambiguous in the absence of other competing ligands. By using various buffer conditions (Cl^- ions and DTT) this effect can be significantly reduced, yet according to most of the literature data, Hg(II) still could effectively inhibit the DNA binding of a ZFP [40,44,51].

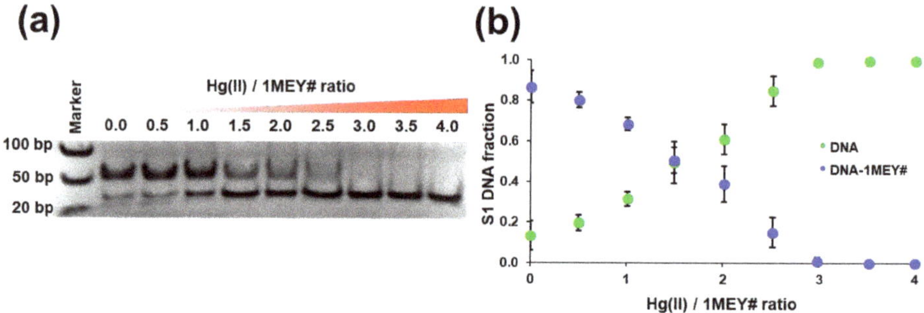

Figure 6. (**a**) Representative electrophoretic gel mobility shift assay of Zn(II)-1MEY# with specific S1 DNA in the presence of increasing equivalents of Hg(II). c_{DNA} = 1 μM, $c_{Zn(II)\text{-}1MEY\#}$ = 1 μM. (**b**) Distribution of S1-DNA based on electrophoretic gel mobility shift assay gel intensities.

3. Materials and Methods

3.1. Protein Expression and Purification

The protein expression and purification steps were described earlier [52]. Briefly, the 1MEY# protein (a consensus peptide-based 1MEY [69,70] derivative) was expressed in *E. coli* BL21 (DE3) cells from a pETM11 vector, with an N-terminal His-SUMO affinity tag. Ni(II) affinity purification was applied, then the N-terminal affinity and SUMO tag were cleaved specifically using the ULP1 protease [71]. Purification was followed by buffer exchange to 10 mM Cl^--free 4-(2-hydroxyethyl)-1-piperazineethanesulfonic acid (HEPES), pH = 7.40, using Amicon 3K 15 mL filters (Merck KGaA, Darmstadt, Germany) at 4000× *g* 8 × 30 min 15 °C and filtration through 0.22 μm, Ø = 13 mm PES filters (Merck KGaA, Darmstadt, Germany). This process provided the Zn(II)-loaded ZFP; thus, for the Cd(II)-1MEY# experiments, Zn(II) was removed from the protein by 0.5 mM EDTA (~25× excess over 1MEY#) and 0.2 mM TCEP (tris(2-carboxyethyl)phosphine) treatment for 10 min at 25 °C. EDTA was washed away during ultra-filtration (Amicon 3K 0.5 mL filters (Merck KGaA, Darmstadt, Germany), 14,000× *g* 5 × 5 min 15 °C) with 10 mM HEPES, 0.2 mM TCEP and 50 mM $NaClO_4$ (pH = 7.40) buffer. The TCEP reducing agent was included in order to protect the free cysteines from potential oxidation. After the calculated EDTA content of the protein sample dropped to below 1 μM, 150 μM final concentration $Cd(ClO_4)_2$ was added to a portion of the sample and incubated for 5 min at 25 °C. Then, additional buffer exchange was performed with 10 mM HEPES and 50 mM $NaClO_4$ (pH = 7.40) buffer by ultra-filtration as described earlier.

3.2. Mass Spectrometric Analysis of the Protein

Intact protein analysis was performed on an LTQ-Orbitrap Elite (Thermo Scientific, Thousand Oaks, CA, USA) mass spectrometer coupled with a TriVersa NanoMate (Advion, Ithaca, NY, USA) chip-based electrospray ion source. Measurements were carried out in positive mode at 120,000 resolution in 8.2 mM ammonium hydrogen carbonate buffer (pH ~7.8), as described previously [72]. Fitting of the ESI-MS data was performed by the Solver add in of Microsoft Excel based on statistical considerations [73,74].

3.3. CD Spectroscopy

A J-1500 Jasco spectrophotometer was used during spectroscopic measurements under constant nitrogen flow in stepwise scanning mode over the range of 180–330 nm. Synchrotron radiation (SR) CD spectra were recorded at the CD1 beamline of the storage ring ASTRID at the Institute for Storage Ring Facilities (ISA), University of Aarhus, Denmark between the 170 and 330 nm wavelength range [75,76].

All spectra were recorded with 1 nm steps and a dwell time of 2 s per step, using l = 0.2 mm cylindrical quartz cells (SUPRA-SIL, Hellma GmbH, Müllheim, Germany). Each sample containing 8–20 μM protein was prepared separately in 6.6–7.5 mM HEPES buffer (pH = 7.40) and was kept at room temperature for at least 5 min prior to measurement. Samples for CD measurements involving DNA contained 33–45 mM $NaClO_4$ as well. If CD data were fitted, the 215–260 nm wavelength range was selected and the PSEQUAD program was used for calculations [58].

3.4. Electrophoretic Mobility Shift Assay (EMSA)

Electrophoretic mobility shift assays were performed using 34 bp DNA probes containing zero (later referred to as S0 DNA) or one (later referred to as S1 DNA) specific 1MEY# target sequence: 5′–GAGGCAGAA–3′. The S0 DNA probe was obtained by the hybridisation of the Forward-S0: 5′–CTAGTTTGCTGAACTGGGGTCACATAGATTAATA–3′ and Reverse-S0: –5′-TATTAATCTATGTGACCCCAGTTCAGCAAACTAG-3′ oligonucleotides. The S1 DNA probe was obtained by the hybridisation of the Forward-S1: 5′–GAATTCCTGCTGA<u>GAGGCAGAA</u>ACATAGGGGTCG–3′ and Reverse-S1: 5′–CGACCCC-TATGTTTCTGCCTCTCAGCAGGAATTC–3′ oligonucleotides (target sequence of 1MEY# is underlined). Oligonucleotides were obtained by solid phase synthesis (Invitrogen—Thermo Scientific, CA, USA). EMSA experiments were performed as described earlier in 10 mM HEPES, 150 mM $NaClO_4$ and 10 m/V% glycerol buffer (pH = 7.40) [77]. The FastRuler Ultra Low Range DNA Ladder (Thermo Scientific, CA, USA) served as a reference (marker) and the ImageJ program was used for the quantification of gel intensities [68]. Calculations were performed by the PSEQUAD program [58].

3.5. Fluorimetry

The competition reactions of the Zn(II)-loaded 1MEY# were monitored using the FluoZin-3 fluorescent probe in a CLARIOstar Plus plate reader (BMG Labtech, Ortenberg, Germany). FluoZin-3 has an absorption maximum at 494 nm and exhibits strong fluorescence at 516 nm when binding Zn(II), with a conditional stability constant of $\log\beta'$ = 8.04 (pH = 7.40) [78]. In a typical measurement, 480–490 nm extinction and 510–520 nm emission filters were used during the titrations of 200 μL protein-FluoZin-3 sample in 96-well, polystyrene, non-binding, flat-bottom, black microplates (Greiner Bio-One, Kremsmünster, Austria) in 10 mM HEPES and 150 mM $NaClO_4$ buffer (pH = 7.40). An amount of 3 μL titrant (or buffer) was injected to the samples at each titration point by the two built-in injectors of the CLARIOstar Plus plate reader. Each injection was followed by 30 s 150 rpm double orbital shaking of the plate and 5 min incubation at 25 °C to reach equilibrium. The titration process was automated using custom built scripts (Table S1). The FluoZin-3 concentration was determined spectrophotometrically (λ_{max} = 491 nm, ε_{max} = 71,143 $M^{-1}cm^{-1}$ (pH = 7.40)). During each measurement, additional control samples were prepared containing only FluoZin-3 or FluoZin-3 and the equivalent amount of Zn(II) which can be released from the 1MEY# ZFP during competition reactions, in order to calculate correct relative fluorescence values (Figure S5). Fitting of the fluorometric data was performed by PSEQUAD [58] and by the Solver add in of Microsoft Excel.

4. Conclusions

The interaction of Zn(II)-loaded Cys2His2 ZFPs with other metal ions can decrease, alter or destroy their DNA binding function. Therefore, a better understanding of these systems is essential. Several research works have aimed at studying the interactions between toxic metal ions and ZFPs, but the available data with model peptides and natural ZFPs cannot be directly compared. In a previous publication, we quantitatively characterised the Zn(II) and DNA binding of a ZFP consisting of three CP1-like model peptide subunits [52]. Based on CD measurements, we could prove that Cd(II) binding of 1MEY# resulted in an almost identical secondary structure to the Zn(II) complex. The protein could bind Cd(II) with the highest affinity determined so far for the Cys2Hi2 binding sites. Yet, the Zn(II) binding affinity of 1MEY# is even higher by 1–2 orders of magnitude. Similar tendencies were observed between the Zn(II) and Cd(II) binding of other investigated ZFs [33,38]. Thus, a large excess of Cd(II) would be necessary to compete with Zn(II), while the Cd(II) substituted ZFP can be reverted into Zn(II)-complex by adding two equivalents of Zn(II) per binding site, based on CD measurements. The metal exchange reaction occurred stepwise, with each ZF subunit behaving independently. Formation of Cd(II)-Cys3 or Cd(II)-Cs4 complexes could not be observed by ESI-MS and CD measurements. The protein bound three Cd(II) ions and no further spectral changes were visible, even when applying 120 equivalents of Cd(II). We could demonstrate that the Cd(II) complex was also capable of specific DNA binding by EMSA and CD measurements, although the affinity decreased by $\log\beta'$ = 0.6 units compared to the Zn(II) complex. The DNA binding ability of Zn(II)-loaded 1MEY# was not influenced even by ~100 eqs of Cd(II), suggesting that either no exchange occurred or the mixed complexes could recognise the DNA target as well.

Hg(II) behaved in a completely different manner. By using perchlorate salt, we could observe the affinity of Hg(II) towards the 1MEY# ZFP to be $\log\beta'_{Hg(II)\text{-}1MEY\#\ bs} \geq 16.7$. During the quantitative exchange of Zn(II) with Hg(II), the well-defined $\beta\beta\alpha$ secondary structure of 1MEY# collapsed. Excess of Hg(II) could also bind to 1MEY# with nanomolar affinity. Even a 13 Hg(II)-bound form was visible in ESI-MS, which could also be followed by CD spectroscopy through the charge transfer bands of Hg(II). In contrast, the Hg(II) affinity of the Ros87 ZFP investigated by Sivo et al. using $HgCl_2$ salt was three orders of magnitude weaker than the Zn(II) affinity of the same protein [51]. Hg(II) not only disrupted the structure of 1MEY#, but also destroyed the DNA binding of the ZFP. An investigation into the multi-metal binding sites in the Hg(II)-bound 1MEY# would be of interest.

Supplementary Materials: The following supporting information can be downloaded at: https://www.mdpi.com/article/10.3390/inorganics11020064/s1, Section S1. Statistical considerations during ESI-MS measurements. Scheme S1. Schematic representation of the Zn(II)/Cd(II) exchange in the 1MEY# ZFP. In the transition states, Cd(II) can replace Zn(II) of any zinc finger subunit with equal probability. Scheme S2. Schematic representation of the Zn(II)/Hg(II) exchange in the 1MEY# ZFP. In the transition states, Hg(II) can replace Zn(II) of any zinc finger subunit with equal probability. Figure S1: Cartoon representation of the crystal structure of (a) the 1st ZF subunit of the 1MEY ZFP and (b) the whole 1MEY ZFP. ZFP: blue, Zn(II): grey sphere, cysteine thiols: yellow (PyMOL representation of 1MEY PDB [4]). (c) Alignment of the amino acid sequence of 1MEY# ZFP (constructed from 1MEY ZFP [4]) with the 26 amino acid-long consensus Cys2His2 model peptide CP1 and CP1 K/S mutant, established and investigated by Berg et al. [5]. The identical amino acids of 1MEY# compared to CP1 are marked with green, while the ones that differ both compared to the CP1 and CP1 K/S mutant are marked with red. The amino acids differing only compared to CP1 are marked with light red. Figure S2: UV–Vis absorption spectra of 1MEY# in either Zn(II) (blue) or Cd(II) (orange) saturated form. $c_{1MEY\#}$ = 13.5 μM in 10 mM HEPES and 50 mM $NaClO_4$ (pH 7.4); l = 1 cm. Figure S3: Measured (separate symbols) and calculated (full lines) relative fluorescence values of Zn(II)–FluoZin-3 systems in the presence of an increasing amount of (**a**) $Cd(ClO_4)_2$; and (**b**) $Hg(ClO_4)_2$. Samples (200 μL) were loaded into the plate wells and titrated with 3 μL aliquots of the titrant at 25 °C. $c_{FluoZin\text{-}3}$ = 3.98 μM, 10 mM HEPES and 150 mM $NaClO_4$ (pH 7.40). The calculations were performed by the PSEQUAD program [6]. Figure S4: Circular dichroism spectra of Zn(II)-loaded 1MEY# (full black line), Hg(II)-loaded (red) and metal-free form using 5 eqs of EDTA per 1MEY# (1.7 eqs per binding site) (dashed black) are also presented. $c_{1MEY\#}$ = 16.4 μM in 7.5 mM HEPES (pH = 7.4) buffer. CD1 beamline of the storage ring ASTRID, Aarhus l = 0.2 mm. Figure S5: Fluorometric titration procedure. Baseline fluorescence was determined by applying a 10-fold excess of EDTA over FluoZin-3. The maximal achievable fluorescence value was determined by applying 0.5 eq Zn(II) to FluoZin-3 ('Max'). A two-fold excess of FluoZin-3 was necessary to make sure 100% of Zn(II) is complexed. A sample containing an identical amount of Zn(II) to the 'Max' reference well was titrated with the titrant. The dilution effect during titration was determined by the injection of buffer (instead of the titrant) to the reference wells and to an additional sample well. Table S1: Automated titration script for CLARIOstar Plus plate reader.

Author Contributions: Conceptualisation, B.H. and B.G.; investigation, B.H. and É.H.-G.; data curation, B.H., É.H.-G. and B.G.; writing—original draft preparation, B.H. and B.G.; writing—review and editing, B.H., É.H.-G. and B.G.; funding acquisition, B.H., É.H.-G. and B.G. All authors have read and agreed to the published version of the manuscript.

Funding: Supported by the ÚNKP-22-4-SZTE-491 New National Excellence Program of the Ministry for Culture and Innovation from the source of the National Research, Development and Innovation Fund and by the Hungarian National Research, Development and Innovation Office (GINOP-2.3.2-15-2016-00038, GINOP-2.3.2-15-2016-00001, GINOP-2.3.2-15-2016-00020, 2019-2.1111-TÉT-2019-00089, and K_16/120130) by the EU Horizon 2020 grant no. 739593. This research was partially funded by the CM_SMP_471080_2021 Campus Mundi Student Mobility Traineeship from the Tempus Public Foundation. The support of SRCD measurements from the CALIPSOplus (EU Framework Programme for Research and Innovation HORIZON 2020, grant no. 730872) is also greatly acknowledged.

Institutional Review Board Statement: Not applicable.

Data Availability Statement: Not applicable.

Acknowledgments: The authors thank Peter Baker for the development and maintenance of the ELKH Cloud (https://science-cloud.hu/ (accessed on 27 December 2022)) for hosting the ProteinProspector search engine. The authors would like to thank Milan Kožíšek for providing the pETM11-SUMO3 plasmid.

Conflicts of Interest: The authors declare no conflict of interest.

References

1. Mackay, J.P.; Crossley, M. Zinc fingers are sticking together. *Trends Biochem. Sci.* **1998**, *23*, 1–4. [CrossRef] [PubMed]
2. Tupler, R.; Perini, G.; Green, M.R. Expressing the human genome. *Nature* **2001**, *409*, 832–833. [CrossRef] [PubMed]
3. Miller, J.; McLachlan, A.; Klug, A. Repetitive zinc-binding domains in the protein transcription factor IIIA from Xenopus oocytes. *EMBO J.* **1985**, *4*, 1609–1614. [CrossRef] [PubMed]
4. Klug, A. The Discovery of Zinc Fingers and Their Applications in Gene Regulation and Genome Manipulation. *Annu. Rev. Biochem.* **2010**, *79*, 213–231. [CrossRef]
5. Cassandri, M.; Smirnov, A.; Novelli, F.; Pitolli, C.; Agostini, M.; Malewicz, M.; Melino, G.; Raschellà, G. Zinc-finger proteins in health and disease. *Cell Death Discov.* **2017**, *3*, 17071. [CrossRef]
6. Berg, J.M.; Shi, Y. The Galvanization of Biology: A Growing Appreciation for the Roles of Zinc. *Science* **1996**, *271*, 1081–1085. [CrossRef]
7. Bulaj, G.; Kortemme, T.; Goldenberg, D.P. Ionization–Reactivity Relationships for Cysteine Thiols in Polypeptides. *Biochemistry* **1998**, *37*, 8965–8972. [CrossRef]
8. Maynard, A.T.; Covell, D.G. Reactivity of Zinc Finger Cores: Analysis of Protein Packing and Electrostatic Screening. *J. Am. Chem. Soc.* **2001**, *123*, 1047–1058. [CrossRef]

9. Smith, J.N.; Hoffman, J.T.; Shirin, Z.; Carrano, C.J. H-Bonding Interactions and Control of Thiolate Nucleophilicity and Specificity in Model Complexes of Zinc Metalloproteins. *Inorg. Chem.* **2005**, *44*, 2012–2017. [CrossRef]
10. Lee, Y.-M.; Lim, C. Factors Controlling the Reactivity of Zinc Finger Cores. *J. Am. Chem. Soc.* **2011**, *133*, 8691–8703. [CrossRef]
11. Quintal, S.M.; Depaula, Q.A.; Farrell, N.P. Zinc finger proteins as templates for metal ion exchange and ligand reactivity. Chemical and biological consequences. *Metallomics* **2011**, *3*, 121–139. [CrossRef] [PubMed]
12. Frankel, A.D.; Berg, J.M.; Pabo, C.O. Metal-dependent folding of a single zinc finger from transcription factor IIIA. *Proc. Natl. Acad. Sci. USA* **1987**, *84*, 4841–4845. [CrossRef] [PubMed]
13. Krishna, S.S.; Majumdar, I.; Grishin, N.V. Structural classification of zinc fingers. *Nucleic Acids Res.* **2003**, *31*, 532–550. [CrossRef] [PubMed]
14. Kellis, J.T.; Nyberg, K., Jr.; Šail, D.; Fersht, A.R. Contribution of hydrophobic interactions to protein stability. *Nature* **1988**, *333*, 784–786. [CrossRef] [PubMed]
15. La Gatta, S.; Leone, L.; Maglio, O.; De Fenza, M.; Nastri, F.; Pavone, V.; Chino, M.; Lombardi, A. Unravelling the Structure of the Tetrahedral Metal-Binding Site in METP3 through an Experimental and Computational Approach. *Molecules* **2021**, *26*, 5221. [CrossRef] [PubMed]
16. Lander, E.S.; Linton, L.M.; Birren, B.; Nusbaum, C. Initial sequencing and analysis of the human genome. *Nature* **2001**, *409*, 860–921. [CrossRef]
17. Kim, Y.G.; Cha, J.; Chandrasegaran, S. Hybrid restriction enzymes: Zinc finger fusions to Fok I cleavage domain. *Proc. Natl. Acad. Sci. USA* **1996**, *93*, 1156–1160. [CrossRef]
18. Urnov, F.D.; Rebar, E.J.; Holmes, M.C.; Zhang, H.S.; Gregory, P.D. Genome editing with engineered zinc finger nucleases. *Nat. Rev. Genet.* **2010**, *11*, 636–646. [CrossRef]
19. Gaj, T.; Gersbach, C.A.; Barbas, C.F., III. ZFN, TALEN, and CRISPR/Cas-based methods for genome engineering. *Trends Biotechnol.* **2014**, *31*, 397–405. [CrossRef]
20. Segal, D.J.; Meckler, J.F.; Carroll, D.; Voytas, D.F.; Porteus, M.; Bickmore, W.A.; Walden, H.; Deans, A.J.; Moldovan, G.-L.; D'Andrea, A.D.; et al. Genome Engineering at the Dawn of the Golden Age. *Annu. Rev. Genom. Hum. Genet.* **2013**, *14*, 135–158. [CrossRef]
21. Urnov, F.D.; Miller, J.C.; Lee, Y.-L.; Beausejour, C.M.; Rock, J.M.; Augustus, S.; Jamieson, A.C.; Porteus, M.H.; Gregory, P.D.; Holmes, M.C. Highly efficient endogenous human gene correction using designed zinc-finger nucleases. *Nature* **2005**, *435*, 646–651. [CrossRef] [PubMed]
22. Passananti, C.; Corbi, N.; Oniri, A.; Di Certo, M.G.; Mattei, E. Transgenic Mice Expressing an Artifical Zinc Finger Regulator Targeting an Endogenous Gene, Engineered Zinc Finger Proteins. *Methods Mol. Biol.* **2010**, *649*, 183–206. [CrossRef] [PubMed]
23. Petolino, J.F. Genome editing in plants via designed zinc finger nucleases. *In Vitro Cell. Dev. Biol. Plant* **2015**, *51*, 1–8. [CrossRef]
24. Ousterout, D.G.; Kabadi, A.M.; Thakore, P.I.; Perez-Pinera, P.; Brown, M.T.; Majoros, W.H.; Reddy, T.E.; Gersbach, C.A. Correction of Dystrophin Expression in Cells from Duchenne Muscular Dystrophy Patients through Genomic Excision of Exon 51 by Zinc Finger Nucleases. *Mol. Ther.* **2015**, *23*, 523–532. [CrossRef]
25. Rodríguez, J.; Mosquera, J.; García-Fandiño, R.; Vázquez, M.E.; Mascareñas, J.L. A designed DNA binding motif that recognizes extended sites and spans two adjacent major grooves. *Chem. Sci.* **2016**, *7*, 3298–3303. [CrossRef]
26. Shi, Y.; Beger, R.D.; Berg, J.M. Metal binding properties of single amino acid deletion mutants of zinc finger peptides: Studies using cobalt(II) as a spectroscopic probe. *Biophys. J.* **1993**, *64*, 749–753. [CrossRef]
27. Hartwig, A.; Asmuss, M.; Ehleben, I.; Herzer, U.; Kostelac, D.; Pelzer, A.; Schwerdtle, T.; Bürkle, A. Interference by toxic metal ions with DNA repair processes and cell cycle control: Molecular mechanisms. *Environ. Health Perspect.* **2002**, *110*, 797–799. [CrossRef]
28. Witkiewiczkucharczyk, A.; Bal, W. Damage of zinc fingers in DNA repair proteins, a novel molecular mechanism in carcinogenesis. *Toxicol. Lett.* **2006**, *162*, 29–42. [CrossRef]
29. Kluska, K.; Adamczyk, J.; Krężel, A. Metal binding properties, stability and reactivity of zinc fingers. *Co-ord. Chem. Rev.* **2018**, *367*, 18–64. [CrossRef]
30. Párraga, G.; Horvath, S.J.; Eisen, A.; Taylor, W.E.; Hood, L.; Young, E.T.; Klevit, R.E. Zinc-dependent structure of a single-finger domain of yeast ADR1. *Science* **1988**, *241*, 1489–1492. [CrossRef] [PubMed]
31. Kopera, E.; Schwerdtle, T.; Hartwig, A.; Bal, W. Co(II) and Cd(II) Substitute for Zn(II) in the Zinc Finger Derived from the DNA Repair Protein XPA, Demonstrating a Variety of Potential Mechanisms of Toxicity. *Chem. Res. Toxicol.* **2004**, *17*, 1452–1458. [CrossRef] [PubMed]
32. Hartwig, A. Zinc Finger Proteins as Potential Targets for Toxic Metal Ions: Differential Effects on Structure and Function. *Antioxidants Redox Signal.* **2001**, *3*, 625–634. [CrossRef] [PubMed]
33. Krizek, B.A.; Merkle, D.L.; Berg, J.M. Ligand variation and metal ion binding specificity in zinc finger peptides. *Inorg. Chem.* **1993**, *32*, 937–940. [CrossRef]
34. Heinz, U.; Hemmingsen, L.; Kiefer, M.; Adolph, H.-W. Structural Adaptability of Zinc Binding Sites: Different Structures in Partially, Fully, and Heavy-Metal Loaded States. *Chem. A Eur. J.* **2009**, *15*, 7350–7358. [CrossRef] [PubMed]
35. Petering, D.H.; Huang, M.; Moteki, S.; Iii, C.F.S. Cadmium and lead interactions with transcription factor IIIA from Xenopus laevis: A model for zinc finger protein reactions with toxic metal ions and metallothionein. *Mar. Environ. Res.* **2000**, *50*, 89–92. [CrossRef]

36. Huang, M.; Krepkiy, D.; Hu, W.; Petering, D.H. Zn-, Cd-, and Pb-transcription factor IIIA: Properties, DNA binding, and comparison with TFIIIA-finger 3 metal complexes. *J. Inorg. Biochem.* **2004**, *98*, 775–785. [CrossRef]
37. Hanas, J.S.; Gunn, C.G. Inhibition of Transcription Factor IIIA-DNA Interactions by Xenobiotic Metal Ions. *Nucleic Acids Res.* **1996**, *24*, 924–930. [CrossRef]
38. Makowski, G.S.; Sunderman, F., Jr. The interactions of zinc, nickel, and cadmium with Xenopus transcription factor IIIA, assessed by equilibrium dialysis. *J. Inorg. Biochem.* **1992**, *48*, 107–119. [CrossRef]
39. Krepkiy, D.; Försterling, F.H.; Petering, D.H. Interaction of Cd^{2+} with Zn Finger 3 of Transcription Factor IIIA: Structures and Binding to Cognate DNA. *Chem. Res. Toxicol.* **2004**, *17*, 863–870. [CrossRef]
40. Razmiafshari, M.; Zawia, N.H. Utilization of a Synthetic Peptide as a Tool to Study the Interaction of Heavy Metals with the Zinc Finger Domain of Proteins Critical for Gene Expression in the Developing Brain. *Toxicol. Appl. Pharmacol.* **2000**, *166*, 1–12. [CrossRef]
41. Kothinti, R.; Blodgett, A.; Tabatabai, N.M.; Petering, D.H. Zinc Finger Transcription Factor Zn_3-Sp1 Reactions with Cd^{2+}. *Chem. Res. Toxicol.* **2010**, *23*, 405–412. [CrossRef] [PubMed]
42. Kothinti, R.K.; Blodgett, A.B.; Petering, D.H.; Tabatabai, N.M. Cadmium down-regulation of kidney Sp1 binding to mouse SGLT1 and SGLT2 gene promoters: Possible reaction of cadmium with the zinc finger domain of Sp1. *Toxicol. Appl. Pharmacol.* **2010**, *244*, 254–262. [CrossRef] [PubMed]
43. Bittel, D.; Dalton, T.; Samson, S.L.-A.; Gedamu, L.; Andrews, G.K. The DNA Binding Activity of Metal Response Element-binding Transcription Factor-1 Is Activated In Vivo and in Vitro by Zinc, but Not by Other Transition Metals. *J. Biol. Chem.* **1998**, *273*, 7127–7133. [CrossRef] [PubMed]
44. Kuwahara, J.; Coleman, J.E. Role of the zinc(II) ions in the structure of the three-finger DNA binding domain of the Sp1 transcription factor. *Biochemistry* **1990**, *29*, 8627–8631. [CrossRef]
45. Malgieri, G.; Palmieri, M.; Esposito, S.; Maione, V.; Russo, L.; Baglivo, I.; de Paola, I.; Milardi, D.; Diana, D.; Zaccaro, L.; et al. Zinc to cadmium replacement in the prokaryotic zinc-finger domain. *Metallomics* **2014**, *6*, 96–104. [CrossRef]
46. Roesijadi, G.; Bogumil, R.; Vasák, M.; Kägi, J.H.R. Modulation of DNA Binding of a Tramtrack Zinc Finger Peptide by the Metallothionein-Thionein Conjugate Pair. *J. Biol. Chem.* **1998**, *273*, 17425–17432. [CrossRef]
47. Zhang, B.; Georgiev, O.; Hagmann, M.; Günes, C.; Cramer, M.; Faller, P.; Vasák, M.; Schaffner, W. Activity of Metal-Responsive Transcription Factor 1 by Toxic Heavy Metals and H_2O_2 In Vitro Is Modulated by Metallothionein. *Mol. Cell. Biol.* **2003**, *23*, 8471–8485. [CrossRef]
48. Giedroc, D.P.; Chen, X.; Pennella, M.A.; LiWang, A.C. Conformational Heterogeneity in the C-terminal Zinc Fingers of Human MTF-1: An Nmr and Zinc-Binding Study. *J. Biol. Chem.* **2001**, *276*, 42322–42332. [CrossRef]
49. Ciavatta, L.; Grimaldi, M. Equilibrium constants of mercury(II) chloride complexes. *J. Inorg. Nucl. Chem.* **1968**, *30*, 197–205. [CrossRef]
50. Makowski, G.S.; Lin, S.-M.; Brennan, S.M.; Smilowitz, H.M.; Hopfer, S.M.; Sunderman, F.W. Detection of two Zn-finger proteins ofXenopus laevis, TFIIIA, and p43, by probing western blots of ovary cytosol with $^{65}Zn^{2+}$, $^{63}Ni^{2+}$, or $^{109}Cd^{2+}$. *Biol. Trace Element Res.* **1991**, *29*, 93–109. [CrossRef]
51. Sivo, V.; D'Abrosca, G.; Baglivo, I.; Iacovino, R.; Pedone, P.V.; Fattorusso, R.; Russo, L.; Malgieri, G.; Isernia, C. Ni(II), Hg(II), and Pb(II) Coordination in the Prokaryotic Zinc-Finger Ros87. *Inorg. Chem.* **2019**, *58*, 1067–1080. [CrossRef]
52. Hajdu, B.; Hunyadi-Gulyás, É.; Kato, K.; Kawaguchi, A.; Nagata, K.; Gyurcsik, B. Zinc binding of a Cys_2His_2 type zinc finger protein is enhanced by the interaction with DNA. *J. Biol. Inorg. Chem.* 2022, *accepted for publication*.
53. Sénèque, O.; Latour, J.-M. Coordination Properties of Zinc Finger Peptides Revisited: Ligand Competition Studies Reveal Higher Affinities for Zinc and Cobalt. *J. Am. Chem. Soc.* **2010**, *132*, 17760–17774. [CrossRef] [PubMed]
54. Kägi, J.H.; Vallee, B.L.; Carlson, J.M. Metallothionein: A Cadmium and Zinc-containing Protein from Equine Renal Cortex: II. Physicochemical properties. *J. Biol. Chem.* **1961**, *236*, 2435–2442. [CrossRef]
55. Willner, H.; Vasak, M.; Kaegi, J.H.R. Cadmium-thiolate clusters in metallothionein: Spectrophotometric and spectropolarimetric features. *Biochemistry* **1987**, *26*, 6287–6292. [CrossRef] [PubMed]
56. Micsonai, A.; Moussong, É.; Wien, F.; Boros, E.; Vadászi, H.; Murvai, N.; Lee, Y.-H.; Molnár, T.; Réfrégiers, M.; Goto, Y.; et al. BeStSel: Webserver for secondary structure and fold prediction for protein CD spectroscopy. *Nucleic Acids Res.* **2022**, *50*, W90–W98. [CrossRef] [PubMed]
57. Basile, L.A.; Coleman, J.E. Optical activity associated with the sulfur to metal charge transfer bands of Zn and Cd GAL4. *Protein Sci.* **1992**, *1*, 617–624. [CrossRef]
58. Zékány, L.; Nagypál, I. A Comprehensive Program for the Evaluation of Potentiometric and/or Spectrophotometric Equilibrium Data Using Analytical Derivatives. In *Computational Methods for the Determination of Formation Constants*; Springer: Boston, MA, USA, 1985; pp. 291–353. [CrossRef]
59. Malaiyandi, L.M.; Sharthiya, H.; Barakat, A.N.; Edwards, J.R.; Dineley, K.E. Using FluoZin-3 and fura-2 to monitor acute accumulation of free intracellular Cd^{2+} in a pancreatic beta cell line. *Biometals* **2019**, *32*, 951–964. [CrossRef]
60. Palmieri, M.; Malgieri, G.; Russo, L.; Baglivo, I.; Esposito, S.; Netti, F.; Del Gatto, A.; de Paola, I.; Zaccaro, L.; Pedone, P.V.; et al. Structural Zn(II) Implies a Switch from Fully Cooperative to Partly Downhill Folding in Highly Homologous Proteins. *J. Am. Chem. Soc.* **2013**, *135*, 5220–5228. [CrossRef]

61. Grazioso, R.; García-Viñuales, S.; Russo, L.; D'Abrosca, G.; Esposito, S.; Zaccaro, L.; Iacovino, R.; Milardi, D.; Fattorusso, R.; Malgieri, G.; et al. Substitution of the Native Zn(II) with Cd(II), Co(II) and Ni(II) Changes the Downhill Unfolding Mechanism of Ros87 to a Completely Different Scenario. *Int. J. Mol. Sci.* **2020**, *21*, 8285. [CrossRef]
62. Szunyogh, D.; Gyurcsik, B.; Larsen, F.H.; Stachura, M.; Thulstrup, P.W.; Hemmingsen, L.; Jancsó, A. Zn^{II} and Hg^{II} binding to a designed peptide that accommodates different coordination geometries. *Dalton Trans.* **2015**, *44*, 12576–12588. [CrossRef] [PubMed]
63. Kägi, J.H.; Vasák, M.; Lerch, K.; Gilg, D.E.; Hunziker, P.; Bernhard, W.R.; Good, M. Structure of mammalian metallothionein. *Environ. Health Perspect.* **1984**, *54*, 93–103. [CrossRef] [PubMed]
64. Stricks, W.; Kolthoff, I.M. Reactions between Mercuric Mercury and Cysteine and Glutathione. Apparent Dissociation Constants, Heats and Entropies of Formation of Various Forms of Mercuric Mercapto-Cysteine and -Glutathione. *J. Am. Chem. Soc.* **1953**, *75*, 5673–5681. [CrossRef]
65. Pires, S.; Habjanič, J.; Sezer, M.; Soares, C.M.; Hemmingsen, L.; Iranzo, O. Design of a Peptidic Turn with High Affinity for HgII. *Inorg. Chem.* **2012**, *51*, 11339–11348. [CrossRef] [PubMed]
66. DeSilva, T.M.; Veglia, G.; Porcelli, F.; Prantner, A.M.; Opella, S.J. Selectivity in heavy metal- binding to peptides and proteins. *Biopolymers* **2002**, *64*, 189–197. [CrossRef] [PubMed]
67. Rousselot-Pailley, P.; Sénèque, O.; Lebrun, C.; Crouzy, S.; Boturyn, D.; Dumy, P.; Ferrand, M.; Delangle, P. Model Peptides Based on the Binding Loop of the Copper Metallochaperone Atx1: Selectivity of the Consensus Sequence MxCxxC for Metal Ions Hg(II), Cu(I), Cd(II), Pb(II), and Zn(II). *Inorg. Chem.* **2006**, *45*, 5510–5520. [CrossRef] [PubMed]
68. Schneider, C.A.; Rasband, W.S.; Eliceiri, K.W. NIH Image to ImageJ: 25 Years of image analysis. *Nat. Methods* **2012**, *9*, 671–675. [CrossRef]
69. Bjerrum, J. *Metal Amine Formation in Aqueous Solution*; Haase: Copenhagen, Denmark, 1941.
70. Beck, M.T.; Nagypál, I. *Chemistry of Complex Equilibria*; Akadémiai Kiadó: Budapest, Hungary; Ellis Horwood Limited: Chichester, UK, 1990.
71. Malakhov, M.P.; Mattern, M.R.; Malakhova, O.A.; Drinker, M.; Weeks, S.D.; Butt, T.R. SUMO fusions and SUMO-specific protease for efficient expression and purification of proteins. *J. Struct. Funct. Genom.* **2004**, *5*, 75–86. [CrossRef]
72. Elhameed, H.A.A.; Hajdu, B.; Balogh, R.K.; Hermann, E.; Hunyadi-Gulyás, É.; Gyurcsik, B. Purification of proteins with native terminal sequences using a Ni(II)-cleavable C-terminal hexahistidine affinity tag. *Protein Expr. Purif.* **2019**, *159*, 53–59. [CrossRef]
73. Kim, C.A.; Berg, J.M. A 2.2 Å resolution crystal structure of a designed zinc finger protein bound to DNA. *Nat. Struct. Biol.* **1996**, *3*, 940–945. [CrossRef]
74. Krizek, B.A.; Amann, B.T.; Kilfoil, V.J.; Merkle, D.L.; Berg, J.M. A consensus zinc finger peptide: Design high-affinity metal binding a pH-dependent structure and a His to Cys sequence variant. *J. Am. Chem. Soc.* **1991**, *113*, 4518–4523. [CrossRef]
75. Miles, A.J.; Hoffmann, S.V.; Wallace, B.A.; Tao, Y.; Janes, R.W. Synchrotron Radiation Circular Dichroism (SRCD) spectroscopy: New beamlines and new applications in biology. *Spectroscopy* **2007**, *21*, 245–255. [CrossRef]
76. Miles, A.J.; Janes, R.W.; Brown, A.; Clarke, D.T.; Sutherland, J.C.; Tao, Y.; Wallace, B.A.; Hoffmann, S.V. Light flux density threshold at which protein denaturation is induced by synchrotron radiation circular dichroism beamlines. *J. Synchrotron Radiat.* **2008**, *15*, 420–422. [CrossRef]
77. Kluska, K.; Veronesi, G.; Deniaud, A.; Hajdu, B.; Gyurcsik, B.; Bal, W.; Krężel, A. Structures of Silver Fingers and a Pathway to Their Genotoxicity. *Angew. Chem. Int. Ed.* **2022**, *61*, e202116621. [CrossRef] [PubMed]
78. Marszałek, I.; Krężel, A.; Goch, W.; Zhukov, I.; Paczkowska, I.; Bal, W. Revised stability constant, spectroscopic properties and binding mode of Zn(II) to FluoZin-3, the most common zinc probe in life sciences. *J. Inorg. Biochem.* **2016**, *161*, 107–114. [CrossRef] [PubMed]

 inorganics

MDPI

Article

Low-Dimensional Compounds Containing Bioactive Ligands. Part XIX: Crystal Structures and Biological Properties of Copper Complexes with Halogen and Nitro Derivatives of 8-Hydroxyquinoline

Martina Kepeňová [1], Martin Kello [2], Romana Smolková [3], Michal Goga [4], Richard Frenák [4], Ľudmila Tkáčiková [5], Miroslava Litecká [6], Jan Šubrt [6] and Ivan Potočňák [1,*]

1 Institute of Chemistry, P. J. Šafárik University in Košice, Moyzesova 11, 041 54 Košice, Slovakia
2 Department of Pharmacology, P. J. Šafárik University in Košice, Trieda SNP 1, 040 11 Košice, Slovakia
3 Department of Ecology, Faculty of Humanities and Natural Sciences, University of Prešov, Ulica 17. Novembra 1, 081 16 Prešov, Slovakia
4 Department of Botany, Faculty of Science, Institute of Biology and Ecology, P. J. Šafárik University in Košice, Mánesova 23, 040 01 Košice, Slovakia
5 Department of Microbiology and Immunology, University of Veterinary Medicine and Pharmacy, Komenského 73, 041 81 Košice, Slovakia
6 Centre of Instrumental Techniques, Institute of Inorganic Chemistry of the CAS, Husinec-Řež č.p. 1001, CZ-25068 Řež, Czech Republic
* Correspondence: ivan.potocnak@upjs.sk; Tel.: +421-55-234-2335

Citation: Kepeňová, M.; Kello, M.; Smolková, R.; Goga, M.; Frenák, R.; Tkáčiková, Ľ.; Litecká, M.; Šubrt, J.; Potočňák, I. Low-Dimensional Compounds Containing Bioactive Ligands. Part XIX: Crystal Structures and Biological Properties of Copper Complexes with Halogen and Nitro Derivatives of 8-Hydroxyquinoline. *Inorganics* **2022**, *10*, 223. https://doi.org/10.3390/inorganics10120223

Academic Editors: Peter Segľa and Ján Pavlik

Received: 7 November 2022
Accepted: 22 November 2022
Published: 25 November 2022

Publisher's Note: MDPI stays neutral with regard to jurisdictional claims in published maps and institutional affiliations.

Abstract: Six new copper(II) complexes were prepared: $[Cu(ClBrQ)_2]$ (**1a**, **1b**), $[Cu(ClBrQ)_2]\cdot 1/2$ diox (**2**) (diox = 1,4-dioxane), $[Cu(BrQ)_2]$ (**3**), $[Cu(dNQ)_2]$ (**4**), $[Cu(dNQ)_2(DMF)_2]$ (**5**) and $[Cu(ClNQ)_2]$ (**6**), where HClBrQ is 5-chloro-7-bromo-8-hydroxyquinoline, HBrQ is 7-bromo-8-hydroxyquinoline, HClNQ is 5-chloro-7-nitro-8-hydroxyquinoline and HdNQ is 5,7-dinitro-8-hydroxyquinoline. Prepared compounds were characterised by infrared spectroscopy, elemental analysis and by X-ray structural analysis. Structural analysis revealed that all complexes are molecular. Square planar coordination of copper atoms in $[Cu(XQ)_2]$ (XQ = ClBrQ (**1a**, **1b**), BrQ (**3**) and ClNQ (**6**)) and tetragonal bipyramidal coordination in $[Cu(dNQ)_2(DMF)_2]$ (**5**) complexes were observed. In these four complexes, bidentate chelate coordination of XQ ligands via oxygen and nitrogen atoms was found. Hydrogen bonds stabilizing the structure were observed in $[Cu(dNQ)_2(DMF)_2]$ (**5**) and $[Cu(ClNQ)_2]$ (**6**), no other nonbonding interactions were noticed in all five structures. The stability of the complexes in DMSO and DMSO/water was evaluated by UV-Vis spectroscopy. Cytotoxic activity of the complexes and ligands was tested against MCF-7, MDA-MB-231, HCT116, CaCo2, HeLa, A549 and Jurkat cancer cell lines. The selectivity of the complexes was verified on a noncancerous Cos-7 cell line. Antiproliferative activity of the prepared complexes was very low in comparison with cisplatin, except complex **3**; however, its activity was not selective and was similar to the activity of its ligand HBrQ. Antibacterial potential was observed only with ligand HClNQ. Radical scavenging experiments revealed relatively high antioxidant activity of complex **3** against ABTS radical.

Keywords: copper complexes; derivatives of 8-hydroxyquinoline; crystal structure; bromination; biological properties

1. Introduction

Cancer treatment by using cisplatin spread widely after discovery of its anti-neoplastic activity [1]. Since then, some derivatives, such as carboplatin, oxaliplatin, nedaplatin, heptaplatin, lobaplatin and miriplatin, were prepared and used as anticancer agents, too [2]. Nowadays, nearly half of all cancer diseases are being treated by platinum-based drugs. These compounds are exceptionally successful against a broad spectrum of cancers, which is caused by their high reactivity [3]. Unfortunately, due this property, some negative phenomena are observed. Treatment with platinum-based drugs negatively influences healthy cells, which causes side-effects, such as neurotoxicity, renal toxicity, vomiting, and damage to the gastrointestinal tract, hair follicles and other tissues [4]. Another observed negative impact is the development of resistance, by which cancer cells try to escape apoptosis [5]. The necessity to reduce the negative impacts widely opened a new research field focused on other metal complexes as biological agents.

The quinoline family, including 8-hydroxyquinoline (8-HQ) and its derivatives, represents compounds with interesting pharmacological properties. For these compounds, anticancer, antibacterial, antifungal, antimalaria, antineurodegenerative and antiHIV effects were described [6–8]. These ligands can be coordinated to different metal atoms by oxygen and nitrogen atoms, and the resulting complexes often show increased anticancer activity. As an example, we can mention several complexes of Pd [9,10], Zn [11,12], Ga [13–15], Ru [16,17] and lanthanides [18–21], among which complexes with halogen- and nitro-derivatives of 8-HQ exhibited the highest activity. However, information on the anticancer activity of copper complexes rarely appears in the literature [12,22–25]. Therefore, we decided to prepare a series of copper complexes with commercially unavailable halogen- and nitro-derivatives of 8-HQ (HClBrQ = 5-chloro-7-bromo-8-hydroxyquinoline, HClNQ = 5-chloro-7-nitro-8-hydroxyquinoline and HdNQ = 5,7-dinitro-8-hydroxyquinoline), as well as with the commercially available, but hitherto unstudied HBrQ (HBrQ = 7-bromo-8-hydroxyquinoline) ligand: $[Cu(ClBrQ)_2]$ (**1a**, **1b**), $[Cu(ClBrQ)_2]\cdot 1/2$ diox (**2**), $[Cu(BrQ)_2]$ (**3**), $[Cu(dNQ)_2]$ (**4**), $[Cu(dNQ)_2(DMF)_2]$ (**5**) and $[Cu(ClNQ)_2]$ (**6**). In this paper, we present synthesis of these complexes and the results of infrared and UV-Vis spectroscopy, and elemental and monocrystal X-ray structural analysis. Moreover, we also discuss their antiproliferative activity against seven cancer cell lines and, using one non-cancerous Cos-7 cell line, we evaluate their selectivity. We also compare their anticancer activity with the activity of the corresponding ligands and cisplatin. Finally, we present the antimicrobial activity of the complexes and ligands against one gram-positive and one gram-negative bacteria, as well as their antioxidant activity.

2. Results and Discussion

2.1. Syntheses

The described copper complexes were synthesised by a simple mixing and stirring of solutions of corresponding ligand and copper(II) salt at laboratory (**1a**) or higher temperature (**1b–6**). While HClQ and HBrQ ligands used in the syntheses of complexes were obtained commercially, the HClNQ, HdNQ and HClBrQ ligands were first synthesised by previously described synthetic routes [26–28]. Interestingly, in the synthesis of **1a**, where $CuBr_2$ was used, in situ bromination of HClQ ligand was observed. The mechanism of bromination of organic substances using $CuBr_2$ was described in the literature [29,30]. This motivated us to prepare HClBrQ ligand and use it for the synthesis of **1b** to compare its structure with the structure of **1a**, because, in the case of in situ bromination, only 85% of the ligand molecules were brominated, as confirmed by semiquantitative EDS analysis (Figure S1) and X-ray structural analysis. Crystals of all prepared complexes suitable for X-ray structural analysis were obtained by slow crystallisation from corresponding solutions.

The composition of the prepared complexes was suggested by elemental (**1b–4**, **6**) and X-ray structural analysis (**1a**, **1b**, **3**, **5**, **6**). Crystals of **5** were unstable in air, due to the releasing of DMF from the structure, and crystals of **1a** were not prepared in sufficient quantity, and, therefore, they could not be characterised by elemental analysis.

2.2. Infrared Spectroscopy

All complexes were first characterised by IR spectroscopy (Figure 1) to confirm the presence of the ligands or solvates in the complexes. The presence of XQ ligands in the prepared compounds was confirmed by several bands, including very weak bands of

ν(C–H)$_{ar}$ vibrations observed at 3075–3098 cm^{-1}. Coordination of the ligands to the copper central atom was supported by the absence of ν(O–H) vibration, from the hydroxyl group, which should be observed in uncoordinated 8-hydroxyquinoline and its derivatives as a broad band in the 3700–3400 cm^{-1} region [31–35]. Characteristic bands of halogen functional groups in positions 5 and 7 presented, in the ranges 973–984 (ν(C$_5$–Cl) vibrations in (**1a**, **1b**, **2**, **6**) and 861–873 cm^{-1} (ν(C$_7$–Br) vibrations in (**1a**–**3**). The presence of a nitro group in **4**–**6** was manifested by bands of ν(N–O)$_{as}$ vibrations observed at 1561–1569 cm^{-1} and ν(N–O)$_{sym}$ vibrations at 1323 cm^{-1} [35].

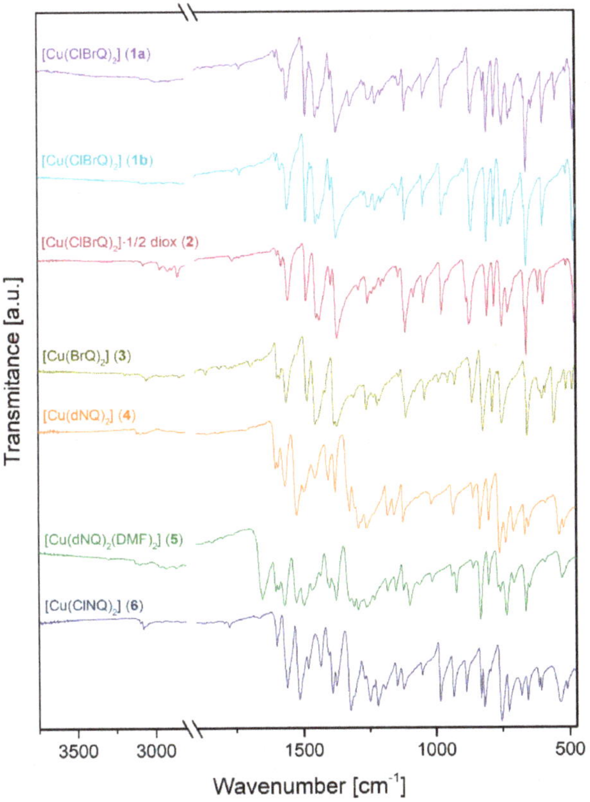

Figure 1. FT–IR spectra of **1**–**6**.

If we compare the IR spectra of **2** and **5** with the spectra of **1b** and **4**, respectively, we can clearly identify characteristic bands of used solvents in the IR spectra of the first two complexes. The bands of ν(CH$_2$) vibrations at 2965, 2912, 2887, 2849 cm^{-1} and ring breathing vibrations at 613 and 1182 cm^{-1} confirmed the presence of 1,4-dioxane in **2** [36–38]. Molecules of DMF in **5** manifested themselves as weak bands of ν(C–H)$_{al}$ vibrations at 2968, 2928, and 2860 cm^{-1}, as a band at 1098 cm^{-1}, which belonged to the deformation vibrations of the methyl group, and as a strong band of ν(C=O) vibrations at 1651 cm^{-1} [39].

2.3. UV-Vis Spectroscopy

A study of the stability of the prepared complexes (**1b**–**4**, **6**) was performed by comparison of the UV-Vis spectra of the complexes freshly suspended in Nujol with the spectra of the complexes in DMSO and DMSO/water (1:1) solutions, which were remeasured every 24 h over 3 days. As can be seen in Figure 2, the spectra of **1b** in DMSO and DMSO/water measured for 3 days were identical and very similar to the spectrum of **1b** prepared in Nujol, which suggested the stability of **1b** in the solutions. Similar results were obtained for **2**, **4** and **6**, but this was not the case for **3**. Figure 2 shows its UV-Vis spectra and it was obvious that the spectra in DMSO/water did not coincide with the spectra in Nujol or DMSO. This might be explained by a very low concentration of **3** in the DMSO/water solution, due to its continuous precipitation from this solution.

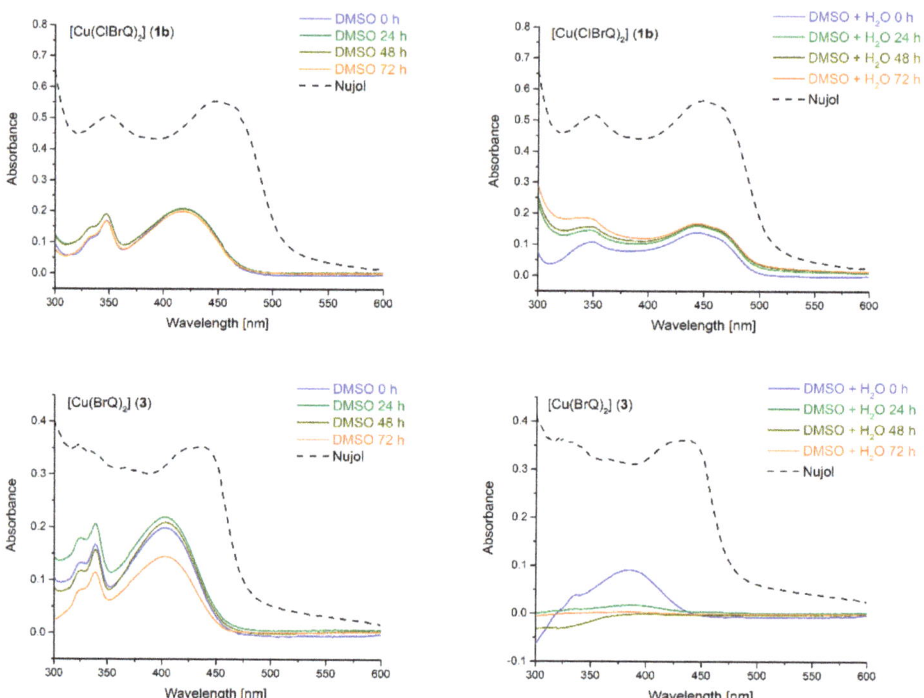

Figure 2. UV-VIS spectra of **1b** (**top**) and **3** (**bottom**).

2.4. X-ray Structure Analysis

The crystals of **1a**, **1b**, **3**, **5** and **6** were suitable for X-ray structural analysis, which confirmed their molecular character. Complexes **1a**, **1b** and **6** crystallised in the monoclinic space group $P2_1/c$, while **3** crystallised in $P2_1/n$ space group. Their unit cells contained two $[Cu(XQ)_2]$ molecules with copper atoms sitting on inversion centres. The central atoms were chelate-coordinated by two deprotonated corresponding XQ ligands (XQ = ClBrQ, BrQ and ClNQ) via oxygen and nitrogen atoms in a distorted square planar fashion (Figures 3 and S2). The shape of their polyhedral coordination was confirmed by bond lengths and angles (Table 1). Cu1–O1 bonds (1.920(2)–1.926(2) Å) were slightly shorter than Cu1–N1 bonds (1.953(2)–1.964(2) Å), due to the smaller covalent radius of the oxygen atom. Similar bond distances and angles were observed in other copper complexes with derivatives of 8-hydroxyquinoline [11,23,40,41].

Even though the crystals of **5** were not stable, being out of the maternal solution, collected X-ray data was sufficient to solve the structure. Complex **5** crystallised in the triclinic space group *P*-1. Cu1 atom, sitting on an inversion centre, was hexacoordinated by pairs of oxygen and nitrogen atoms in *trans*-positions from two molecules of dNQ, and axial positions were occupied by two oxygen atoms from two molecules of DMF (Figure 4). As can be seen in Table 2, the shape of the coordination polyhedron could be described as elongated tetragonal bipyramid, due to the Jahn–Teller effect. Nevertheless, the Cu1–O1 and Cu1–N1 bond lengths were close to those observed in **1a**, **1b**, **3** and **6**. All carbon atoms of DMF were disordered over two positions with site occupation factors being 0.62 and 0.38.

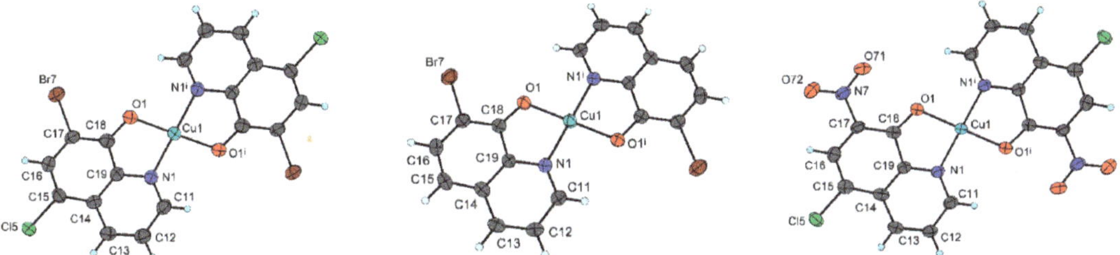

Figure 3. Molecular structure of **1b**, **3** and **6** (left to right). Displacement ellipsoids are drawn at the 80% or 50% (**3**) probability levels. Symmetry code: i = $-x$, $-y + 1$, $-z + 1$.

Table 1. Selected bond lengths [Å] and angles [°] for **1a**, **1b**, **3** and **6**.

	1a	1b	3	6
Cu1–O1	1.922(3)	1.920(2)	1.922(2)	1.926(2)
Cu1–N1	1.964(3)	1.965(2)	1.958(2)	1.953(2)
O1–Cu1–N1^{i}	94.90(11)	94.76(8)	94.57(9)	95.81(9)
O1–Cu1–N1	85.10(11)	85.24(8)	85.43(9)	84.19(9)

Symmetry code: i = $-x$, $-y + 1$, $-z + 1$.

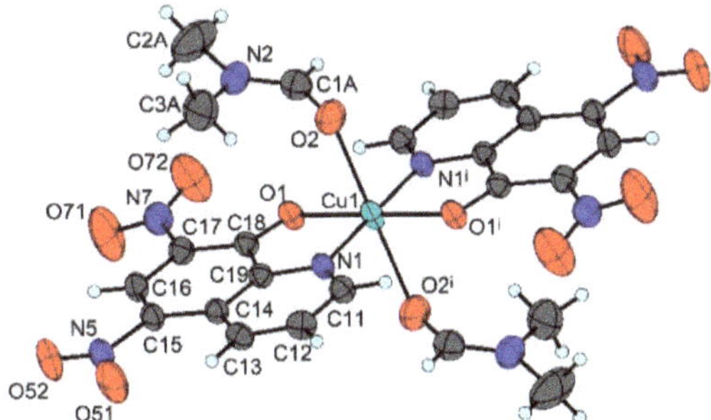

Figure 4. Molecular structure of **5**. Displacement ellipsoids are drawn at the 50% probability level. Only one position of carbon atoms of DMF is shown because of clarity. Symmetry code: i = $-x + 1$, $-y + 1$, $-z$.

Table 2. Selected bond lengths [Å] and angles [°] for **5**.

Bonds		Angles	
Cu1–O1	1.9508(14)	O1–Cu1–N1^{i}	96.82(7)
Cu1–N1	1.9706(17)	O1–Cu1–N1	83.18(7)
Cu1–O2	2.4920(19)	O1–Cu1–O2	94.14(6)
		N1–Cu1–O2	91.91(7)
		O1^{i}–Cu1–O2	85.86(6)
		N1^{i}–Cu1–O2	88.09(7)

Symmetry code: i = $-x + 1$, $-y + 1$, $-z$.

From the above-described structures, only structures with nitro groups (**5** and **6**) were stabilised by hydrogen bonds. Two hydrogen bonds presented in **5** (Table S1) created a layer parallel with the (01-1) plane (Figure 5). Only one hydrogen bond in the structure of **6** (Table S1) created a layer parallel with the (100) plane (Figure 5). No other significant intermolecular interactions were present in the structures of the complexes.

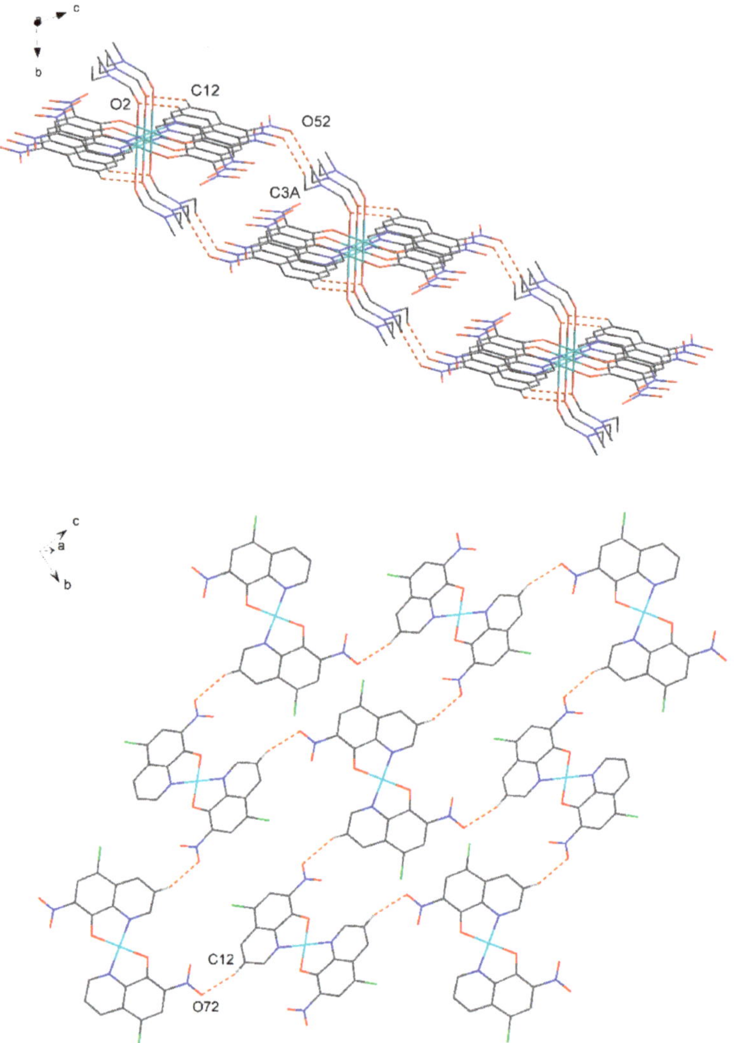

Figure 5. Layers formed by hydrogen bonds (dashed lines) in **5** (**top**) and **6** (**bottom**). Hydrogen atoms not involved in hydrogen bonds are omitted because of clarity.

2.5. Antiproliferative Activity

In the present work, four copper(II) complexes (**1b**, **3**, **4** and **6**) and their ligands were screened for potential antiproliferative activity. Furthermore, cisplatin, as a common chemotherapy agent, was used as a standard. There is evidence that copper complexes should be studied for their pro-apoptotic potential, also due to the fact that cancer cells take up larger amounts of copper than normal cells, as reviewed in [42]. Moreover, it was published that copper complexes with quinoline induced apoptosis and had cytotoxic effects on cancer cell lines [43–45]. In our study, the most potent novel copper(II) complex was **3** with IC_{50} values in the range 5.3–6.0 μM on all tested cancer cell lines (Table 3). However, we did not observe selectivity towards non-cancerous Cos-7 cells (IC_{50} = 5.8 μM). This complex was more efficient than cisplatin and its ligand HBrQ. Other tested complexes (**1b**, **4** and **6**) showed IC_{50} values above 200 μM on all tested cell lines, except for Caco-2 cells, where complex **1b** showed IC_{50} 106.8 μM and complex **4** around 46.4 μM with a selectivity towards Cos-7 cells, but lower effectivity than cisplatin. Based on the literature [46–51] the antiproliferative activity of copper(II) chloride against the seven cancer cell lines under study was not tested, because $CuCl_2$ displayed inconsequential in vitro toxicity.

Table 3. IC_{50} values of tested copper(II) complexes and their ligands.

	Cell Lines (IC50 in μM)							
	MCF-7	**MDA-MB-231**	**HCT116**	**Caco-2**	**HeLa**	**A549**	**Jurkat**	**Cos-7**
1a	NT	NT	NT	NT	NT	NT	NT	NT
1b	>200	>200	>200	106.8	>200	>200	>200	>200
2	NT	NT	NT	NT	NT	NT	NT	NT
3	5.8	6.0	5.3	5.4	5.4	5.6	5.7	5.8
4	204.8	>200	>200	46.4	>200	>200	>200	>200
5	NT	NT	NT	NT	NT	NT	NT	NT
6	>200	>200	>200	>200	>200	>200	>200	>200
HClBrQ	36.4	6.2	23.1	19.7	40.2	24.4	5.7	65.3
HBrQ	9.1	6.1	5.6	6.5	20.1	6.4	5.9	6.5
HdNQ	78.1	79.5	81.5	>200	>200	>200	72.8	>200
HClNQ	6.8	13.0	8.1	75.8	47.1	28.0	34.2	112.2
cisplatin	29.7	7.1	7.4	23.2	35.4	13.5	6.3	18.3

NT: not tested.

2.6. Antibacterial Activity

The antibacterial activity of the prepared complexes and their ligands were tested against gram-positive (*S. aureus*) and gram-negative (*E. coli*) bacteria ($CuCl_2$ was not tested, due to its inactivity against bacteria [52]). The RIZD, as well as MIC, were performed. Only ligand HClNQ in RIZD exhibited test inhibition against gram-positive bacteria *S. aureus* as well as gram-negative bacteria *E. coli*. The RIZD for *S. aureus* was 153% and for *E. coli* 123% (Table 4), whereas the inhibition of gentamicin sulfate as positive control was 100%. Other complexes were not suitable for antibacterial activity in concentration of 33.6 μM. This concentration was still possible, due to the solubility of the tested compounds.

Table 4. Antibacterial activity of tested complexes. RIZD (%) means percentage of relative inhibition zone diameter. All complexes were tested against *E. coli* and *S. aureus* in 3 replicates (n = 3, ±SD).

RIZD (%)	1b	3	4	6	HClBrQ	HBrQ	HdNQ	HClNQ
E. coli	NA	NA	NA	NA	NA	NA	NA	123.51
S. aureus	NA	NA	NA	NA	NA	NA	NA	153.31

NA: no activity.

MIC was tested with ligand HClNQ in all dilutions (1:1; 1:2; 1:4; 1:8). For *E. coli*, as well as *S. aureus*, the absorbance in dilutions 1:1 and 1:2 was lower and comparable with negative control (average 0.042 ± 0.005). Dilutions 1:4 and 1:8 showed higher absorbance, due to turbidity caused by the growing bacteria (Table 5).

Table 5. MIC (minimal inhibition concentration) of tested complex HClNQ with dilution ratio. Values mean absorbance. Absorbance was based on the cloudiness of the sample. Higher value means that bacteria were growing. Positive control represents bacterial growth without any treatment.

	1:1	**1:2**	**1:4**	**1:8**
E. coli				
HClNQ	0.047 ± 0.001	0.051 ± 0.001	0.105 ± 0.007	0.166 ± 0.043
Positive control	0.363 ± 0.007	0.363 ± 0.007	0.363 ± 0.007	0.363 ± 0.007
S. aureus				
HClNQ	0.046 ± 0.001	0.063 ± 0.003	0.107 ± 0.001	0.224 ± 0.012
Positive control	0.341 ± 0.035	0.341 ± 0.035	0.341 ± 0.035	0.341 ± 0.035

2.7. Radical Scavenging Activity

Radical scavenging activity was tested for complexes **1b**, **3**, **4** and **6**, along with free parental ligands against ABTS and DPPH radicals. Ligands HClBrQ and HdNQ showed low radical scavenging activity, while HClNQ was completely inactive within the measured range. Among the prepared compounds only **3** was active against both radicals (Table 6); however, its parental ligand HBrQ showed even stronger antioxidant activity [53]. The lower activity of studied complexes in comparison with free ligand molecules suggested that the radical scavenging mechanism involved reaction of radicals with 8-hydroxyquinoline derivatives, rather than Cu(II) central atoms. Antioxidant activity was more pronounced against ABTS radical than DPPH radical for all active compounds. This trend was previously observed for analogous metal complexes with the derivatives of 8-quinolinol [54,55].

Table 6. ABTS and DPPH radical scavenging activity (IC_{50} in μM; and SC% for 200 μM antioxidant concentration) for complex **3**, free ligands and L-ascorbic acid.

	ABTS		DPPH	
	IC_{50} (μM)	SC (%)	IC_{50} (μM)	SC (%)
3	33.33 ± 0.31	100	-	6.09 ± 2.08
HClBrQ	109.29 ± 1.50	82.80 ± 1.87	-	24.08 ± 0.78
HBrQ [a]	8.23 ± 0.52	100	46.03 ± 1.09	100
HdNQ	-	0	-	13.489 ± 0.87
L-ascorbic acid	21.03 ± 0.30	100	35.24 ± 1.25	100

Complexes **1b**, **4** and **6** and ligand HClNQ were inactive in the measured range 0–200 μM. [a] Data from [53].

3. Materials and Methods

3.1. Materials and Chemicals

The coordination compounds to be investigated were prepared using copper(II) chloride dihydrate, p.a. (Lachema, Neratovice, Czech Republic), copper(II) bromide, 99% (Sigma Aldrich, Bratislava, Slovakia), 5-chloro-8-hydroxyquinoline (HClQ), 95% (Sigma Aldrich), 7-bromo-8-hydroxyquinoline (HBrQ), 97% (Sigma Aldrich), N,N-dimethylformamide, 99% (Merck KGaA, Darmstadt, Germany), 1,2-dimethoxyethane, 99% (Alfa Aesar, Karlsruhe, Germany), ethanol, 96% (BGV, Hniezdne, Slovakia), methanol, p.a. (Centralchem, Bratislava, Slovakia), 1,4-dioxane, 99% (Centralchem), dimethyl sulfoxide, ≥99.9% (Sigma Aldrich). All commercially available chemicals were used without further purification. HdNQ, HClNQ and HClBrQ ligands were synthesised.

3.2. Syntheses

3.2.1. Synthesis of $[Cu(ClBrQ)_2]$ (**1a**)

The HClQ (35.9 mg, 0.2 mmol) was dissolved in DMF (10 mL). While continuously stirring, 10 mL of DMF solution of $CuBr_2$ (44.7 mg, 0.2 mmol) was added. After 30 min of stirring, the beaker was laid down at room temperature. After three months, yellow needles of **1a** had formed, and were filtered off, and dried in the air.

$[Cu(ClBrQ)_2]$ (**1a**)—Calc. for $C_{18}H_{8.30}Br_{1.70}Cl_2N_2O_2Cu$ (554.85 $g{\cdot}mol^{-1}$): C, 38.96; H, 1.51; N, 5.05%. Found: not measured. IR (ATR, cm^{-1}): $\nu(C–H)_{ar}$ 3075 (vw), $\nu(C=C)_{ar}$ 1595 (w), 1578 (w), 1555 (m), 1485 (m), ν(C=N) 1448 (m), ν(C–C) 1370 (s), 1223 (m), 1135 (m), ν(C–O) 1114 (m), β(C–H) 1045 (m), $\nu(C_5–Cl)$ 973 (m), $\nu(C_7–Br)$ 864 (m), γ(C–H) 806 (m), Ring breathing 779 (m), 748 (m), β(CCC) 723 (m), 654 (s), β(CNC) 596 (m), $\beta(C_5–Cl)$ 508 (w), γ(CCC) 485 (m).

3.2.2. Synthesis of $[Cu(ClBrQ)_2]$ (**1b**)

HClBrQ (25.9 mg, 0.1 mmol) was dissolved in ethanol (10 mL) and warmed to 60 °C. While continuously stirring, 10 mL of DMF solution of $CuCl_2$ (8.5 mg of $CuCl_2{\cdot}2H_2O$, 0.05 mmol) (warmed to 60 °C) was added. After 30 min of stirring, the beaker was laid down at room temperature. After five days, yellow needles of **1b** had formed, and were filtered off, and dried in the air.

[Cu(ClBrQ)$_2$] (**1b**)—Calc. for $C_{18}H_8Br_2Cl_2N_2O_2Cu$ (578.73 g·mol^{-1}): C, 37.37; H, 1.39; N, 4.84%. Found: C, 37.72; H, 1.58; N, 4.65%. IR (ATR, cm^{-1}): ν(C–H)$_{ar}$ 3074 (vw), ν(C=C)$_{ar}$ 1594 (w), 1578 (w), 1555 (m), 1485 (m), ν(C=N) 1448 (m), ν(C–C) 1370 (s), 1223 (m), 1136 (m), ν(C–O) 1114 (m), β(C–H) 1045 (m), ν(C$_5$–Cl) 974 (m), ν(C$_7$–Br) 863 (m), γ(C–H) 806 (m), Ring breathing 779 (m), 750 (m), β(CCC) 723 (m), 654 (s), β(CNC) 596 (m), β(C$_5$–Cl) 509 (w), γ(CCC) 484 (m).

3.2.3. Synthesis of [Cu(ClBrQ)$_2$]·1/2 Diox (**2**)

HClBrQ (25.9 mg, 0.1 mmol) was dissolved in 1,4-dioxane (10 mL) and warmed to 60 °C. While continuously stirring, 10 mL ethanol solution of $CuCl_2$ (8.5 mg of $CuCl_2{\cdot}2H_2O$, 0.05 mmol) (warmed to 60 °C), was added. After 30 min of stirring, the beaker was laid down and the precipitate of **2** formed during stirring was filtered off. Mother liquor was laid to crystallise at room temperature. After five days, yellow needles of **2** had formed, and were filtered off, and dried in the air. The identity of the powder and crystals was verified by IR spectroscopy.

[Cu(ClBrQ)$_2$]·1/2 diox (**2**)—Calc. for $C_{20}H_{12}Br_2Cl_2N_2O_3Cu$ (622,58 g·mol^{-1}): C, 38.58; H, 1.94; N, 4.50%. Found: C, 38.34; H, 1.92; N, 4.42%. IR (ATR, cm^{-1}): ν(C–H)$_{ar}$ 3079 (vw), ν(CH$_2$) 2965 (w), 2912 (w), 2887 (w), 2849 (m), ν(C=C)$_{ar}$ 1592 (w), 1577 (w), 1553 (m), 1486 (m), ν(C=N) 1448 (m), ν(C–C) 1369 (s), 1253 (m), 1139 (m), ν(C–O) 1113 (m), β(C–H) 1045 (m), ν(C$_5$–Cl) 975 (m), ν(C$_7$–Br) 873 (m), γ(C–H) 805 (m), Ring breathing 779 (m), 613 (w), 750 (m), 728 (m), β(CCC) 657 (s), β(CNC) 594 (m), β(C$_5$–Cl) 509 (w), γ(CCC) 482 (m).

3.2.4. Synthesis of [Cu(BrQ)$_2$] (**3**)

HBrQ (22.4 mg, 0.1 mmol) was dissolved in 10 mL of ethanol and warmed to 60 °C. While continuously stirring, 10 mL of 1,4-dioxane solution of $CuCl_2$ ($CuCl_2{\cdot}2H_2O$ of 8.5 mg, 0.05 mmol) (warmed to 60 °C) was added. After 30 min of stirring, the beaker was laid down at room temperature. After five days, yellow needles of **3** had formed, and were filtered off, and dried in the air.

[Cu(BrQ)$_2$] (**3**)—Calc. for $C_{18}H_{10}Br_2Cl_2N_2O_2Cu$ (509.64 g·mol^{-1}): C, 42.42; H, 1.98; N, 5.50%. Found: C, 42.57; H, 2.10; N, 5.37%. IR (ATR, cm^{-1}): ν(C–H)$_{ar}$ 3054 (vw), ν(C=C)$_{ar}$ 1584 (m), 1561 (s), 1483 (s), ν(C=N) 1453(s), ν(C–C) 1371 (s), 1259 (m), 1231 (m), 1219 (m), ν(C–O) 1112 (m), β(C–H) 1041 (m), ν(C$_7$–Br) 861 (m), γ(C–H) 821 (m), Ring breathing 786 (m), 751 (s), β(CCC) 654 (s), β(CNC) 599 (w), 587(w), β(C–O) 555 (m), γ(CCC) 482 (m).

3.2.5. Synthesis of [Cu(dNQ)$_2$] (**4**)

HdNQ (23.5 mg, 0.1 mmol) was dissolved in methanol (10 mL) and warmed to 60 °C. While continuously stirring, solution of $CuCl_2$ (8.5 mg of $CuCl_2{\cdot}2H_2O$, 0.05 mmol) in dimethoxyethane (10 mL, warmed to 60 °C) was added. After 30 min of stirring, the precipitate of **4** was filtered off and dried in the air.

[Cu(dNQ)$_2$] (**4**)—Calc. for $C_{18}H_8N_6O_{10}Cu$ (531,84 g·mol^{-1}): C, 40.65; H, 1.52; N, 15.80%. Found: C, 41.04; H, 1.88; N, 15.48%. IR (ATR, cm^{-1}): ν(C–H)$_{ar}$ 3056 (vw), ν(C=C)$_{ar}$ 1594 (m), 1518 (m), 1492 (m), ν(N–O)$_{as}$ 1567 (m), ν(C=N) 1453 (m), ν(C–C) 1405 (w), 1379 (m), 1255 (m), ν(N–O)$_{sym}$ 1321 (m), β(C–H) 1169 (m), 1155 (m), 1016 (w), ν(C–O) 1113 (m), δ(NO$_2$) 928 (m), 699 (m), γ(C–H) 831 (m), 797 (m), Ring breathing 749 (s), 733 (w), β(CCC) 655 (m), β(CNC) 592 (w), β(C–O) 530 (m), γ(CCC) 457 (m).

3.2.6. Synthesis of [Cu(dNQ)$_2$(DMF)$_2$] (**5**)

HdNQ (23.5 mg, 0.1 mmol) was dissolved in DMF (10 mL) and warmed to 60 °C. While continuously stirring, 10 mL of DMF solution of $CuCl_2$ (8.5 mg of $CuCl_2{\cdot}2H_2O$, 0.05 mmol) (warmed to 60 °C) was added. After 30 min of stirring, the beaker was laid down at room temperature. After two days, green prisms of **5** had formed, and were filtered off, and dried in the air.

$[Cu(dNQ)_2DMF_2]$ (**5**)—Calc. for $C_{18}H_8N_6O_{10}Cu$ (531.84 g·mol^{-1}): C, 40.65; H, 1.52; N, 15.80%. Found: not measured, IR (ATR, cm^{-1}): ν(C–H)$_{ar}$ 3058 (vw), ν(C–H)$_{al}$ 2968 (vw), 2928 (vw), 2860 (vw), ν(C=O) 1651 (s), ν(C=C)$_{ar}$ 1601 (m), 1591 (m), 1525 (m), 1497 (m), ν(N–O)$_{as}$ 1569 (m), ν(C=N) 1454 (m), ν(C–C)$_{ar}$ 1399 (m), 1374 (m), 1260 (m), ν(N–O)$_{sym}$ 1323 (m), β(C–H) 1182 (m), 1151 (m), ν(C–O) 1122 (m), $\delta(CH_3)$ 1098 (m), $\delta(NO_2)$ 922 (m), 705 (m), γ(C–H) 831 (m), 801 (m), Ring breathing 756 (m), 733 (s), β(CCC) 660 (m), β(CNC) 591 (w), β(C–O) 527 (m), γ(CCC) 470 (w), 450 (w).

3.2.7. Synthesis of $[Cu(ClNQ)_2]$ (**6**)

HClNQ (22.5 mg, 0.1 mmol) was dissolved in 10 mL of DMF and warmed to 60 °C. While continuously stirring, 10 mL of DMF solution of $CuCl_2$ (8.5 mg of $CuCl_2·2H_2O$, 0.05 mmol) (warmed to 60 °C), was added. After 30 min of stirring, the beaker was laid down and the precipitate of **6**, that formed during stirring, was filtered off. The mother liquor was laid to crystallise at room temperature. After five days, green needles of **6** had formed, and were filtered off, and dried in the air. The identity of the powder and crystals was verified by IR spectroscopy.

$[Cu(ClNQ)_2]$ (**6**)—Calc. for $C_{18}H_8Cl_2N_4O_6Cu$ (510.73 g·mol^{-1}): C, 42.50; H, 1.78; N, 10.97%. Found: C, 42.33; H, 1.58; N, 10.97%. IR (ATR, cm^{-1}): ν(C–H)$_{ar}$ 3078 (vw), ν(C=C)$_{ar}$ 1599 (m), 1482 (s) ν(N–O)$_{as}$ 1561 (m), ν(C=N) 1434 (s), ν(C–C) 1384 (m), 1374 (m), 1249 (m), 1220 (s), ν(N–O)$_{sym}$ 1323 (s), ν(C–O) 1123 (m), β(C–H) 1148 (m), 1052 (m), ν(C_5–Cl) 984 (m), $\delta(NO_2)$ 934 (m), 679 (m), γ(C–H) 831 (m), 818 (s), Ring breathing 754 (s), β(CCC) 726 (m), 679 (m), 655 (s), β(CNC) 613 (w), β(C–O) 535 (m), β(C_5–Cl) 510 (w), γ(CCC) 476 (m).

3.3. *Physical Measurements*

Infrared spectra of prepared complexes were recorded on a Nicolet 6700 FT-IR spectrometer from Thermo Scientific with a diamond crystal Smart OrbitTM, in the range 4000–400 cm^{-1}. Elemental analyses of C, H, and N were with a CHNS Elemental Analyzer varioMICRO from Elementar Analysensysteme GmbH. Absorption spectra were measured with a SPECORD 250 spectrophotometer (Analytik Jena, Jena, Germany), from 300 to 600 nm, in Nujol, and DMSO and DMSO/water (1:1) solutions at 24, 48 and 72 h intervals. IR and UV-Vis spectra were described in Origin 2022b [56]. The morphological characteristics of the samples were studied using a JEOL JSM 6510 scanning electron microscope (W cathode, 20 nm resolution at 1 kV). Semi-quantitative chemical analysis of the samples was determined using the attached Oxford Instruments EDS analyser INCA X act. Measurements were performed on native samples without any conductive coating.

3.4. *X-ray Structure Analysis*

The crystal structures of **3** and **5** were determined using an Oxford Diffraction Xcalibur2 diffractometer equipped with a Sapphire2 CCD detector, while the structures of **1a**, **1b** and **6** were determined using a Rigaku XtaLAB Synergy S diffractometer with Hybrid Pixel Array detector (HyPix-6000HE). CrysAlis Pro software was used for data collection and cell refinement, data reduction and absorption correction [57]. Structures of prepared complexes were solved by SHELXT [58] and refined by SHELXL [59], implemented in the WinGX program [60]. For all non-H atoms, anisotropic displacement parameters were refined. Hydrogen atoms of XQ and DMF molecules were placed in calculated positions and refined riding on carbon atoms. Presence of hydrogen bonds was analysed by using SHELXL, while PLATON [61], running under WinGX, was used to analyse π–π interaction. Diamond was used for molecular graphics [62]. The summary of crystal data and structure refinements for **1a**, **1b**, **3**, **5** and **6** is presented in Table 7.

Table 7. Crystal data and structural refinement of **1a**, **1b**, **3**, **5** and **6**.

Compound	1a	1b	3	5	6
Empirical formula	$C_{18}H_{8.30}Br_{1.70}Cl_2CuN_2O_2$	$C_{18}H_8Br_2Cl_2CuN_2O_2$	$C_{18}H_{10}Br_2CuN_2O_2$	$C_{24}H_{22}CuN_8O_{12}$	$C_{18}H_8Cl_2CuN_4O_6$
Formula weight [g·mol^{-1}]	554.85	578.52	509.64	678.03	510.72
Temperature [K]	100(2)	100(2)	173(2)	173(2)	100(2)
Wavelength [Å]	1.54184	1.54184	0.71073	0.71073	1.54184
Crystal system	monoclinic	monoclinic	monoclinic	triclinic	monoclinic
Space group	$P2_1/c$	$P2_1/c$	$P2_1/n$	$P-1$	$P2_1/c$
Unit cell dimensions [Å, °]	a = 4.89650(10) b = 10.32300(10) c = 17.5282(2) β = 90.182(1)	a = 4.88650(10) b = 10.3474(3) c = 17.5221(4) β = 90.201(2)	a = 4.9383(2) b = 10.1335(4) c = 16.2276(6) β = 90.581(3)	a = 6.5539(6) b = 10.1953(7) c = 11.1513(7) α = 103.833(5) β = 100.445(6) γ = 103.446(7)	a = 3.7670(2) b = 12.4567(6) c = 18.1614(5) β = 93.452(3)
Volume [Å^3]	885.99(2)	885.96(4)	812.02(5)	681.18(9)	850.67(7)
Z; density (calculated) [g·cm^{-3}]	2; 2.080	2; 2.169	2; 2.084	1; 1.653	2; 1.994
Absorption coefficient [mm^{-1}]	9.162	9.962	6.280	0.883	5.193
F(000)	538	558	494	347	510
Crystal shape, colour	needle, yellow	needle, yellow	needle, yellow	prism, green	needle, green
Crystal size [mm]	0.161 × 0.078 × 0.031	0.050 × 0.040 × 0.020	0.560 × 0.228 × 0.064	0.654 × 0.216 × 0.152	0.048 × 0.015 × 0.009
θ range for data collection [°]	4.972–77.225	4.964–76.881	3.216–28.707	3.295–28.514	4.306–77.467
Index ranges	$-5 \leq h \leq 6$, $-13 \leq k \leq 13$, $-21 \leq l \leq 22$	$-5 \leq h \leq 6$, $-12 \leq k \leq 11$, $-21 \leq l \leq 21$	$-6 \leq h \leq 6$, $-13 \leq k \leq 11$, $-20 \leq l \leq 21$	$-8 \leq h \leq 8$, $-13 \leq k \leq 13$, $-11 \leq l \leq 14$	$-2 \leq h \leq 4$, $-11 \leq k \leq 15$, $-22 \leq l \leq 22$
Reflections collected/independent	30,356/1833 [R(int) = 0.0494]	8341/1769 [R(int) = 0.0323]	5221/1863 [R(int) = 0.0257]	7356/3093 [R(int) = 0.0261]	6520/1709 [R(int) = 0.0335]
Data/restrains/parameters	1833/0/124	1769/0/124	1863/0/115	3093/0/237	1709/0/142
Goodness-of-fit on F^2	1.161	1.062	1.070	1.089	1.161
Final R indices [$I > 2\sigma(I)$]	R1 = 0.0347, $wR2$ = 0.0768	R1 = 0.0271, $wR2$ = 0.0776	R1 = 0.0312, $wR2$ = 0.0646	R1 = 0.0428, $wR2$ = 0.0905	R1 = 0.0367, $wR2$ = 0.1057
R indices (all data)	R1 = 0.0371, $wR2$ = 0.0777	R1 = 0.0296, $wR2$ = 0.0788	R1 = 0.0441, $wR2$ = 0.0718	R1 = 0.0549, $wR2$ = 0.0975	R1 = 0.0452, $wR2$ = 0.1093
Largest diff. peak and hole [e·Å^{-3}]	0.435 and −0.476	0.458 and −0.573	0.424 and −0.384	0.402 and −0.452	0.461 and −0.643

3.5. Cell Cultures

The human cancer cell lines were purchased from ATCC (American Type Culture Collection; Manassas, VA, USA). HCT116 (human colorectal carcinoma), HeLa (human cervical adenocarcinoma) and Jurkat (human leukemic T cell lymphoma) were cultured in RPMI 1640 medium (Biosera, Kansas City, MO, USA) while A549 (human alveolar adenocarcinoma), MCF-7 (human Caucasian breast adenocarcinoma), Caco-2 (human colorectal adenocarcinoma) and MDA-MB-231 (human breast cancer cell line) were maintained in a growth medium consisting of high glucose Dulbecco's Modified Eagle Medium (DMEM) + sodium pyruvate (Biosera). The human kidney fibroblasts (Cos-7) were cultured in DMEM medium (Biosera). All media were supplemented with a 10% fetal bovine serum (FBS; Invitrogen, Carlsbad, CA, USA), Antibiotic/Antimycotic Solution (Sigma, St. Louis, MO, USA) and maintained in an atmosphere containing 5% CO_2 in humidified air at 37 °C. Prior to each experiment, cell viability was greater than 95%.

Screening of Antiproliferative/Cytotoxic Activity

The antiproliferative/cytotoxic effect of copper complexes (concentrations of 10, 50 and 100 μM) was determined by resazurin assay in HCT116, Caco-2, A549, MCF-7, MDA-MB-231, HeLa, Jurkat and Cos-7 cells. Tested cells (1×10^4/well) were seeded in 96-well plates. After 24 h, final copper complex concentrations, prepared from DMSO stock solution, were added, and incubation proceeded for the next 72 h. Ten microliters of resazurin solution (Merck, Darmstadt, Germany), at a final concentration of 40 μM, was added to each well at the end-point (72 h). After a minimum of 1 h incubation, the fluorescence of the metabolic product resorufin was measured by the automated Cytation^{TM} 3 cell imaging multi-mode reader (Biotek, Winooski, VT, USA) at 560 nm excitation/590 nm emission filter. The results were expressed as a fold of the control, where control fluorescence was taken as 100%. All experiments were performed in triplicate. The IC_{50} values were calculated from these data.

3.6. Antibacterial Activity

3.6.1. Microorganisms Used

The tested bacteria (S. aureus CCM 4223 and E. coli CCM 3988) were obtained from the Czech collection of microorganisms (CCM, Brno, Czech Republic).

3.6.2. Agar Well-Diffusion Method

The antibacterial properties of the four complexes, **1b**, **3**, **4** and **6**, and their ligands, HClBrQ, HBrQ, HdNQ and HClNQ, were evaluated by the agar well diffusion method using a slightly modified process compared to [63]. Firstly, each compound was dissolved in a small amount of 100% DMSO and then dissolved to 33.6 μM solution. As a positive control, gentamicin sulfate (Biosera, Nuaille, France), with concentration 50 μg/mL, was used.

Bacteria were cultured overnight, aerobically, at 37 °C in LB medium (Sigma-Aldrich, Saint-Louis, MO, USA), with agitation. The inoculum from these overnight cultures was prepared by adjusting the density of the culture to equal that of the 0.5 McFarland standard ($1–2 \times 10^8$ CFU/mL), by adding a sterile saline solution. These bacterial suspensions were diluted 1:300 in liquid plate count agar (HIMEDIA, Mumbai, India) resulting in a final concentration of bacteria approximately 5×10^5 CFU/mL, and 20 mL of this inoculated agar was poured into a Petri dish (diameter 90 mm). Once the agar was solidified, five mm diameter wells were punched in the agar and filled with 50 μL of samples. Gentamicin sulfate, with a concentration of 50 μg/mL, was used as a positive control. The plates were incubated for 18–20 h at 37 °C. Afterwards, the plates were photographed, and the inhibition zones were measured by ImageJ 1.53e software (U. S. National Institutes of Health, Bethesda, MD, USA). The values used for the calculation were mean values calculated from 3 replicate tests.

The antibacterial activity was calculated by applying the formula reported in [63]:

$$\%RIZD = [(IZD\ sample - IZD\ negative\ control)/IZD\ gentamicin] \times 100$$

where RIZD is the relative inhibition zone diameter (%) and IZD is the inhibition zone diameter (mm).

As a negative control, the inhibition zones of 5% DMSO equal to 0 were taken. The inhibition zone diameter (IZD) was obtained by measuring the diameter of the transparent zone.

3.6.3. Determination of the Minimum Inhibitory Concentration (MIC) by the Microdilution Method

The minimum inhibitory concentration was determined by the microdilution method with a slight modification of the procedure described in [64].

Stock solutions of the test samples prepared in 5% DMSO (Sigma-Aldrich, USA) were two-fold diluted (1:1 to 1:8) in wells of a 96-well plate (Greiner Bio-One, Germany): the wells of the microtiter plate were filled with 50 μL of Mueller–Hinton Broth (MHB, HIMEDIA, Mumbai, India), and 50 μL of stock solution of the test substance was added to the first well (1:1 dilution). After mixing, 50 μL of this solution was transferred to the next well (1:2 dilution), etc. Each sample was tested in triplicate.

Bacteria were cultured overnight, aerobically, at 37 °C in MHB, with agitation. The inoculum from the overnight cultures was prepared by adjusting the density of bacterial suspension to 0.5 McFarland standard (1–2 $\times$ 10^8 CFU/mL) by adding sterile saline and then diluting 1:150 in MHB. Subsequently, 50 μL of this inoculum was added to each well with 50 μL of diluted test samples (final concentration of bacteria in the well ca. 5 $\times$ 10^5 CFU/mL). Wells filled only with MHB medium and bacterial suspension were used as a positive control, and wells filled with sterile MHB alone were used as a negative control. The plates were covered with a lid and incubated for 18–20 h at 37 °C. The evaluation was made by measuring the absorbance at 600 nm on a microplate reader (Synergy HT, Biotek, Santa Clara, CA, USA).

3.7. *Radical Scavenging Experiments*

Radical scavenging activity of complexes and free ligands was estimated by DPPH (2,2-diphenyl-1-picrylhydrazyl) and ABTS (diammonium 2,2′-azino-bis(3-ethylbenzothiazoline-6-sulfonate) radicals, according to slightly modified methods described in literature [65,66]. Compounds were dissolved in DMSO and the resulting solutions were diluted with MeOH in 1:3 ratio. The prepared solutions were mixed with methanolic solutions of the respective radicals and incubated in the dark at room temperature (7 min ABTS and 30 min DPPH). The absorbance was recorded at 734 nm (ABTS) and 517 nm (DPPH), respectively. L-ascorbic acid was used as a standard, and experiments were performed in triplicate. The IC_{50} parameters were calculated from a linear plot of the inhibition percentage against the concentration of compounds.

4. Conclusions

In this work, we present six new copper(II) complexes: $[Cu(ClBrQ)_2]$ (**1a**, **1b**), $[Cu(ClBrQ)_2]\cdot 1/2$ diox (**2**) (diox = 1,4-dioxane), $[Cu(BrQ)_2]$ (**3**), $[Cu(dNQ)_2]$ (**4**), $[Cu(dNQ)_2(DMF)_2]$ (**5**) and $[Cu(ClNQ)_2]$ (**6**), where HClBrQ is 5-chloro-7-bromo-8-hydroxyquinoline, HBrQ is 7-bromo-8-hydroxyquinoline, HClNQ is 5-chloro-7-nitro-8-hydroxyquinoline and HdNQ is 5,7-dinitro-8-hydroxyquinoline. Prepared complexes were characterised by IR spectroscopy (**1–6**), and elemental (**1b–4** and **6**) and X-ray structural (**1a**, **1b**, **3**, **5** and **6**) analysis. The sability of **1b–4** and **6** in DMSO and DMSO/water solutions was verified by UV-Vis spectroscopy.

Complexes **1a**, **1b**, **3**, **5** and **6** were molecular compounds. Crystal structures of **1a**, **1b**, **3** and **6** formed by square planar $[Cu(XQ)_2]$ complexes, in which copper atoms were coordinated by two pairs of oxygen and nitrogen atoms from two deprotonated XQ ligands. On the other hand, a tetragonal bipyramidal coordination of the copper atom in **5** was observed. The equatorial plane was occupied by the same atoms as in the mentioned $[Cu(XQ)_2]$ complexes, and oxygen atoms from DMF molecules occupied axial positions. Interestingly, despite similar structures, intramolecular interactions were observed only in

5 and **6**, where hydrogen bonds existed. They created layers parallel with (01-1) (**5**) and (100) (**6**) plane.

After cytotoxic evaluation, only complex **3** showed promising potential as an anticancer agent, but, due to the low selectivity towards non-cancerous cells, modification of the complex was required.

Among the prepared compounds only **3** was active against both tested radicals (ABTS and DPPH). Antioxidant activity could be involved in the cytotoxic effects of **3** against the tested cell line; however, the free ligand showed even better antiradical properties. Apparently, the coordination of HBrQ ligand to Cu(II) decreased its antioxidant activity.

Only the HClNQ ligand showed antibacterial potential with concentration 33.6 μM.

Supplementary Materials: The following supporting information can be downloaded at: https://www.mdpi.com/article/10.3390/inorganics10120223/s1, Figure S1: SEM micrograph (a) of **1a** along with EDS elemental analysis (b); Figure S2: Molecular structure of **1a**; Table S1: Hydrogen bonds [Å and °] for **5** and **6**.

Author Contributions: Conceptualization, I.P. and M.K. (Martina Kepeňová); investigation, M.K. (Martina Kepeňová), M.K. (Martin Kello), R.S., M.G., R.F., Ľ.T., M.L., J.Š. and I.P.; resources, M.K. (Martin Kello), M.G., M.L. and I.P.; writing—original draft preparation, M.K. (Martina Kepeňová), M.K. (Martin Kello), R.S., M.G. and I.P.; Writing—Review & Editing, M.K. (Martina Kepeňová), R.S. and I.P.; visualization, M.K. (Martina Kepeňová) and J.Š.; supervision, M.K. (Martin Kello), M.G. and I.P.; project administration, M.K. (Martin Kello), M.G. and I.P.; funding acquisition, M.K. (Martin Kello), M.L. and I.P. All authors have read and agreed to the published version of the manuscript.

Funding: This work was supported by VEGA 1/0148/19, VEGA 1/0653/19, APVV-18-0016 and VVGS-PF-2022-2134. Moreover, this publication is the result of the project implementation, "Open scientific community for modern interdisciplinary research in medicine (OPENMED)", ITMS2014+: 313011V455, supported by the Operational Programme Integrated Infrastructure, funded by the ERDF. We also thank the Research Infrastructure NanoEnviCz project, supported by the Ministry of Education, Youth and Sports of the Czech Republic, Project No. LM2018124 for instrumentation.

Data Availability Statement: CCDC 2216977-2216981 contain the supplementary crystallographic data for **1a**, **1b**, **3**, **5**, and **6**. These data can be obtained free of charge via http://www.ccdc.cam.ac.uk/conts/retrieving.html, (accessed on 5 November 2022) or from the Cambridge Crystallographic Data Centre, 12 Union Road, Cambridge CB2 1EZ, UK; Fax: (+44) 1223-336-033; or e-mail: deposit@ccdc.cam.ac.uk.

Acknowledgments: Special thanks go to Petra Ecorchard. (Centre of Instrumental Techniques, Institute of Inorganic Chemistry of the CAS) for the preparing of SEM samples.

Conflicts of Interest: The authors declare no conflict of interest. The funders had no role in the design of the study; in the collection, analyses, or interpretation of data; in the writing of the manuscript; or in the decision to publish the results.

References

1. Rosenberg, B.; van Camp, L.; Krigas, T. Platinum compounds: A new class of potent antitumour agents. *Nature* **1969**, *222*, 385–386. [CrossRef]
2. Peng, K.; Liang, B.B.; Liu, W.; Mao, Z.W. What blocks more anticancer platinum complexes from experiment to clinic: Major problems and potential strategies from drug design perspectives. *Coord. Chem. Rev.* **2021**, *449*, 214210. [CrossRef]
3. Casini, A.; Vessières, A.; Meier-Menches, S.M. *Metal-Based Anticancer Agents*; Royal Society of Chemistry: Cambridge, UK, 2019; pp. 3–30.
4. Duan, Z.Y.; Liu, J.Q.; Yin, P.; Li, J.J.; Cai, G.Y.; Chen, X.M. Impact of aging on the risk of platinum-related renal toxicity: A systematic review and meta-analysis. *Cancer Treat. Rev.* **2018**, *69*, 243–253. [CrossRef] [PubMed]
5. Nedeljković, M.; Damjanović, A. Mechanisms of chemotherapy resistance in triple-negative breast cancer-how we can rise to the challenge. *Cells* **2019**, *8*, 957. [CrossRef] [PubMed]
6. Gupta, R.; Luxami, V.; Paul, K. Insights of 8-hydroxyquinolines: A novel target in medicinal chemistry. *Bioorganic Chem.* **2021**, *108*, 104663. [CrossRef] [PubMed]
7. Cherdtrakulkiat, R.; Boonpangrak, S.; Sinthupoom, N.; Prachayasittikul, S.; Ruchirawat, S.; Prachayasittikul, V. Derivatives (halogen, nitro and amino) of 8-hydroxyquinoline with highly potent antimicrobial and antioxidant activities. *Biochem. Biophys. Rep.* **2016**, *6*, 135–141. [CrossRef]

8. Saadeh, H.A.; Sweidan, K.A.; Mubarak, M.S. Recent advances in the synthesis and biological activity of 8-hydroxyquinolines. *Molecules* **2020**, *25*, 4321. [CrossRef]
9. Low, K.H.; Xu, Z.X.; Xiang, H.F.; Chui, S.S.Y.; Roy, V.A.L.; Che, C.M. Bis(5,7-dimethyl-8-hydroxyquinolinato)platinum(II) complex for efficient organic heterojunction solar cells. *Chem. Asian J.* **2011**, *6*, 3223–3229. [CrossRef]
10. Vranec, P.; Potočňák, I. Low-dimensional compounds containing bioactive ligands. Part III: Palladium(II) complexes with halogenated quinolin-8-ol derivatives. *J. Mol. Struct.* **2013**, *1041*, 219–226. [CrossRef]
11. Liu, Y.C.; Wei, J.H.; Chen, Z.F.; Liu, M.; Gu, Y.Q.; Huang, K.B.; Li, Z.Q.; Liang, H. The antitumor activity of zinc(II) and copper(II) complexes with 5,7-dihalo-substituted-8-quinolinoline. *Eur. J. Med. Chem.* **2013**, *69*, 554–563. [CrossRef]
12. Zhang, H.R.; Liu, Y.C.; Meng, T.; Qin, Q.P.; Tang, S.F.; Chen, Z.F.; Zou, B.Q.; Liu, Y.N.; Liang, H. Cytotoxicity, DNA binding and cell apoptosis induction of a zinc(II) complex of HBrQ. *MedChemComm* **2015**, *6*, 2224–2231. [CrossRef]
13. Valiahdi, S.M.; Heffeter, P.; Jakupec, M.A.; Marculescu, R.; Berger, W.; Rappersberger, K.; Keppler, B.K. The gallium complex KP46 exerts strong activity against primary explanted melanoma cells and induces apoptosis in melanoma cell lines. *Melanoma Res.* **2009**, *19*, 283–293. [CrossRef]
14. Hreusova, M.; Novohradsky, V.; Markova, L.; Kostrhunova, H.; Potočňák, I.; Brabec, V.; Kasparkova, J. Gallium(III) Complex with Cloxyquin Ligands Induces Ferroptosis in Cancer Cells and Is a Potent Agent against Both Differentiated and Tumorigenic Cancer Stem Rhabdomyosarcoma Cells. *Bioinorg. Chem. Appl.* **2022**, *2022*, 3095749. [CrossRef] [PubMed]
15. Kubista, B.; Schoefl, T.; Mayr, L.; van Schoonhoven, S.; Heffeter, P.; Windhager, R.; Keppler, B.K.; Berger, W. Distinct activity of the bone-targeted gallium compound KP46 against osteosarcoma cells—Synergism with autophagy inhibition. *J. Exp. Clin. Cancer Res.* **2017**, *36*, 5658–5668. [CrossRef] [PubMed]
16. Gobec, M.; Kljun, J.; Sosič, I.; Mlinarič-Raščan, I.; Uršič, M.; Gobec, S.; Turel, I. Structural characterization and biological evaluation of a clioquinol-ruthenium complex with copper-independent antileukaemic activity. *Dalton Trans.* **2014**, *43*, 9045–9051. [CrossRef]
17. Kubanik, M.; Holtkamp, H.; Söhnel, T.; Jamieson, S.M.F.; Hartinger, C.G. Impact of the Halogen Substitution Pattern on the Biological Activity of Organoruthenium 8-Hydroxyquinoline Anticancer Agents. *Organometallics* **2015**, *34*, 5658–5668. [CrossRef]
18. Chen, Z.F.; Gu, Y.Q.; Song, X.Y.; Liu, Y.C.; Peng, Y.; Liang, H. Synthesis, crystal structure, cytotoxicity and DNA interaction of 5,7-dichloro-8-quinolinolato-lanthanides. *Eur. J. Med. Chem.* **2013**, *59*, 194–202. [CrossRef]
19. Wang, R.; Zou, B.Q.; Qin, Q.P.; Wang, Z.F.; Tan, M.X.; Liang, H. Synthesis, characterization and the anticancer activity of six lanthanides(III) complexes with 5,7-dihalogenated-8-quinolinol and 2,2′-bipyridine derivatives. *Transit. Met. Chem.* **2020**, *45*, 477–483. [CrossRef]
20. Meng, T.; Qin, Q.P.; Chen, Z.L.; Zou, H.H.; Wang, K.; Liang, F.P. High in vitro and in vivo antitumor activities of Ln(III) complexes with mixed 5,7-dichloro-2-methyl-8-quinolinol and 4,4′-dimethyl-2,2′-bipyridyl chelating ligands. *Eur. J. Med. Chem.* **2019**, *169*, 103–110. [CrossRef]
21. Liu, Y.C.; Chen, Z.F.; Song, X.Y.; Peng, Y.; Qin, Q.P.; Liang, H. Synthesis, crystal structure, cytotoxicity and DNA interaction of 5,7-dibromo-8-quinolinolato-lanthanides. *Eur. J. Med. Chem.* **2013**, *59*, 168–175. [CrossRef]
22. Tardito, S.; Barilli, A.; Bassanetti, I.; Tegoni, M.; Bussolati, O.; Franchi-Gazzola, R.; Mucchino, C.; Marchiò, L. Copper-dependent cytotoxicity of 8-hydroxyquinoline derivatives correlates with their hydrophobicity and does not require caspase activation. *J. Med. Chem.* **2012**, *55*, 10448–10459. [CrossRef]
23. Kuchárová, V.; Kuchár, J.; Zaric, M.; Canovic, P.; Arsenijevic, N.; Volarevic, V.; Misirkic, M.; Trajkovic, V.; Radojević, I.D.; Čomić, L.R.; et al. Low-dimensional compounds containing bioactive ligands. Part XI: Synthesis, structures, spectra, in vitro anti-tumor and antimicrobial activities of 3d metal complexes with 8-hydroxyquinoline-5-sulfonic acid. *Inorg. Chim. Acta* **2019**, *497*, 119062. [CrossRef]
24. Lüköová, A.; Baran, P.; Volarevic, V.; Ilic, A.; Vilková, M.; Litecká, M.; Harmošová, M.; Potočňák, I. Low-dimensional compounds containing bioactive ligands. Part XVI: Halogenated derivatives of 8-quinolinol N-oxides and their copper(II) complexes. *J. Mol. Struct.* **2021**, *1246*, 131144. [CrossRef]
25. Potočňák, I.; Vranec, P.; Farkasová, V.; Sabolová, D.; Vataščinová, M.; Kudláčová, J.; Radojević, I.D.; Čomić, L.R.; Markovic, B.S.; Volarevic, V.; et al. Low-dimensional compounds containing bioactive ligands. Part VI: Synthesis, structures, in vitro DNA binding, antimicrobial and anticancer properties of first row transition metal complexes with 5-chloro-quinolin-8-ol. *J. Inorg. Biochem.* **2016**, *154*, 67–77. [CrossRef]
26. Nikolaides, N.; Bogdan, S.E.; Szalma, J.S. Preparation of 8-alkyl 7-(2-imidazolinylamino)quinolines via palladium mediated alkylations. *Synth. Commun.* **2002**, *32*, 2027–2033. [CrossRef]
27. Clavier, S.; Rist, Ø.; Hansen, S.; Gerlach, L.O.; Högberg, T.; Bergman, J. Preparation and evaluation of sulfur-containing metal chelators. *Org. Biomol. Chem.* **2003**, *1*, 4248–4253. [CrossRef] [PubMed]
28. Basak, A.; Abouelhassan, Y.; Norwood, V.M.; Bai, F.; Nguyen, M.T.; Jin, S.; Huigens, R.W. Synthetically Tuning the 2-Position of Halogenated Quinolines: Optimizing Antibacterial and Biofilm Eradication Activities via Alkylation and Reductive Amination Pathways. *Chem. A Eur. J.* **2016**, *22*, 9181–9189. [CrossRef]
29. Bhatt, S.; Nayak, S.K. Copper(II) bromide: A simple and selective monobromination reagent for electron-rich aromatic compounds. *Synth. Commun.* **2007**, *37*, 1381–1388. [CrossRef]
30. Hao, W.; Liu, Y. C-H bond halogenation catalyzed or mediated by copper: An overview. *Beilstein J. Org. Chem.* **2015**, *11*, 2132–2144. [CrossRef]

31. Arjunan, V.; Mohan, S.; Ravindran, P.; Mythili, C.V. Vibrational spectroscopic investigations, ab initio and DFT studies on 7-bromo-5-chloro-8-hydroxyquinoline. *Spectrochim. Acta Part A Mol. Biomol. Spectrosc.* **2009**, *72*, 783–788. [CrossRef]
32. Krishnakumar, V.; Ramasamy, R. DFT studies and vibrational spectra of isoquinoline and 8-hydroxyquinoline. *Spectrochim. Acta Part A Mol. Biomol. Spectrosc.* **2005**, *61*, 673–683. [CrossRef] [PubMed]
33. Wagner, C.C.; Calvo, S.; Torre, M.H.; Baran, E.J. Vibrational spectra of clioquinol and its Cu(II) complex. *J. Raman Spectrosc.* **2007**, *38*, 373–376. [CrossRef]
34. Vranec, P.; Potočňák, I.; Sabolová, D.; Farkasová, V.; Ipóthová, Z.; Pisarčíková, J.; Paulíková, H. Low-dimensional compounds containing bioactive ligands. V: Synthesis and characterization of novel anticancer Pd(II) ionic compounds with quinolin-8-ol halogen derivatives. *J. Inorg. Biochem.* **2014**, *131*, 37–46. [CrossRef] [PubMed]
35. Bahgat, K.; Ragheb, A.G. Analysis of vibratinal spectra of 8-hydroxyquinoline and its 5,7-dichloro, 5,7-dibromo, 5,7-diiodo and 5,7-dinitro derivatives based on density functional theory calculations. *Cent. Eur. J. Chem.* **2007**, *5*, 201–220. [CrossRef]
36. Malherbe, F.E.; Bernstein, H.J. The infrared and Raman spectra of p-dioxane. *J. Am. Chem. Soc.* **1952**, *74*, 4408–4410. [CrossRef]
37. Alver, Ö.; Parlak, C. Spectroscopic investigations of 1,4-dioxane adsorbed on bentonite from anatolia. *Anadolu Univ. J. Sci. Technol. A Appl. Sci. Eng.* **2016**, *17*, 273–277. [CrossRef]
38. Borowski, P.; Gac, W.; Pulay, P.; Woliński, K. The vibrational spectrum of 1,4-dioxane in aqueous solution-theory and experiment. *New J. Chem.* **2016**, *40*, 7663–7670. [CrossRef]
39. Shastri, A.; Das, A.K.; Krishnakumar, S.; Singh, P.J.; Raja Sekhar, B.N. Spectroscopy of N, N-dimethylformamide in the VUV and IR regions: Experimental and computational studies. *J. Chem. Phys.* **2017**, *147*, 224305. [CrossRef]
40. Wang, C.; Niu, J.; Li, J.; Ma, X. Synthesis, characterization, structures and Suzuki coupling reaction of Cu(II) complexes derived from N and O-containing organic ligand. *Inorg. Chim. Acta* **2017**, *464*, 81–87. [CrossRef]
41. di Vaira, M.; Bazzicalupi, C.; Orioli, P.; Messori, L.; Bruni, B.; Zatta, P. Clioquinol, a drug for Alzheimer's disease specifically interfering with brain metal metabolism: Structural characterization of its zinc(II) and copper(II) complexes. *Inorg. Chem.* **2004**, *43*, 3795–3797. [CrossRef]
42. Denoyer, D.; Clatworthy, S.A.S.; Cater, M.A. Copper Complexes in Cancer Therapy. *Met. Ions Life Sci.* **2018**, *18*, 469–506. [CrossRef]
43. Shen, W.Y.; Jia, C.P.; Mo, A.N.; Liang, H.; Chen, Z.F. Chemodynamic therapy agents Cu(II) complexes of quinoline derivatives induced ER stress and mitochondria-mediated apoptosis in SK-OV-3 cells. *Eur. J. Med. Chem.* **2021**, *223*, 113636. [CrossRef] [PubMed]
44. Radhakrishnan, K.; Khamrang, T.; Sambantham, K.; Sali, V.K.; Chitgupi, U.; Lovell, J.F.; Mohammad, A.A.; Venugopal, R. Identification of cytotoxic copper(II) complexes with phenanthroline and quinoline, quinoxaline or quinazoline-derived mixed ligands. *Polyhedron* **2021**, *194*, 114886. [CrossRef]
45. Hu, K.; Liu, C.; Li, J.; Liang, F. Copper(II) complexes based on quinoline-derived Schiff-base ligands: Synthesis, characterization, HSA/DNA binding ability, and anticancer activity. *MedChemComm* **2018**, *9*, 1663–1672. [CrossRef]
46. Firmino, G.S.S.; de Souza, M.V.N.; Pessoa, C.; Lourenco, M.C.S.; Resende, J.A.L.C.; Lessa, J.A. Synthesis and evaluation of copper(II) complexes with isoniazid-derived hydrazones as anticancer and antitubercular agents. *BioMetals* **2016**, *29*, 953–963. [CrossRef]
47. Hindo, S.S.; Frezza, M.; Tomco, D.; Heeg, M.J.; Hryhorczuk, L.; McGarvey, B.R.; Dou, Q.P.; Verani, C.N. Metals in anticancer therapy: Copper(II) complexes as inhibitors of the 20S proteasome. *Eur. J. Med. Chem.* **2009**, *44*, 4353–4361. [CrossRef]
48. Martínez Medina, J.J.; Naso, L.G.; Pérez, A.L.; Rizzi, A.; Okulik, N.B.; Valcarcel, M.; Salado, C.; Ferrer, E.G.; Williams, P.A.M. Synthesis, characterization, theoretical studies and biological (antioxidant, anticancer, toxicity and neuroprotective) determinations of a copper(II) complex with 5-hydroxytryptophan. *Biomed. Pharmacother.* **2019**, *111*, 414–426. [CrossRef]
49. Rossi, A.; Poverini, R.; di Lullo, G.; Modesti, A.; Modicat, A.; Scarinos, M.L. Heavy metal toxicity following apical and basolateral exposure in the human intestinal cell line Caco-2. *Toxicol. Vitr.* **1996**, *10*, 27–36. [CrossRef]
50. Panjehpour, M.; Taher, M.-A.; Bayesteh, M. The growth inhibitory effects of cadmium and copper on the MDA-MB468 human breast cancer cells. *J. Res. Med. Sci.* **2010**, *15*, 279–286.
51. Peña, Q.; Sciortino, G.; Maréchal, J.D.; Bertaina, S.; Simaan, A.J.; Lorenzo, J.; Capdevila, M.; Bayón, P.; Iranzo, O.; Palacios, Ò. Copper(II) N, N, O-chelating complexes as potential anticancer agents. *Inorg. Chem.* **2021**, *60*, 2939–2952. [CrossRef]
52. Beeton, M.L.; Aldrich-Wright, J.R.; Bolhuis, A. The antimicrobial and antibiofilm activities of copper(II) complexes. *J. Inorg. Biochem.* **2014**, *140*, 167–172. [CrossRef] [PubMed]
53. Litecká, M.; Hreusová, M.; Kašpárková, J.; Gyepes, R.; Smolková, R.; Obuch, J.; David, T.; Potočňák, I. Low-dimensional compounds containing bioactive ligands. Part XIV: High selective antiproliferative activity of tris(5-chloro-8-quinolinolato)gallium(III) complex against human cancer cell lines. *Bioorganic Med. Chem. Lett.* **2020**, *30*, 127206. [CrossRef] [PubMed]
54. Lüköová, A.; Drweesh, E.A.; Volarevic, V.; Miloradovic, D.; Markovic, B.S.; Smolková, R.; Samoľová, E.; Kuchár, J.; Vilková, M.; Potočňák, I. Low-dimensional compounds containing bioactive ligands. Part XIII: Square planar anti-cancer Pd(II) complexes with halogenderivatives of 8-quinolinol and dimethylamine. *Polyhedron* **2020**, *184*, 114535. [CrossRef]
55. Litecká, M.; Prachařová, J.; Kašpárková, J.; Brabec, V.; Smolková, R.; Gyepes, R.; Obuch, J.; Kubíček, V.; Potočňák, I. Low-dimensional compounds containing bioactive ligands. Part XV: Antiproliferative activity of tris(5-nitro-8-quinolinolato)gallium(III) complex with noticeable selectivity against the cancerous cells. *Polyhedron* **2020**, *187*, 114672. [CrossRef]
56. *Origin*; Version 2022B; OriginLab Corporation: Northampton, MA, USA, 2022.
57. *CrysAlisPRO*; Version 1.0.43; Oxford Diffraction/Agilent Technologies UK Ltd.: Yarnton, UK, 2020.

58. Sheldrick, G.M. SHELXT—Integrated space-group and crystal-structure determination. *Acta Crystallogr. Sect. A Found. Adv.* **2015**, *71*, 3–8. [CrossRef]
59. Sheldrick, G.M. Crystal structure refinement with SHELXL. *Acta Crystallogr. Sect. C Struct. Chem.* **2015**, *71*, 3–8. [CrossRef]
60. Farrugia, L.J. WinGX and ORTEP for Windows: An update. *J. Appl. Crystallogr.* **2012**, *45*, 849–885. [CrossRef]
61. Spek, A.L. Structure validation in chemical crystallography. *Acta Crystallogr. Sect. D Biol. Crystallogr.* **2009**, *65*, 148–155. [CrossRef]
62. Brandenburg, K. *Diamond*; Version 3.2K; Crystal Impact GbR: Bonn, Germany, 2014.
63. Rojas, J.J.; Ochoa, V.J.; Ocampo, S.A.; Muñoz, J.F. Screening for antimicrobial activity of ten medicinal plants used in Colombian folkloric medicine: A possible alternative in the treatment of non-nosocomial infections. *BMC Complement. Altern. Med.* **2006**, *6*, 2. [CrossRef]
64. Andrews, J.M. Determination of minimum inhibitory concentrations. *J. Antimicrob. Chemother.* **2001**, *48* (Suppl. S1), 5–16. [CrossRef]
65. Re, R.; Pellegrini, N.; Proteggente, A.; Pannala, A.; Yang, M.; Rice-Evans, C. Antioxidant activity applying an improved ABTS radical cation decolorization assay. *Free. Radic. Biol. Med.* **1999**, *26*, 1231–1237. [CrossRef] [PubMed]
66. Brand-Williams, W.; Cuvelier, M.E.; Berset, C. Use of a free radical method to evaluate antioxidant activity. *LWT Food Sci. Technol.* **1995**, *28*, 25–30. [CrossRef]

Article

Low-Dimensional Compounds Containing Bioactive Ligands. Part XX: Crystal Structures, Cytotoxic, Antimicrobial Activities and DNA/BSA Binding of Oligonuclear Zinc Complexes with Halogen Derivatives of 8-Hydroxyquinoline

Michaela Harmošová [1], Martin Kello [2], Michal Goga [3], Ľudmila Tkáčiková [4], Mária Vilková [1], Danica Sabolová [1], Simona Sovová [1], Erika Samoľová [5], Miroslava Litecká [6], Veronika Kuchárová [7], Juraj Kuchár [1] and Ivan Potočňák [1,*]

[1] Institute of Chemistry, P. J. Šafárik University in Košice, Moyzesova 11, SK-04154 Košice, Slovakia
[2] Department of Pharmacology, P. J. Šafárik University in Košice, Trieda SNP 1, 040 11 Košice, Slovakia
[3] Department of Botany, Institute of Biology and Ecology, Faculty of Science, P. J. Šafárik University in Košice, Mánesova 23, 040 01 Košice, Slovakia
[4] Department of Microbiology and Immunology, University of Veterinary Medicine and Pharmacy, Komenského 73, 041 81 Košice, Slovakia
[5] Institute of Physics of the Czech Academy of Science, Na Slovance 2, CZ-182 21 Prague, Czech Republic
[6] Centre of Instrumental Techniques, Institute of Inorganic Chemistry of the CAS, Husinec-Řež č.p. 1001, CZ-25068 Řež, Czech Republic
[7] Institute of Experimental Physics, Watsonova 47, SK-04001 Košice, Slovakia
* Correspondence: ivan.potocnak@upjs.sk; Tel.: +421-55-234-2335

Citation: Harmošová, M.; Kello, M.; Goga, M.; Tkáčiková, Ľ.; Vilková, M.; Sabolová, D.; Sovová, S.; Samoľová, E.; Litecká, M.; Kuchárová, V.; et al. Low-Dimensional Compounds Containing Bioactive Ligands. Part XX: Crystal Structures, Cytotoxic, Antimicrobial Activities and DNA/BSA Binding of Oligonuclear Zinc Complexes with Halogen Derivatives of 8-Hydroxyquinoline. *Inorganics* **2023**, *11*, 60. https://doi.org/10.3390/inorganics11020060

Academic Editors: Peter Segľa and Ján Pavlik

Received: 4 January 2023
Revised: 19 January 2023
Accepted: 24 January 2023
Published: 26 January 2023

Abstract: Two tetranuclear $[Zn_4Cl_2(ClQ)_6]\cdot 2DMF$ (**1**) and $[Zn_4Cl_2(ClQ)_6(H_2O)_2]\cdot 4DMF$ (**2**), as well as three dinuclear $[Zn_2(ClQ)_3(HClQ)_3]I_3$ (**3**), $[Zn_2(dClQ)_2(H_2O)_6(SO_4)]$ (**4**) and $[Zn_2(dBrQ)_2(H_2O)_6(SO_4)]$ (**5**), complexes (HClQ = 5-chloro-8-hydroxyquinoline, HdClQ = 5,7-dichloro-8-hydroxyquinoline and HdBrQ = 5,7-dibromo-8-hydroxyquinoline) were prepared as possible anticancer or antimicrobial agents and characterized by IR spectroscopy, elemental analysis and single crystal X-ray structure analysis. The stability of the complexes in solution was verified by NMR spectroscopy. Antiproliferative activity and selectivity of the prepared complexes were studied using in vitro MTT assay against the HeLa, A549, MCF-7, MDA-MB-231, HCT116 and Caco-2 cancer cell lines and on the Cos-7 non-cancerous cell line. The most sensitive to the tested complexes was Caco-2 cell line. Among the tested complexes, complex **3** showed the highest cytotoxicity against all cell lines. Unfortunately, all complexes showed only poor selectivity to normal cells, except for complex **5**, which showed a certain level of selectivity. Antibacterial potential was observed for complex **5** only. Moreover, the DNA/BSA binding potential of complexes **1–3** was investigated by UV-vis and fluorescence spectroscopic methods.

Keywords: crystal structure; Zn complexes; 8-hydroxyquinoline; cytotoxicity; antimicrobial activity; DNA/BSA binding

1. Introduction

Cancer diseases represent one of the greatest challenges for society. Cancer is responsible of around 10.0 million deaths worldwide and 19.3 million new cases each year [1]. There are seven platinum-containing drugs with a great effect in a cancer treatment. Three of them (cisplatin, carboplatin and oxaliplatin) are approved worldwide while other four (nedaplatin, lobaplatin, heptaplatin and miriplatin) are approved in specific Asian countries [2]. Despite their success in a cancer treatment, there are not only several side effects, such as allergic reactions, kidney problems, ototoxicity and decreased immunity associated with the use of these drugs, but also platinum drugs cell resistance can be developed.

Therefore, new drugs need to be synthesized to overcome the drug-resistance and unwanted side effects [3]. New transition metal complexes could be a suitable choice because of their bioactivity, biocompatibility, varying coordination numbers and geometries; they can also bind with DNA through covalent, noncovalent and electrostatic interactions, which can lead to cellular death [4]. There are many classes of organic compounds, including heterocyclic compounds, which have been synthesized as suitable bioactive ligands, and their complexes were tested as possible anticancer drugs. Growing interest is noted for 8-hydroxyquinoline (H8-HQ) and its derivatives (HXQ) [5,6]. H8-HQ, as well as HXQ, are known to exhibit anti-SARS-CoV-2 [7], antimicrobial [8,9], antifungal [10], anti-inflammatory [11], anticancer [12–14] and antioxidant activity [15,16]. They also have significant effects in the treatment of malaria, HIV and neurological diseases [6]. The lipophilic derivative of H8-HQ, clioquinol (HCQ = 5-chloro-7-iodo-8-hydroxyquinoline), has been studied as a treatment for Alzheimer's disease [17]. The halogenation of H8-HQ at various positions makes it more lipophilic, which leads to better absorption and better activity. In addition, the complexation of HXQ with metal ions shows better anticancer activity [6 and citations therein]. Recently, we published several papers describing the biological activities of both transition and non-transition metal complexes with HXQ. Some of them, such as $K[PdCl_2(dClQ)]$, $Cs[PdCl_2(dClQ)]$ (HdClQ = 5,7-dichloro-8-hydroxyquinoline) [18] and $[Ga(ClQ)_3]$ (HClQ = 5-chloro-8-hydroxyquinoline) [19], exhibited strong and selective cytotoxic effects against tested cancer cell lines. On the other hand, although mononuclear zinc complexes with HXQ, such as $K[Zn(dClQ)_3]\cdot 2DMF\cdot H_2O$, $(HdClQ)_2[ZnCl_4]\cdot 2H_2O$ or $[Zn(dBrQ)_2(H_2O)]_2\cdot DMF\cdot H_2O$ (HdBrQ = 5,7-dibromo-8-hydroxyquinoline), have shown significant cytotoxic activity against colon cancer HCT116 cell lines, they were not selective. However, we have observed strong antimicrobial and antifungal activity of these complexes [20]. These results motivated us to continue our work, and we have prepared a series of di- and tetranuclear zinc complexes with mono and dihalogen derivatives of H8-HQ with the aim of studying their cytotoxic and antimicrobial activity. In this work, we describe preparation of five new zinc(II) complexes, $[Zn_4Cl_2(ClQ)_6]\cdot 2DMF$ (**1**), $[Zn_4Cl_2(ClQ)_6(H_2O)_2]\cdot 4DMF$ (**2**), $[Zn_2(ClQ)_3(HClQ)_3]I_3$ (**3**), $[Zn_2(dClQ)_2(H_2O)_6(SO_4)]$ (**4**) and $[Zn_2(dBrQ)_2(H_2O)_6(SO_4)]$ (**5**), which were characterized by IR and NMR spectroscopy, as well as elemental and single crystal X-ray structure analysis. Moreover, the results of cytotoxic and antimicrobial assays and their DNA/BSA binding properties are presented.

2. Results and Discussion

2.1. Syntheses

In this work, we have focused on the preparation of zinc complexes with halogen derivatives of H8-HQ. We have used several zinc salts, solvents, different synthetic strategies and different procedures of crystallization. Regarding solvents, DMF was used to dissolve respective HXQ ligands, while ethanol was used to dissolve zinc(II) chloride or iodide. Because of the better solubility of $ZnSO_4\cdot 7H_2O$ in methanol, for the preparations of sulphato complexes **4** and **5**, methanol was used instead of ethanol. The preparation of **1**, **3**, **4** and **5** was performed at room temperature by a simple one-pot synthesis. After the mixing of reactants, complexes were isolated after several months of crystallization at room temperature. On the other hand, the preparation of **2** was performed at low temperatures; moreover, KOH was used to improve deprotonation of the HClQ hydroxyl group and crystallization was carried out at a low temperature in a fridge. It has to be noted that we attempted to prepare complexes with all three HXQ derivatives at room temperature and also under reflux either with or without added KOH; however, no other products were obtained. Figure 1 schematically describes the synthetic procedure used for the synthesis of **1**–**5**.

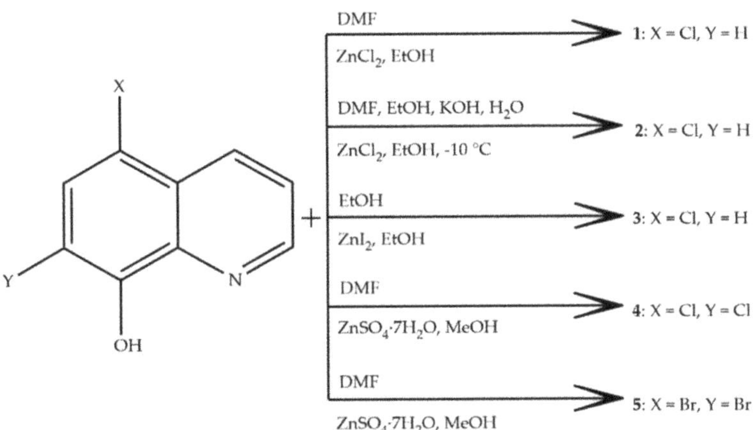

Figure 1. Synthesis and formulae of **1**–**5**. **1**: bis(μ$_3$-5-chloro-8-hydroxyquinolinato)-tetrakis(μ$_2$-5-chloro-8-hydroxyquinolinato)-dichloride-tetra-zinc(II) bis-dimethylformamide solvate, $[Zn_4Cl_2(ClQ)_6]\cdot 2DMF$; **2**: bis(μ$_3$-5-chloro-8-hydroxyquinolinato)-tetrakis(μ$_2$-5-chloro-8-hydroxyquinolinato)-diaqua-dichloride-tetra-zinc(II) tetra-dimethylformamide solvate, $[Zn_4Cl_2(ClQ)_6(H_2O)_2]\cdot 4DMF$; **3**: tris(5-chloro-8-hydroxyquinolinato)-tris(5-chloro-8-hydroxyquinoline)-di-zinc(II) triiodide, $[Zn_2(ClQ)_3(HClQ)_3]I_3$; **4**: catena (μ$_2$-sulphato)-haxaaqua-bis(5,7-dichloro-8-hydroxyquinolinato)-di-zinc (II), $[Zn_2(dClQ)_2(H_2O)_6(SO_4)]$; **5**: catena-(μ$_2$-sulphato)-haxaaqua-bis(5,7-dibromo-8-hydroxyquinolinato)-di-zinc(II), $[Zn_2(dBrQ)_2(H_2O)_6(SO_4)]$.

2.2. Infrared Spectroscopy

The presence of HXQ ligands in **1**–**5** was first proven by IR spectroscopy. The IR spectra of pure HdClQ, HdBrQ and HClQ ligands have already been described [21,22]. The characteristic vibrations of the free ligands include the ν(O–H) vibration, which manifests as a broad band starting at 3360 cm^{-1} and ending with a vibration band of $\nu(C–H)_{ar}$ at 3060 cm^{-1}, and its absence in **1**–**5** suggests the coordination of the respective ligand to the zinc atom through the oxygen atom after the deprotonation of a hydroxyl group. The coordination of the ligands to the zinc atom through pyridine nitrogen atom is supported by the characteristic shift of ν(C=N) bands to lower frequencies in the spectra of **1**–**5** (1462–1454 cm^{-1}) compared to the free ligands (1470–1460 cm^{-1}). The remaining ligands bands are still present in the spectra of the complexes. However, they are often shifted compared to the IR spectra of free ligands.

Tetranuclear complexes **1** and **2** have a similar composition with two extra water molecules in **2** and thus their IR spectra will be discussed together. Despite the presence of water molecules, in the spectrum of **2** the valence vibration of the OH group is not clearly visible (Figure 2). We attribute it to a wide band starting at 3305 cm^{-1}, which smoothly passes into the $\nu(C–H)_{ar}$ vibration observed at 3076 cm^{-1}. On the other hand, the planar deformation vibration of H_2O is clearly visible (1650 cm^{-1}); however, the bands of the rocking, twisting and wagging vibrational modes of coordinated water molecules, which should appear in the spectrum in the ranges of 970–930 cm^{-1} and 660–600 cm^{-1} [23], are again missing. DMF molecules in both complexes are proven by the weak bands of $\nu(C–H)_{al}$ vibrations between 2990 and 2850 cm^{-1} and by strong absorption bands of ν(C=O) vibration at 1672 cm^{-1}. The absorption bands of ClQ ligands are at the same wavenumbers in both spectra and are close to those of free ligands with the exception of the above-mentioned ν(O–H) and ν(C=N) bands.

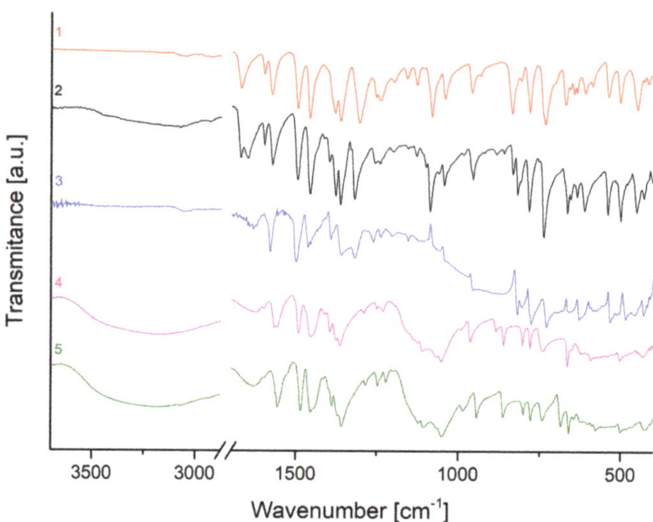

Figure 2. FT–IR spectra of **1–5**.

IR spectrum of **3**, which contains the same ClQ ligand, is similar to those of **1** and **2**; of course, the absorption bands of water and DMF molecules are missing.

The composition of **4** and **5** is again very similar; both complexes differ only by HXQ ligands. The presence of water molecules in these complexes is proven by strong and wide bands of ν(O–H) vibration starting at 3630 cm^{-1} and ending by $\nu(C–H)_{ar}$ vibrations observed at 3072 cm^{-1}, as well as by deformation vibrations around 1630 cm^{-1}.

We observed huge bands around 1110 and 1050 cm^{-1} in the spectra of **4** and **5**, which can be attributed to the ν_{as}(S–O) and ν_s(S–O) valence vibrations, respectively, originating from the sulphato ligand ($ZnSO_4 \cdot 7H_2O$ was used in the synthesis). The weak bands of deformation vibrations of the SO_4 group around 624 cm^{-1} are also observed in the spectra [24]. The main difference in the bands of HdClQ and HdBrQ ligands present in **4** and **5**, respectively, are the bands of ν(C5–X) vibrations. In **4**, the ν(C5–Cl) vibration is observed at 964 cm^{-1} (in the spectra of complexes **1–3** with HClQ ligand, corresponding bands are between 961 and 955 cm^{-1}) while ν(C5–Br) vibration is observed at 943 cm^{-1} in the spectrum of **5**. On the other hand, ν(C7–X) vibrations that were absent in the spectra of **1–3** are in the spectra of **4** and **5** at the same wavenumber, 863 cm^{-1} [21,25,26].

2.3. NMR Characterization and Stability Studies

The assignment of resonances for individual proton and carbon atoms within all complexes were made based on ^{1}H, ^{13}C and ^{1}H, ^{13}C-COSY, ^{1}H, ^{13}C-HSQC, ^{1}H and ^{13}C-HMBC spectra. These studies were performed at 298 K. The data obtained from these spectra are listed in Table 1 and Table S1. It should be noted that the same sets of proton and carbon signals were detected for each quinoline ligand of **1** and **2** because their composition is almost the same. They only differ by the presence of two water molecules in the coordination sphere of complex **2** whose signals are overlapped by the signals of water from DMSO-d_6. On the other hand, the separate and broadened signals were detected for complexes **3, 4** and **5**.

Based on the crystal structures, we assume that complexes **1** and **2** should have symmetric structures in solution and ^{1}H NMR spectra display only one set of chemical shifts (Table 1), corresponding to the appropriate units, and no fluxional phenomena were observed for these complexes.

Table 1. ^{1}H NMR (600 MHz, DMSO-d$_6$) chemical shifts δ_H [ppm] for **1–5**.

		δ_H (mult., *J* [Hz])				
		H-2	**H-3**	**H-4**	**H-6**	**H-7**
1 *	ClQ	8.74 (br s)	7.75 (dd, 8.5, 4.5)	8.54 (d, 8.5)	7.50 (d, 8.5)	6.73 (br s)
2 *	ClQ	8.74 (br s)	7.75 (dd, 8.4, 4.4)	8.54 (dd, 8.4, 1.5)	7.50 (d, 8.4)	6.74 (d, 8.4)
3	ClQ	8.66 (br s)	7.74 (br s)	8.52 (br s)	7.51 (br s)	6.72 (br s)
	HClQ	8.95 (br s)	7.74 (br s)	8.52 (br s)	7.60 (br s)	7.08 (br s)
4 **	dClQ	8.55 (br s) 8.79 (br s)	7.70 (m)	8.50 (d, 8.4)	7.70 (m)	-
5 **	dBrQ	8.47 (br s) 8.76 (br s)	7.70 (br s) 7.75 (br s)	8.41 (d, 8.5) 8.44 (br s)	7.92 (s) 7.92 (s)	-

* 7.95 (s, HDMF), 2.89 (s, MeDMF) and 2.73 (s, MeDMF); ** chemical shifts for major form are in the first line and for the minor form in the second line.

Figure S1 displays the ^{1}H NMR spectrum for **3**. At 25 °C, the ^{1}H NMR spectrum consists of broadened signals attributable to the protons of ClQ and HClQ ligands. At first sight, one could think that this effect arises from a dynamic within a system. However, it is most likely that broadened signals of **3** arise from its low solubility and the presence of some particulate matter in the solution. For the same reason, it was not possible to measure ^{13}C and 2D NMR spectra and to assign carbon chemical shifts.

We assumed that in the solution, **4** and **5** would give well-resolved signals like the above-described complexes **1** and **2**. ^{1}H NMR experiments confirmed the opposite to be true. ^{1}H NMR spectra of complexes **4** and **5** contained broadened peaks consistent with the presence of a dynamic exchange process. As illustrated by Figure 3 and Figure S2, in the ^{1}H NMR spectrum of **4** only, signals belonging to proton H-2 are broadened, and in the ^{1}H NMR spectrum of **5**, almost all proton signals H-2, H-3 and H-4 are broadened. All broadened proton signals are shifted to the low field. This observation indicated that protons described as major must exchange positions with protons signed as minor. It is a consequence of the slow proton exchange of **4** and **5** within the NMR time scale. Similarly, ^{13}C signals of **4** and **5** (Figure 3 and Figure S2) exhibited very interesting changes. At ambient temperature, signals belonging to carbons C-8, C-2, C-8a, C-4a and C-5 of **4** are doubled. We expected a similar pattern for the complex **5**, but the very low solubility of the complex in DMSO-d$_6$ caused the minor form signals to not be well distinguished (Figure S2). Only the signals of carbons C-4, C-6 and C-4a are visibly doubled.

Because of the different structures of all studied complexes, complexes **4** and **5** are fluxional and are characterized by the proton position exchange without the dissociation of the complex. If the process occurred with dissociation, the process should be associated with the presence of free ligand peaks in NMR spectra, which is not the case.

The stability of the complexes **1–5** in DMSO-d$_6$ solution was tested using ^{1}H NMR time-dependent spectra to evaluate their suitability for biological testing.

As shown in Figure 4 and Figure S3, at ambient temperature, the time-dependent ^{1}H NMR spectra of complexes **1** and **2** show peaks corresponding to the one set of chemical shifts, an indication of the presence of complexes that are structurally the same for 72 h. However, as shown in Figure 4 and Figure S3, the ^{1}H NMR spectra after 48 h show new peaks at δ_H 10.20 (s, OH), 9.03 (dd, J 4.7, 1.7, H-2), 8.31 (dd, J 7.9, 1.7, H-4) and 7.86 (dd, J 7.9, 4.7, H-3). Whereas the signals belonging to protons H-6 and H-7 are not detectable. By comparing chemical shifts with free ligand shifts (^{1}H NMR (600 MHz, DMSO-d$_6$): δ_H 10.18 (s, OH), 8.94 (dd, J 4.1, 1.6, H-2), 8.48 (dd, J 8.6, 1.6, H-4), 7.71 (dd, J 8.6, 4.1, H-3), 7.60 (d, J 8.3, H-6) and 7.08 (d, J 8.3, H-7)), we cannot completely exclude complex decomposition. The ratio of the complex and newly formed structure is 90:1 after 48 h and 70:1 after 96 h for **1** and 56:1 after 48 h and 46:1 after 72 h for **2** (based on integral values of H-4 signals). Nevertheless, we believe that the low concentration of the newly formed structures will not significantly affect the results of biological activity tests. Contrary to our findings, Zhang et al. [27] prepared tetranuclear $[Zn_4(8\text{-}HQ)_6Ac_2]$ (Ac = acetate) and $[Zn_4(MeQ)_6Ac_2]$ (MeQ = 2-methyl-8-hydroxylquinoline) complexes, which are very similar to complexes **1** and **2** in our work, and tested them for their stability in the physiological conditions within 48 h. Based on the results of UV-Vis spectroscopy, they concluded that both complexes were stable.

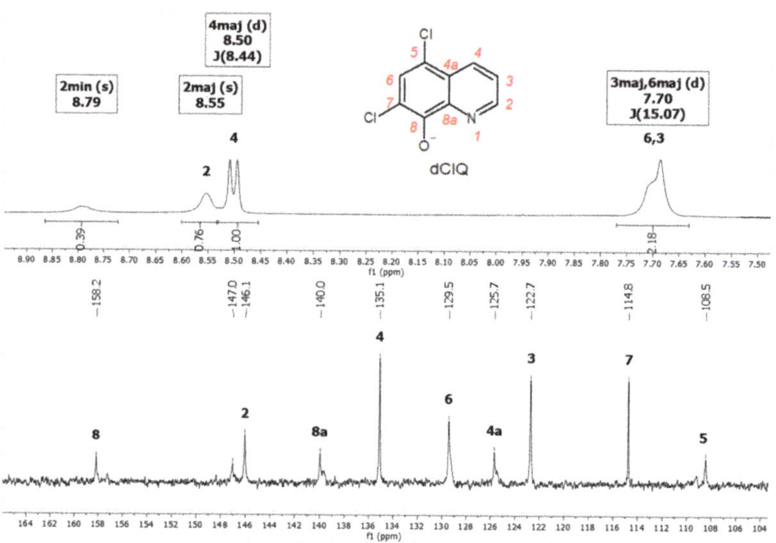

Figure 3. ^1H (600 MHz, DMSO-d_6) and ^{13}C (150 MHz, DMSO-d_6) NMR spectrum of complex **4**.

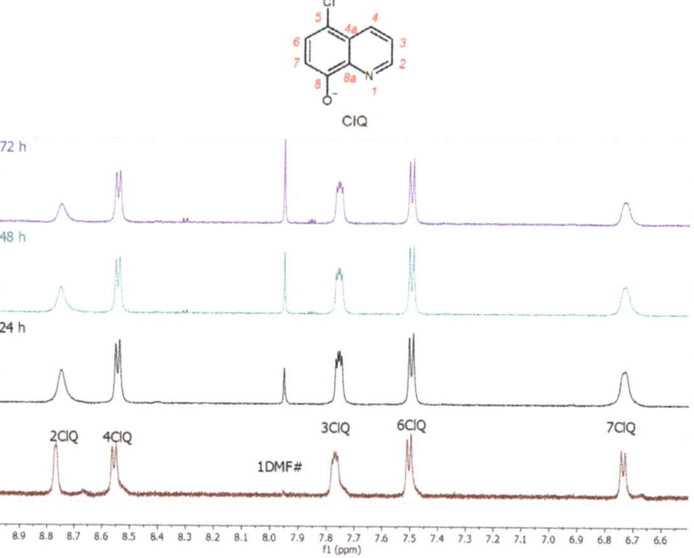

Figure 4. Time-dependent ^1H NMR (600 MHz, DMSO-d6) spectra of complex **1**.

As shown in Figure S3, the ^{1}H NMR spectrum measured the after dissolving of the complex in DMSO-d$_6$ shows broad peaks corresponding to complex **3**. After 24 h, spectrum becomes somewhat better resolved but remains broad. It can be confirmed that all time-dependent spectra (24–72 h) are consistent with the structure of complex **3** observed immediately after dissolving. Likewise, complexes **4** (Figure S5) and **5** (Figure S6) are stable in the DMSO-d$_6$ solution for 72 h. Minor peaks at δ_H 8.75 and 8.29 ppm appeared in the spectra (Figure S6) of the complex **5** and can be attributed to the dynamics rather than the decomposition of complex **5**.

2.4. X-ray Structure Analysis

The $[Zn_4Cl_2(ClQ)_6]\cdot 2DMF$ (**1**) and $[Zn_4Cl_2(ClQ)_6(H_2O)_2]\cdot 4DMF$ (**2**) complexes are similar tetranuclear molecular compounds with a complicated type of structure. They crystallize in the triclinic *P*1 space group. In both complexes, there are two pairs of crystallographically independent zinc atoms. Zn1 atoms are pentacoordinated in both structures by N2 and O2 atoms of one chelate ClQ ligand, with O2 atom being a bridge between Zn1 and Zn2 atoms. Other two coordination places are occupied by bridging O1 and O3 atoms of the other two ClQ chelate-bridging ligands, while the fifth place is occupied by the chloride ligand, Cl1. Zn2 atoms are hexacoordinated in both structures; however, coordination spheres around the Zn2 atoms in both structures are different. In both structures, two chelate-bridging ClQ ligands coordinate to the Zn2 atoms through N1 and N3, and O1 and O3 atoms, which are bridges to the Zn1 atoms. The last two coordination places in **1** are occupied by two bridging oxygen atoms of another two ClQ ligands (Figure 5), while only one place is occupied by the bridging oxygen atom of another ClQ ligand in **2** and the last coordination place is occupied by the O4 atom of a water molecule (Figure 5). In summary, there are two tridentate and four bidentate oxygen atoms of the six ClQ ligands in **1** and each Zn1-Zn2 pair is bridged by two oxygen atoms of two different ClQ ligands. Thus, there are four double-oxygen bridges between Zn1 and Zn2 atoms in **1** (Figure 5). On the other hand, all six ClQ oxygen atoms are bidentate in **2** and thus there are only two double-oxygen bridges between Zn1 and Zn2 atoms, while the remaining two bridges are single-oxygen bridges only.

Based on bond angles (Table 2), the shapes of coordination polyhedra around Zn2 atoms in **1** and **2** are distorted octahedrally. On the other hand, Zn1 atoms are coordinated in a trigonal bipyramidal or square pyramidal shape, respectively, as suggested by the τ parameter [28] and program SHAPE [29] (Table S2).

Bond distances around the zinc atoms in both structures are listed in Table 2. It can be seen that the Zn-O and Zn-N distances within the chelate rings are in the range of 2.0789(12)–2.1525(16) Å, while distances between Zn atoms and bridging oxygen atoms are shorter, 2.0134(12)–2.0463(13) Å, except for the Zn1^i-O2 bond (2.2779(13) Å; i = $-x + 1, -y + 1, -z + 1$) in the structure of **1** which length is comparable to Zn1-Cl1 in **1** and **2** (2.2766(5) and 2.3149(5) Å, respectively). In addition, a water molecule in **2** is coordinated with a Zn2-O4 bond length being 2.2044(13) Å. All mentioned distances and angles are similar to those described in the literature for other tetranuclear zinc complexes [27,30].

Outside the tetranuclear complex species there are two or four solvated molecules of DMF in **1** and **2**, respectively, which are joined with the complexes by hydrogen bonds stabilizing both structures. These and other hydrogen bonds present in the structure of both complexes are gathered in Table 3 and are shown in Figures 6 and 7. Due to these hydrogen bonds, a layered structure parallel with the (011) plane in **1** and a chain-like structure along the *c* axis in **2** is formed.

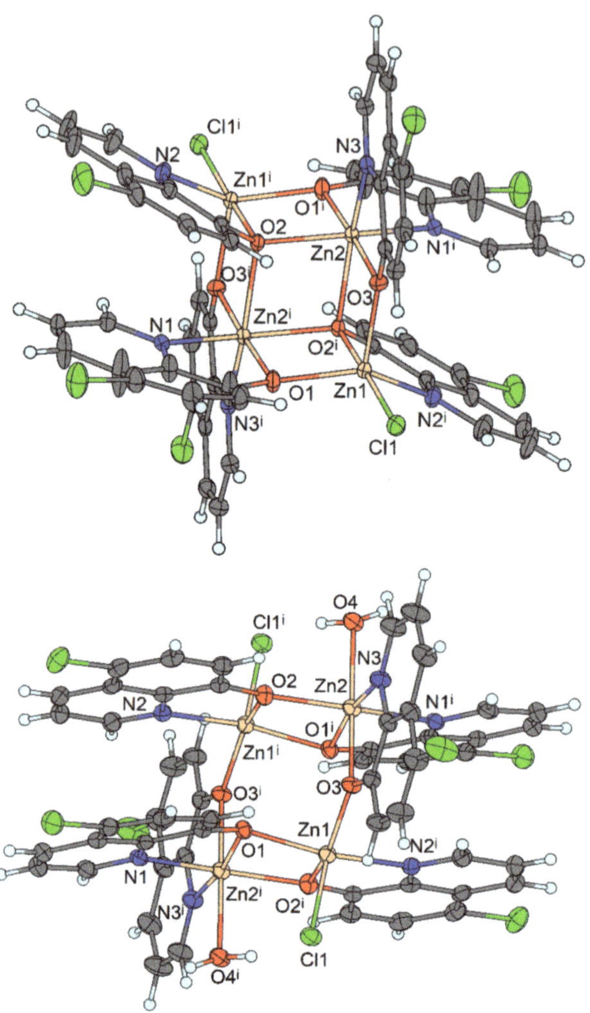

Figure 5. Molecular structure of **1** (up) and **2** (down). Solvated DMF molecules are omitted, symmetry code: i = $-x$, $-y + 1$, $-z + 1$. Displacement ellipsoids are drawn at 50% probability.

The further stabilization of both structures arises from π-π interactions between aromatic rings. Data characterizing π-π interactions are given in Table S3.

Complex **3** crystallizes in the monoclinic space group $I2/c$. In its structure, there is a binuclear $[Zn_2(ClQ)_3(HClQ)_3]^+$ cation (Figure 8, Table 4) in which both Zn atoms are hexacoordinated by three N,O-chelate bonded ligands. Three of them (ClQ) are anionic with deprotonated hydroxyl groups, while remaining three ligands (HClQ) contain protonated hydroxyl groups and are neutral. The hydrogen atoms of these hydroxyl groups are involved in strong hydrogen bonds, which link the two mononuclear Zn complex units into the binuclear cation. One of the hydrogen atoms lies on a two-fold rotation axis and is thus shared equally by O1 and $O1^i$ oxygen atoms (i = $-x + 1$, y, $-z + 1.5$). The remaining two hydrogen atoms involved in hydrogen bonds are shared unequally, and corresponding covalent bonds are elongated and are thus only slightly shorter than hydrogen bonds (Table 5). The positive charge of the binuclear cation is balanced by a triiodide anion.

Selected bond lengths and angles are gathered in Table 4 and are similar to those in **1**, **2** and **5** (below), as well as in other similar zinc complexes [31,32].

Table 2. Selected bond distances and angles [Å, °] for **1** and **2**.

	1	2		1	2
Zn1-Cl1	2.2766 (5)	2.3149 (5)	O3-Zn1-O1	106.79 (6)	95.66 (5)
Zn1-O1	2.0153 (13)	2.0388 (13)	O3-Zn1-N2^{i}	106.90 (6)	93.30 (5)
Zn1^{i}-O2	2.2779 (13)	2.0789 (12)	O1-Zn1-N2^{i}	128.77 (6)	148.68 (5)
Zn1-O3	2.0134 (12)	2.0194 (12)	O3-Zn1-Cl1	109.17 (4)	116.24 (4)
Zn1^{i}-N2	2.1101 (17)	2.1126 (16)	O1-Zn1-Cl1	106.28 (4)	98.52 (4)
Zn2^{i}-O1	2.0926 (13)	2.0893 (12)	N2^{i}-Zn1-Cl1	97.67 (5)	104.23 (4)
Zn2-O2	2.0809 (12)	2.0463 (13)	O3-Zn1-O2^{i}	78.82 (5)	143.62 (5)
Zn2^{i}-O2/O4$^{i\,a}$	2.2058 (12)	2.2044 (13)	O1-Zn1-O2^{i}	75.39 (5)	77.12 (5)
Zn2-O3	2.1016 (13)	2.1054 (12)	N2^{i}-Zn1-O2^{i}	74.63 (6)	77.92 (5)
Zn2^{i}-N1	2.0983 (15)	2.1525 (16)	Cl1-Zn1-O2^{i}	170.49 (4)	100.13 (4)
Zn2-N3	2.0791 (15)	2.0791 (15)	O2-Zn2-N3	100.07 (5)	108.62 (6)
			O2-Zn2-O1^{i}	78.25 (5)	76.73 (5)
			N3-Zn2-O1^{i}	102.75 (6)	170.93 (6)
			O2-Zn2-N1^{i}	153.47 (6)	150.34 (5)
			N3-Zn2-N1^{i}	98.11 (6)	99.11 (6)
			O1^{i}-Zn2-N1^{i}	79.01 (5)	77.27 (5)
			O2-Zn2-O3	99.92 (5)	99.00 (5)
			N3-Zn2-O3	78.85 (6)	79.48 (5)
			O1^{i}-Zn2-O3	177.72 (5)	92.56 (5)
			N1^{i}-Zn2-O3	102.44 (5)	96.15 (5)
			O2-Zn2-O2^{i}/O4$^{i\,a}$	78.62 (5)	81.48 (5)
			N3-Zn2-O2^{i}/O4	156.91 (6)	92.26 (5)
			O1^{i}-Zn2-O2^{i}/O4	99.55 (5)	95.84 (5)
			N1^{i}-Zn2-O2^{i}/O4	91.90 (5)	87.24 (5)
			O3-Zn2-O2^{i}/O4	78.69 (5)	171.46 (5)

Symmetry code: ^{i} = $-x+1, -y+1, -z+1$. ^{a} Atom O4^{i} from **2**.

Table 3. Hydrogen bond interactions (Å, °) in **1** and **2**.

Complex	D-H···A	*d*(D-H)	*d*(H···A)	*d*(D···A)	<(DHA)
1	C22-H22···O40 ^{i}	0.95	2.32	3.127	149.9
	C26-H26···Cl1A ^{ii}	0.95	2.77	3.633	151.6
2	C22-H22···Cl1 ^{iii}	0.95	2.76	3.592	146.2
	C37-H37···Cl1	0.95	2.80	3.661	151.2
	C11-H11···O5	0.95	2.38	3.158	138.3
	C36-H36···Cl25 ^{i}	0.95	2.71	3.631	162.3
	O4-H1O···O6	0.88	1.81	2.685	171.2
	O4-H2O···Cl1 ^{iv}	0.86	2.44	3.247	155.6

Symmetry codes: ^{i} = $x+1, y, z$; ^{ii} = $-x+1, -y+1, -z$ (**1**); ^{i} = $x+1, y, z$; ^{iii} = $x-1, y, z$; ^{iv} = $-x+1, -y+1, -z+1$ (**2**).

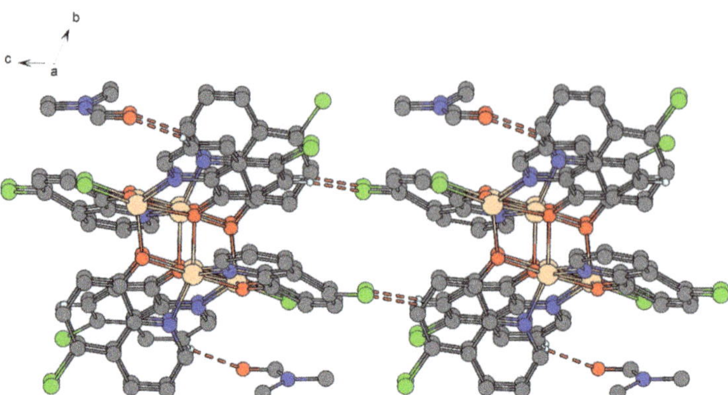

Figure 6. Part of the two-dimensional structure of **1** viewed along the *a* axis with hydrogen bonds (red dashed lines). Only hydrogen atoms involved in the hydrogen bonds are shown for clarity.

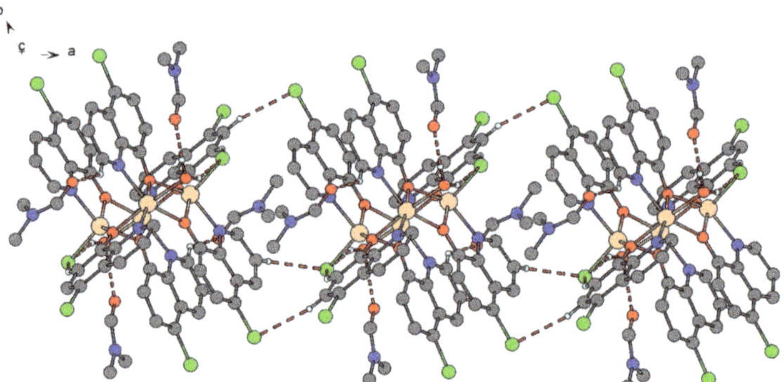

Figure 7. Part of the one-dimensional structure of **2** viewed along the *c* axis with hydrogen bonds (red dashed lines). Only hydrogen atoms involved in the hydrogen bonds are shown for clarity.

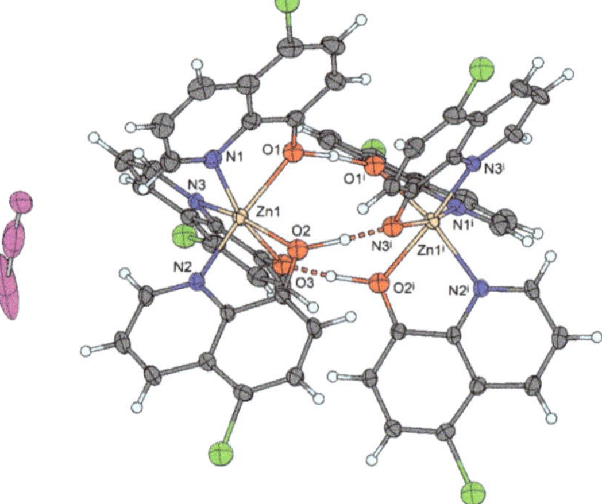

Figure 8. Molecular structure of **3**, symmetry code: i = $-x + 1$, y, $-z + 1.5$. Displacement ellipsoids are drawn at 50% probability.

Table 4. Selected bond distances and angles [Å, °] for **3**.

Zn1-O1	2.097 (3)	O1-Zn1-O2	88.54 (11)
Zn1-O2	2.101 (3)	O1-Zn1-O3	91.48 (10)
Zn1-O3	2.171 (3)	O1-Zn1-N1	77.03 (12)
Zn1-N1	2.142 (3)	O1-Zn1-N2	165.42 (12)
Zn1-N2	2.098 (3)	O1-Zn1-N3	93.46 (12)
Zn1-N3	2.089 (3)	O2-Zn1-N1	93.25 (12)
		O2-Zn1-N2	78.93 (12)
		O2-Zn1-N3	161.62 (12)
		O3-Zn1-O2	84.24 (11)
		O3-Zn1-N1	168.32 (12)
		O3-Zn1-N2	94.61 (11)
		O3-Zn1-N3	77.46 (11)
		N1-Zn1-N2	88.3 (3)
		N3-Zn1-N1	169.4 (3)
		N3-Zn1-N2	84.5 (3)

Table 5. Hydrogen bond interactions [Å, °] for **3**.

D-H···A	*d* (D-H)	*d* (H···A)	*d* (D···A)	<(DHA)
O1-H1-O1 [i]	1.202	1.202	2.399	172
O2-H2···O3 [i]	1.20	1.24	2.435	174

Symmetry code: [i] = $-x + 1, y, -z + 1.5$.

The further stabilization of the structure arises from π-π interactions between parallel aromatic rings of neighboring complexes. Due to these interactions a *zig-zag* chain parallel with the *a* axis is formed (Figure S7). Data characterizing π-π interactions are given in Table S4.

The complexes $[Zn_2(dClQ)_2(H_2O)_6(SO_4)]$ (**4**) and $[Zn_2(dBrQ)_2(H_2O)_6(SO_4)]$ (**5**) crystallize in the triclinic *P*1 space group and their structures are very similar. They are formed by binuclear complexes in which two Zn-species are bridged by a sulphate group (Figure 9). Both Zn atoms in the binuclear complexes are hexacoordinated by a corresponding N,O-bidentate chelate XQ ligand, three oxygen atoms of water molecules and one oxygen atom of the sulphate group. Interestingly, in **4** there are four crystallographically independent binuclear complexes. However, due to the extremely small crystals of **4**, their diffraction power was low, and thus the quality of the obtained data enabled us to present only an isotropic model of its structure without hydrogen atoms. Therefore, we discuss neither bond distances and angles nor nonbonding interactions in this complex.

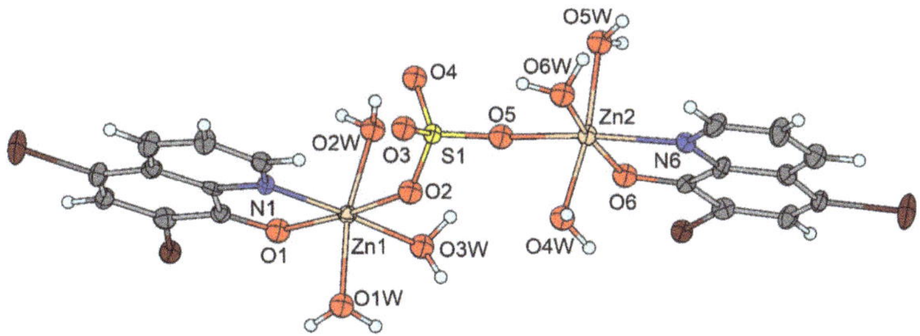

Figure 9. Molecular structure of **5**. Displacement ellipsoids are drawn at 50% probability.

Bond distances around both Zn atoms in **5** are normal (Table 6) and are close to the bond lengths observed in the above-described complexes and also in other Zn–HXQ complexes described in the literature [20,31,32]. Coordination polyhedra around Zn1 and Zn2 atoms are distorted octahedra, as represented by N-Zn-O and O-Zn-O *trans*-angles spanning from 165.6(3) to 173.8(3)°, while *cis*-angles are in a range 79.1(3) to 97.8(3)°. The shape of the coordination polyhedra is also confirmed by octahedral distortion parameter Σ and the results of the program SHAPE (Table S2).

Table 6. Selected bond distances and angles [Å, °] for **5**.

Zn1-O1w	2.079 (6)	O1-Zn1-O2	172.5 (2)
Zn1-O2w	2.053 (6)	N1-Zn1-O1	79.5 (3)
Zn1-O3w	2.078 (6)	N1-Zn1-O2	93.0 (3)
Zn1-N1	2.094 (7)	O1w-Zn1-N1	93.5 (3)
Zn1-O1	2.118 (6)	O1w-Zn1-O1	93.1 (2)
Zn1-O2	2.214 (6)	O1w-Zn1-O2	87.9 (2)
Zn2-O5	2.060 (6)	O2w-Zn1-O2	87.4 (2)
Zn2-O4w	2.061 (7)	O2w-Zn1-O3w	82.8 (2)
Zn2-O6	2.088 (6)	O2w-Zn1-O1w	167.9 (2)
Zn2-O6w	2.095 (7)	O2w-Zn1-N1	97.8 (3)
Zn2-O5w	2.112 (7)	O2w-Zn1-O1	93.1 (3)
Zn2-N6	2.135 (7)	O3w-Zn1-O1w	86.4 (3)
		O3w-Zn1-N1	173.8 (3)
		O3w-Zn1-O1	94.3 (2)
		O3w-Zn1-O2	93.2 (2)
		O4w-Zn2-O6	96.0 (3)
		O4w-Zn2-N6	94.2 (3)
		O4w-Zn2-O5w	169.4 (3)
		O4w-Zn2-O6w	84.5 (3)
		O5-Zn2-O4w	88.3 (3)
		O5-Zn2-O5w	87.6 (3)
		O5-Zn2-O6	86.6 (2)
		O5-Zn2-O6w	101.3 (3)
		O5-Zn2-N6	165.6 (3)
		O5w-Zn2-N6	92.1 (3)
		O6-Zn2-O5w	93.5 (3)
		O6w-Zn2-O5w	86.7 (3)
		O6-Zn2-O6w	172.1 (3)
		O6-Zn2-N6	79.1 (3)
		O6w-Zn2-N6	93.0 (3)

Water molecules are involved in hydrogen bonds (Table 7), forming a hydrophilic layer parallel with the (001) plane (Figure 10). On the borders of this layer there are hydrophobic dBrQ ligands, and between neighboring ligands π-π interactions stabilize the two-dimensional structure of **5** (Figure 10, Table S5).

Table 7. Hydrogen bond interactions [Å, °] in **5**.

D-H···A	*d* (D-H)	*d* (H···A)	*d* (D···A)	<(DHA)
O1w-H1O2···O6 ^{ii}	0.87	1.89	2.740	165.6
O1w-H2O2···O2 ^{ii}	0.87	2.16	3.021	173.0
O2w-H2O4···O6 ^{i}	0.88	1.96	2.788	157.2
O2w-H1O4···O4	0.88	1.97	2.738	146.1
O3w-H1O3···O2 ^{ii}	0.87	2.05	2.879	159.3
O3w-H2O3···O4 ^{i}	0.87	2.21	2.932	139.3
O4w-H2O8···O1 ^{ii}	0.87	2.05	2.789	141.7
O4w-H1O8···O3 ^{iv}	0.87	2.08	2.886	153.6
O5w-H1O7···O1 ^{i}	0.87	1.87	2.707	160.7
O5w-H2O7···O3 ^{iii}	0.87	1.99	2.787	152.3
O6-H1O9···O5w ^{iii}	0.87	2.01	2.855	161.2
O6w-H2O9···O4 ^{iii}	0.87	2.03	2.825	151.5

Symmetry codes: ^{i} = $-x + 2, -y + 1, -z + 1$; ^{ii} = $-x + 1, -y + 1, -z + 1$; ^{iii} = $-x + 2, -y + 2, -z + 1$; ^{iv} = $-x + 1, -y + 2, -z + 1$.

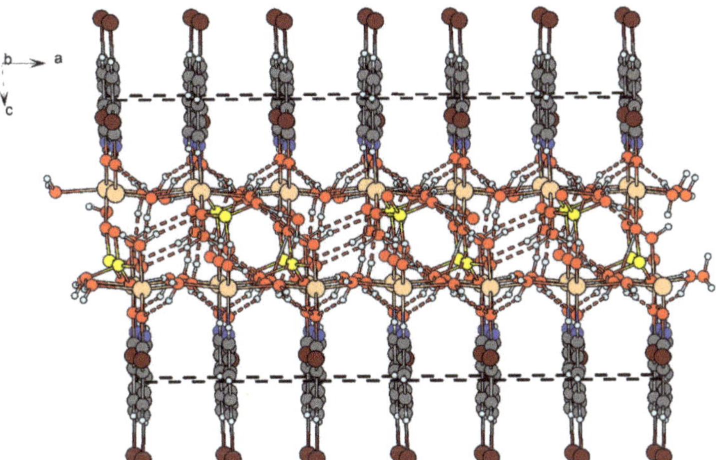

Figure 10. Part of the two-dimensional structure of **5** viewed along the *b* axis with hydrogen bonds (red dashed lines) and π-π interactions (black dashed lines).

2.5. The DNA Binding Potency

The UV-vis spectrometry is one of the basic techniques for analyzing small molecule–DNA interactions. The UV-vis titrations were realized at a fixed concentration of investigated zinc complexes with HClQ ligand (**1–3**) and varying concentrations of calf thymus DNA (ctDNA). The absorbance maxima of complexes **1–3** show reduced absorbance intensity (hypochromism) upon titration with ctDNA (Figures S8 and S9 and Figure 11). Hypochromic effect is noticed when molecules bind into DNA-supporting helix stabilization by the insertion of the flat aromatic part between base pairs [2]. This determined hypochromicity is associated with the intercalative binding mode of studied complexes [33]. The most obvious hypochromicity was recorded for complex **3** (about 46%) (Table 8). This means that the dinuclear complex **3**, which shows a promising cytotoxic effect (see Section 2.7), binds better into DNA compared to the tetranuclear complexes **1** and **2**.

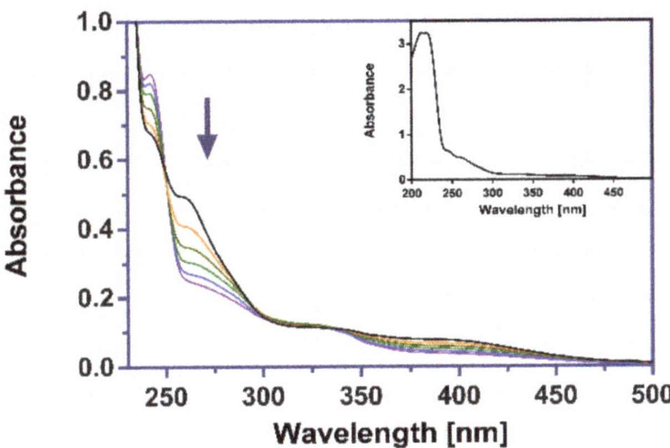

Figure 11. UV-vis spectrum of complex **3** (6.14×10^{-6} M) with ctDNA. The arrow indicates changes in absorbance upon increasing DNA concentration. Inset: UV-vis absorption spectrum of **3**.

Table 8. Spectral binding parameters of complexes **1–3**.

Complex	λ_{max1} [nm]	λ_{max2} [nm]	$\Delta\lambda_{max2}$ [nm]	hypochromism at λ_{max2} [%]	$K_{sv}{}^{EB}$ $\times 10^4$ [M^{-1}]	$K_{sv}{}^{BSA}$ $\times 10^6$ [M^{-1}]
1	215	275	2	16.32	1.06	2.04
2	218	275	1	21.72	0.62	0.68
3	219	259	4	46.54	0.34	5.44

We tried to examine the actual binding mode of the Zn(II) complexes with DNA by competitive fluorometric dye displacement assay with Ethidium bromide (EB) also. EB is often used as a DNA intercalating agent, which displays maximum emission at about 602 nm upon binding to double helical DNA [34]. The fluorescence emission spectra of the DNA-EB system was measured from 565 nm to 700 nm with incremental amounts of complexes **1–3** (Figures S10 and S11 and Figure 12).

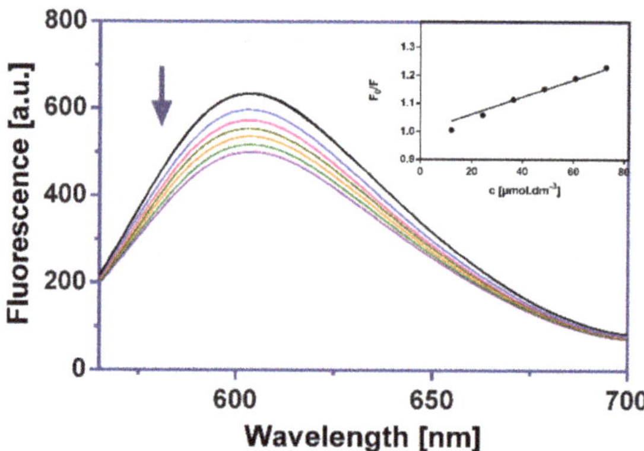

Figure 12. Fluorescence spectrum of DNA–EB complex in the absence (black line) and presence of complex **3**. Inset: the corresponding Stern–Volmer plot for quenching process of EB by **3**.

Table 8 shows Stern–Volmer quenching constants (K_{sv}), which are estimated by the linear regression of the F_o/F against quencher concentration. Obtained K_{sv} are in the range from 3.4×10^3 M^{-1} to 1.06×10^4 M^{-1}. Similar K_{sv} constants were found for complexes [Zn(bpy)(Gly)]NO_3 and [Zn(phen)(Gly)]NO_3 [35]. Since the calculated K_{sv} values are quite low, we assume that the complexes are only weak intercalators.

2.6. BSA Interaction Study

The bovine serum albumin (BSA) works as a model protein to explore drug–protein interactions due to its structural similarities with human serum albumin (HSA). In vivo, the BSA serves as a transporter for biologically efficient drugs and also for endogenous molecules. The binding of biologically effective drugs to serum albumins is an important criterion for pharmacokinetics [36].

The changes noticed in the fluorescence emission spectra of BSA with various concentrations of Zn(II) complexes are presented in Figures S12 and S13 and Figure 13.

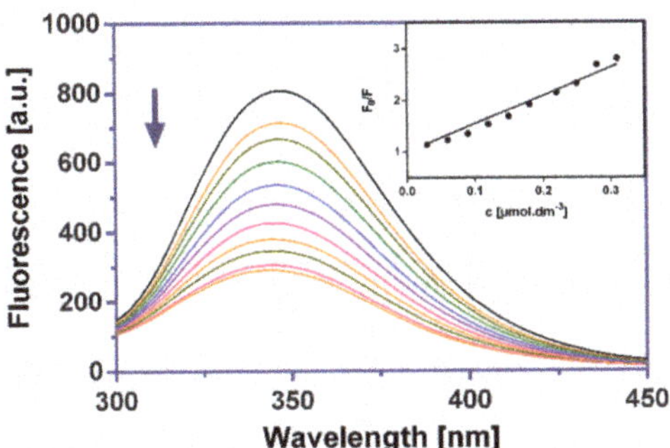

Figure 13. Fluorescence quenching spectra of BSA in presence of complex **3**. Inset: the corresponding Stern–Volmer plot for **3** at 25 °C.

Upon the addition of **1–3,** we have observed a substantial quenching of the fluorescence intensity of BSA at 348 nm. The decrease in fluorescence intensity indicates the formation of certain complexes between the Zn(II) complexes and BSA. The fluorescence quenching is also described by the Stern–Volmer relationship. The values of the parameters K_{sv} for BSA binding are listed in Table 8. The highest K_{sv} value (5.44×10^6 M^{-1}) was found for the complex **3**, which shows the highest cytotoxicity against all cell lines. All estimated K_{sv} values were about 100 times higher than that found by Butkus et al. for small zinc complexes [37].

2.7. Antiproliferative Activity

The robust screening test was used to analyze antiproliferative activity of the zinc complexes in various cancer cell lines (Table 9). As the data clearly showed, based on IC_{50} values, the most cytotoxic complexes were **3** and **4** but unfortunately with poor selectivity towards normal kidney fibroblasts. Good cytotoxic potential and higher selectivity was shown by complex **5** on all cancer cell lines except HeLa. Complexes **1** and **2** showed the lowest cytotoxic IC_{50} values on MCF-7 and HCT116 cells and average IC_{50} concentration on other cell lines. As can be seen from the IC_{50} values, **3** and **4** exhibited higher cytotoxic effect than cisplatin against all cancer cell lines except HCT116. In general, HeLa cells were most resistant to the tested complexes. The ligands used in synthesis showed cell-dependent cytotoxicity in the range of 5 to 103 μM concentration. In general, binuclear complexes **3–5** showed higher cytotoxic activity compared to their free ligands against all cancer cell lines. Tetranuclear complexes **1** and **2** showed the lowest efficiency from the complexes and ligands, probably due to their bigger size preventing their entrance to the cells. As the results showed, changing the structure of the complexes did not lead to an improvement in their cytotoxicity or selectivity in comparison with our mononuclear zinc complexes [20].

Table 9. IC_{50} values [μM] for complexes **1–5**, corresponding ligands and cisplatin in the various cell lines.

	HeLa	A549	MCF-7	MDA-MB-231	HCT116	Caco-2	Cos-7
1	71.2	39.5	8.6	29.9	7.9	33.9	26.5
2	33.6	30.3	7.1	25.5	6.9	28.6	25.2
3	8.3	5.8	5.9	8.1	5.5	6.2	6.9
4	14.88	5.81	5.93	6.92	5.49	6.30	6.53
5	31.83	25.70	5.93	8.10	6.89	17.63	36.76
HClQ	39.60	18.17	6.56	9.51	5.93	103.68	20.82
HdClQ	33.12	31.23	6.57	9.85	7.10	38.86	62.95
HdBrQ	59.91	79.72	12.03	47.15	28.66	85.05	88.17
cisplatin	18.82	13.50	21.60	73.40	3.28	23.20	18.30

2.8. Antimicrobial Activity

The antibacterial activity of the prepared complexes was tested against Gram-positive (*S. aureus*) and Gram-negative (*E. coli*) bacteria. The RIZD (percentage of relative inhibition zone diameter) was performed. Only **5** showed in RIZD test inhibition against Gram-positive bacteria *S. aureus*. The inhibition was 66.97% (Table 10), whereas the inhibition of gentamicin sulphate as positive control was 100%. Oher complexes are not suitable for antibacterial activity in concentrations of 33.6 μM. Concentration was the same as gentamicin sulphate. In the case of **2** and **3**, we observed problems with dissolution and after a few minutes it started flocculation.

Table 10. Antibacterial activity of tested complexes. RIZD (%) means percentage of relative inhibition zone diameter. All complexes were tested against *E. coli* and *S. aureus* in three replicates (n = 3, ±SD). NA means no activity.

RIZD (%)	1	2	3	4	5
E. coli	NA	NA	NA	NA	NA
S. aureus	NA	NA	NA	NA	66.97

3. Materials and Methods

3.1. Materials and Chemicals

Reagents were obtained from the following commercial sources: HClQ—95% from Sigma-Aldrich (Darmstadt, Germany); HdClQ—99%, HdBrQ—98%, and N,N-dimethylform amide—99% from Alfa Aesar (Kandel, Germany); ethanol—96% from BGV (Hniezdne, Slovakia), zinc(II) iodide—98% from Fisher Chemical (Loughborough, UK); zinc sulphate heptahydrate— p.a. from Lachema (Neratovice, Czech Republic); KOH—p.a. from ITES (Vranov, Slovakia); methanol—p.a.; and zinc(II) chloride—98% from Centralchem (Bratislava, Slovakia). All mentioned chemicals were used as received.

3.2. Syntheses

3.2.1. Synthesis of $[Zn_4Cl_2(ClQ)_6]\cdot 2DMF$ (1)

HClQ (179 mg, 1 mmol) was dissolved in DMF (10 mL). Soon afterwards a solution of zinc chloride (136 mg, 1 mmol $ZnCl_2$ in 10 mL of ethanol) was added and mixed for a while at room temperature. After one month at room temperature, orange crystals of **1** were formed, filtered off and dried on air.

$[Zn_4Cl_2(ClQ)_6]\cdot 2DMF$ (**1**): Anal. Calc. for $C_{60}H_{44}Cl_8N_8O_8Zn_4$ (1550.11 g·mol^{-1}): C, 46.35; H, 2.53; N, 6.64%. Found: C, 46.04; H, 2.56; N, 6.91%. Yield: 111.09 mg (43%). ^{1}H NMR (DMSO-d_6): δ = 8.74 (1H, br s, H-2), 8.54 (1H, d, *J* 8.5 Hz, H-4), 7.95 (1H, s, H_{DMF}-1), 7.75 (1H, dd, *J* 8.5, 4.5 Hz, H-3), 7.50 (1H, d, *J* 8.5 Hz, H-6), 6.73 (1H, br s, H-7), 2.89 (3H, s, H_{DMF}-4), 2.73 (3H, s, H_{DMF}-3) ppm. ^{13}C NMR (DMSO-d_6): δ = 162.3 (C-8, C_{DMF}-1), 145.6 (C-2), 139.7 (C-8a), 135.0 (C-4), 130.0 (C-6), 126.6 (C-4a), 122.8 (C-3), 111.6 (C-7), 109.0 (C-5), 35.8 (C-4DMF), 30.8 (C-3DMF) ppm. IR (ATR, cm^{-1}): 3047(vw), 2921(w), 2853(w), 1671(s), 1599(m), 1575(m), 1495(m), 1458(m), 1381(m), 1364(m), 1305(m), 1253(m), 1240(m), 1199(w), 1159(w), 1129(w), 1083(m), 1043(m), 961(m), 935(w), 838(m), 810(w), 783(m), 735(s), 672(m), 647(w), 638(w), 613(w), 590(w), 541(m), 504(m), 450(m).

3.2.2. Synthesis of $[Zn_4Cl_2(ClQ)_6(H_2O)_2]\cdot 4DMF$ (2)

HClQ (263 mg, 1.46 mmol) was dissolved in a mixture of DMF (7.5 mL) and ethanol (15 mL). KOH (29 mg, 1 mmol KOH in 1 mL water) was added afterwards, and the formed solution was cooled down to approximately −10 °C. Additionally, zinc chloride solution, (136 mg, 1 mmol $ZnCl_2$ in 10 mL of ethanol) cooled down to approximately −10 °C, was added and the reaction mixture was mixed for couple of minutes. After keeping the solution in the fridge for around four months, yellow crystals of **2** were formed, filtered off and dried in air.

$[Zn_4Cl_2(ClQ)_6(H_2O)_2]\cdot 4DMF$ (**2**): Anal. Calc. for $C_{66}H_{62}Cl_8N_{10}O_{12}Zn_4$ (1732.42 g·mol^{-1}): C, 45.76; H, 3.61; N, 8.08%. Found: C, 45.91; H, 3.13; N, 7.88%. Yield: 88.53 mg (21%). ^{1}H NMR (DMSO-d$_6$): δ = 8.74 (1H, br s, H-2), 8.54 (1H, dd, *J* 8.4, 1.5 Hz, H-4), 7.95 (1H, s, H_{DMF}-1), 7.75 (1H, dd, *J* 8.4, 4.4 Hz, H-3), 7.50 (1H, d, *J* 8.4 Hz, H-6), 6.74 (1H, d, *J* 8.4 Hz, H-7), 2.89 (3H, s, H_{DMF}-4), 2.73 (3H, s, H_{DMF}-3) ppm. ^{13}C NMR (DMSO-d$_6$): δ = 162.3 (C-8, C_{DMF}-1), 145.7 (C-2), 140.2 (C-8a), 135.0 (C-4), 130.0 (C-6), 126.6 (C-4a), 122.8 (C-3), 111.7 (C-7), 109.0 (C-5), 35.8 (C_{DMF}-4), 30.8 (C_{DMF}-3) ppm. IR (ATR, cm^{-1}): 3075(vw), 2923(w), 2847(w), 1672(m), 1649(m), 1598(m), 1573(m), 1495(m), 1457(m), 1398(w), 1378(m), 1364(m), 1320(m), 1257(w) 1241(w), 1200(w), 1155(w), 1129(w), 1100(w), 1087(m), 1044(m), 956(m), 885(w), 862(w), 834(w), 820(m), 784(m), 739(m), 666(m), 656(w), 636(m), 613(m), 541(m), 501(m), 451(m), 429(m).

3.2.3. Synthesis of $[Zn_2(ClQ)_3(HClQ)_3]I_3$ (3)

ZnI_2 (159 mg, 0.5 mmol) in ethanol (10 mL) was added drop by drop to the solution of HClQ (89 mg, 0.5 mmol) in ethanol (10 mL) under stirring. After one month at room temperature, dark brown crystals of **3** were formed, filtered off and dried in air.

$[Zn_2(ClQ)_3(HClQ)_3]I_3$ (**3**): Anal. Calc. for $C_{54}H_{33}Cl_6I_3N_6O_6Zn_2$ (1586.09 g·mol^{-1}): C, 40.89; H, 2.10; N, 5.30%. Found: C, 40.72; H, 2.29; N, 4.98%. Yield: 87.23 mg (66%). ^{1}H NMR (DMSO-d$_6$): δ = 10.19 (1H, br s, OH_{HClQ}), 8.95 (1H, br s, H-2_{HClQ}), 8.66 (1H, br s, H-2_{ClQ}), 8.52 (2H, br s, H-4_{HClQ}, H-4_{ClQ}), 7.74 (2H, br s, H-3_{HClQ}, H-3_{ClQ}), 7.60 (1H, br s, H-6_{HClQ}), 7.51 (1H, br s, H-6_{ClQ}), 7.08 (1H, br s, H-7_{HClQ}), 6.72 (1H, br s, H-7_{ClQ}) ppm. IR (ATR, cm^{-1}): 3062(w), 1633(w), 1580(m), 1500(m), 1463(m), 1453(w), 1393(m), 1361(m), 1319(m), 1261(m), 1239(w), 1206(w), 1155(w), 1107(w), 1059(w), 955(w), 868(w), 820(m), 807(w), 777(m), 730(m), 660(m), 629(m), 575(w), 534(m), 510(w), 485(m), 458(w), 430(m).

3.2.4. Synthesis of $[Zn_2(dClQ)_2(H_2O)_6(SO_4)]$ (4)

$ZnSO_4\cdot 7H_2O$ (144 mg, 0.5 mmol) in methanol (10 mL) was added drop by drop to the solution of HdClQ (107 mg, 0.5 mmol) in DMF (10 mL) under stirring. After one month at room temperature, yellow crystals of **4** were formed, filtered off and dried in air.

$[Zn_2(dClQ)_2(H_2O)_6(SO_4)]$ (**4**): Anal. Calc. for $C_{18}H_{20}Cl_4N_2O_{12}SZn_2$ (761.01 g·mol^{-1}): C, 28.41; H, 2.65; N, 3.68; S, 4.21%. Found: C, 28.22; H, 2.48; N, 3.39; S, 3.90%. Yield: 62.78 mg (33%). ^{1}H NMR (DMSO-d$_6$): δ = 8.79 (1H, br s, H-2 of minor form), 8.55 (1H, br s, H-2), 8.50 (d, *J* 8.4 Hz, H-4), 7.70 (2H, m, H-3, H-6). ^{13}C NMR (DMSO-d$_6$): δ = 158.2 (C-8), 146.1 (C-2), 140.0 (C-8a), 135.1 (C-4), 129.5 (C-6), 125.7 (C-4a), 122.7 (C-3), 114.8 (C-7), 108.5 (C-5). IR (ATR, cm^{-1}): 3056(vs), 1624(m), 1566(m), 1491(m), 1455(m), 1395(m), 1377(w), 1364(m), 1290(w), 1249(w), 1231(w), 1110(w), 1052(m), 963(m), 885(w), 862(m), 803(m), 780(m), 740(m), 665(m), 593(w), 503(w), 432(w).

3.2.5. Synthesis of $[Zn_2(dBrQ)_2(H_2O)_6(SO_4)]$ (5)

$ZnSO_4\cdot 7H_2O$ (144 mg, 0.5 mmol) in methanol (10 mL) was added drop by drop to the solution of HdBrQ (151 mg, 0.5 mmol) in DMF (10 mL) under stirring. After one month at room temperature, yellow crystals of **5** were formed, filtered off and dried in air.

$[Zn_2(dBrQ)_2(H_2O)_6(SO_4)]$ (**5**): Anal. Calc. for $C_{18}H_{20}Br_4N_2O_{12}S_1Zn_2$ (938.82 g·mol^{-1}): C, 23.03; H, 2.15; N, 2.98; S, 3,42%. Found: C, 23.45; H, 2.13; N, 2,97; S, 3.31%. Yield: 110.31 mg (47%). ^{1}H NMR (DMSO-d$_6$): minor form: δ = 8.76 (1H, br s, H-2), 8.44 (1H, br s, H-4), 7.92 (1H, s, H-6), 7.75 (1H, br s, H-3) ppm. ^{1}H NMR (DMSO-d$_6$): major form: δ = 8.47 (1H, br s, H-2), 8.41 (1H, d, *J* 8.5 Hz, H-4), 7.92 (1H, s, H-6), 7.70 (1H, br s, H-3) ppm. ^{13}C NMR (DMSO-d$_6$): minor form: δ = 159.8 (C-8), 145.9 (C-2), 140.0 (C-8a), 137.6 (C-4), 134.6 (C-6), 127.3 (C-4a), 123.2 (C-3), 105.3 (C-7), 97.7 (C-5) ppm. ^{13}C NMR (DMSO-d$_6$): major form: δ = 159.8 (C-8), 145.9 (C-2), 140.0 (C-8a), 137.4 (C-4), 134.7 (C-6), 127.3 (C-4a), 123.2 (C-3), 105.3 (C-7), 97.7 (C-5) ppm. IR (ATR, cm^{-1}): 3186(m), 3072(vw), 1632(m), 1556(m), 1484(m), 1455(m), 1389(m), 1359(s), 1284(w), 1247(w), 1221(w), 1121(w), 1106(w), 1050(m), 985(w), 943(m), 863(m), 800(w), 777(m), 741(m), 684(m), 660(m), 577(w), 500(w), 428(w).

3.3. Physical Measurements

The infrared spectra of the prepared complexes were recorded on a Nicolet 6700 FT-IR spectrophotometer from Thermo Scientific equipped with a diamond crystal Smart Orbit™ in the range 4000–400 cm^{-1}. Elemental analyses of C, H, N and S were obtained on CHNOS Elemental Analyzer Vario MICRO from Elementar Analysensysteme GmbH. NMR spectra were recorded at room temperature on a Varian VNMRS spectrometer operating at 599.87 MHz for ^1H and 150.84 MHz for ^{13}C. Spectra were recorded in DMSO-d_6 and the chemical shifts were referenced to the residual solvent signal (^1H NMR 2.50 ppm, ^{13}C NMR 39.5 ppm). The 2D gCOSY, gHSQC and gHMBC (optimized for a long-range coupling of 8 Hz) methods were employed.

3.4. X-ray Structure Analysis

The data collection for **1**, **2**, **3** and **4** was carried out on SuperNova diffractometer from Rigaku OD equipped with Atlas2 CCD detector; for **5** a Rigaku XtaLAB Synergy, Dualflex diffractometer equipped with Hybrid Pixel Array Detector (HyPix-6000HE) was used. CrysAlisPro software was used for data collection and cell refinement, data reduction and absorption correction [38]. Crystal structure of **3** was solved and refined by JANA2020 [39], other structures were solved and refined by SHELXT [40] and subsequent Fourier syntheses using SHELXL-2018 [41], respectively, implemented in WinGX program suit [42]. Anisotropic displacement parameters were refined for all non-H atoms, except for heavily disordered triiodide anion in the canals formed in the crystal structure of **3**. The maxima found in the difference Fourier map were assigned as iodine and their thermal movement is described by refinement of the 4th order anharmonic ADP for each atom using JANA2020 program. Hydrogen atoms of HXQ and DMF molecules were placed in calculated positions and refined riding on their parent C atoms. H atoms of hydroxyl groups in **3** and water molecules in **2** and **5** involved in hydrogen bonds were found in a Fourier difference map and refined as riding model. A geometric analysis was performed using SHELXL-2018, PLATON [43] was used to analyze π-π interaction, while DIAMOND [44] was used for molecular graphics. A summary of crystal data and structure refinement for all complexes is presented in Table 11.

Table 11. Crystal data and structural refinement for **1–5**.

Compound	1	2	3	4 [a]	5
Empirical formula	$C_{60}H_{44}Cl_8N_8O_8Zn_4$	$C_{66}H_{62}Cl_8N_{10}O_{12}Zn_4$	$C_{54}H_{33}Cl_6I_3N_6O_6Zn_2$	$C_{18}H_{20}Cl_4N_2O_{12}S_1Zn_2$	$C_{18}H_{20}Br_4N_2O_{12}S_1Zn_2$
Formula weight [$g \cdot mol^{-1}$]	1550.11	1732.33	1586.1	761.01	938.8
Temperature [K]	95 (2)	95 (2)	95 (2)	95 (2)	100 (2)
Wavelength [Å]	1.54184	1.54184	1.54184	1.54184	1.54184
Crystal system	Triclinic	Triclinic	Monoclinic	Triclinic	Triclinic
Space group	*P*1	*P*1	*I*2/*c*	*P*1	*P*1
Unit cell dimensions [Å, °]	a = 10.5958 (2) b = 11.9402 (2) c = 13.4286 (2) α = 110.555 (1) β = 106.214 (1) γ = 97.530 (1)	a = 11.5637 (1) b = 12.8889 (1) c = 13.9496 (2) α = 108.460 (1) β = 103.392 (1) γ = 107.652 (1)	a = 16.1712 (2) b = 15.5676 (2) c = 22.0518 (3) β = 99.646 (1)	a = 10.2395 (2) b = 18.4516 (5) c = 27.6232 (11) α = 75.983 (3) β = 86.842 (2) γ = 89.267 (2)	a = 7.1417 (2) b = 10.3496 (4) c = 17.6888 (8) α = 89.915 (3) β = 86.591 (3) γ = 88.077 (2)
Volume [$Å^3$]	1477.08 (4)	1751.46 (4)	5472.98 (12)	5055.9 (3)	1304.40 (9)
Z; density (calculated) [$g\ cm^{-3}$]	1; 1.743	1; 1.642	4; 1.9249	8; 1.999	2; 2.390
Absorption coefficient [mm^{-1}]	5.708	4.935	17.53	7.616	10.790
F(000)	780	880	3072	3056	908
Crystal shape, color	prism, orange	prism, yellow	prism, dark brown	plate, yellow	plate, yellow

Table 11. *Cont.*

Compound	1	2	3	4 [a]	5
Crystal size [mm^3]	0.187 × 0.119 × 0.058	0.154 × 0.113 × 0.088	0.20 × 0.18 × 0.12	0.016 × 0.029 × 0.204	0.042 × 0.056 × 0.086
θ range for data collection [°]	3.753 to 76.298	3.584 to 76.466	3.45 to 74.56	3.285 to 74.115	2.502 to 77.357
Index ranges	$-12 \leq h \leq 13$, $-15 \leq k \leq 15$, $-16 \leq l \leq 16$	$-14 \leq h \leq 14$, $-16 \leq k \leq 16$, $-17 \leq l \leq 17$	$-20 \leq h \leq 20$, $-18 \leq k \leq 19$, $-27 \leq l \leq 27$	$-12 \leq h \leq 10$, $-23 \leq k \leq 21$, $-30 \leq l \leq 34$	$-8 \leq h \leq 8$, $-13 \leq k \leq 13$, $-22 \leq l \leq 22$
Reflections collected/independent	54196/6102 [*R* (int) = 0.0240]	64538/7280 [*R* (int) = 0.0348]	52620/5599 [*R* (int) = 0.0319]	30212/19479 [*R* (int) = 0.0375]	12316/12316 [*R* (int) = 0.0705]
Data/restraints/parameters	6102/0/449	7280/0/455	5590/0/390	19479/0/745	12316/0/361
Goodness-of-fit on F^2	1.022	1.047	3.579	1.078	1.091
Final *R* indices [$I>2\sigma(I)$]	R1 = 0.0276, *w*R2 = 0.0683	R1 = 0.0282, *w*R2 = 0.0721	R1 = 0.0567, *w*R2 = 0.1826	R1 = 0.1206, *w*R2 = 0.3232	R1 = 0.0527, *w*R2 = 0.1657
R indices (all data)	R1 = 0.0292, *w*R2 = 0.0692	R1 = 0.0316, *w*R2 = 0.0748	R1 = 0.0601, *w*R2 = 0.1854	R1 = 0.1362, *w*R2 = 0.3334	R1 = 0.0567, *w*R2 = 0.1696
Largest diff. peak and hole [e Å^{-3}]	0.528 and −0.779	0.491 and −0.565	2.64 and −1.26	3.824 and −1.557	1.502 and −0.960

[a] Only isotropic model.

3.5. DNA/BSA Binding Studies

3.5.1. UV-vis Absorption Study

UV-vis spectrophotometric measurements were carried out on a Varian Cary 100 UV-vis spectrophotometer at room temperature (24 °C) in 10 mM Tris buffer (pH 7.4) using quartz cuvettes with 10 mm light path in the range of 230-500 nm. The complexes were dissolved in DMSO, while the stock solution of ctDNA was prepared in Tris–EDTA buffer (pH 7.4). The DNA concentration per nucleotide was determined by absorption spectroscopy using the molar absorption coefficient (6600 M^{-1} cm^{-1}) at 260 nm. An equal volume of nucleic acid was added to both the sample and reference cuvettes to eliminate any interference due to the absorbance of DNA itself.

3.5.2. Ethidium Bromide Displacement Assay

The fluorescence emission spectra were measured using a Varian Cary Eclipse spectrofluorometer at room temperature (24 °C) in 10 mM Tris buffer (pH 7.4). The fluorescence spectra were measured at an excitation wavelength of 510 nm and slit width 10 nm for the excitation and emission beams. The emission spectra were recorded in the range 565–700 nm and analyzed according to the classical Stern–Volmer equation:

$$\frac{F_0}{F} = 1 + K_{SV} \cdot [Q]$$

where F_0 and F represent the fluorescence intensities in the absence and presence of the quencher, K_{SV} is the Stern–Volmer quenching constant and $[Q]$ is the concentration of the quencher [33].

3.5.3. BSA Binding Experiments

The fluorescence spectra were recorded using a Varian Cary Eclipse spectrofluorometer at a temperature of 25 °C in 0.01 M phosphate-buffered saline (pH 7.4). The excitation wavelength (λ_{ex}) for BSA was 280 nm and slit width 10 nm for the excitation and emission beams. The spectrofluorometric titrations with increasing concentrations of the zinc complexes were recorded in the range 300–450 nm.

3.6. In Vitro Antitumor Activity

3.6.1. Cell Lines and Cell Culture

The human cancer cell lines purchased from ATCC (American Type Culture Collection; Manassas, VA, USA): HCT116 (human colorectal carcinoma) and HeLa (human cervical adenocarcinoma) were cultured in RPMI 1640 medium (Biosera, Kansas City, MO, United States), and A549 (human alveoral adenocarcinoma), MCF-7 (human Caucasian breast adenocarcinoma), Caco-2 (human colorectal adenocarcinoma) and MDA-MB-231 (human breast adenocarcinoma) were maintained in growth medium consisting of high glucose Dulbecco´s Modified Eagle Medium (DMEM) + sodium pyruvate (Biosera). The human kidney fibroblasts (Cos-7) were cultured in DMEM medium (Biosera). All media were supplemented with a 10% fetal bovine serum (FBS; Invitrogen, Carlsbad, CA, USA), 1X HyClone™ antibiotic/antimycotic solution (GE Healthcare, Little Chalfont, UK) and maintained in an atmosphere containing 5% CO_2 in humidified air at 37 °C. Prior to each experiment, cell viability was greater than 95%.

3.6.2. Assessment of Cytotoxicity

The antiproliferative effect of zinc complexes (concentrations of 10, 50, 100 μM) was determined by resazurin assay in HCT116, Caco-2, A549, MCF-7, MDA-MB-231, HeLa and Cos-7 cells. Tested cells (1×10^4/well) were seeded in 96-well plates. After 24 h, final zinc complex concentrations prepared from DMSO stock solution were added, and incubation proceeded for the next 72 h. Ten microliters of resazurin solution (Merck, Darmstadt, Germany) at a final concentration of 40 μM was added to each well at the end-point (72 h). After a minimum of 1 h incubation, the fluorescence of the metabolic product resorufin was measured by the automated Cytation™ 3 cell imaging multi-mode reader (Biotek, Winooski, VT, USA) at 560 nm excitation/590 nm emission filter. The results were expressed as a fold of the control, where control fluorescence was taken as 100%. All experiments were performed in triplicate. The IC_{50} values were calculated from these data.

3.7. *Antimicrobial Activity*

3.7.1. Microorganisms Used

The tested bacteria (*S. aureus* CCM 4223 and *E. coli* CCM 3988) were obtained from the Czech collection of microorganisms (CCM, Brno, Czech Republic).

3.7.2. Agar Well-Diffusion Method

The antibacterial properties of the complexes **1**–**5** and their ligands HClQ, HdClQ and HdBrQ were evaluated by the agar well diffusion method by slightly modified process compared to [45]. Firstly, each compound was dissolved in a small amount of 100% DMSO and then dissolved to 33.6 μM solution. As a positive control, gentamicin sulphate (Biosera, Nuaille, France) with a concentration of 50 μg/mL was used.

Bacteria were cultured overnight, aerobically at 37 °C in LB medium (Sigma-Aldrich, Saint-Louis, MO, USA) with agitation. The inoculum from these overnight cultures was prepared by adjusting the density of culture to equal that of the 0.5 McFarland standard ($1–2 \times 10^8$ CFU/mL) by adding a sterile saline solution. These bacterial suspensions were diluted 1:300 in liquid plate count agar (HIMEDIA, Mumbai, India), resulting in a final concentration of bacteria approximately 5×10^5 CFU/mL, and 20 mL of this inoculated agar was poured into a Petri dish (diameter 90 mm). Once the agar was solidified, five mm diameter wells were punched in the agar and filled with 50 μL of samples. Gentamicin sulphate with a concentration of 50 μg/mL was used as a positive control. The plates were incubated for 18–20 h at 37 °C. Afterward, the plates were photographed, and the inhibition zones were measured by the ImageJ 1.53e software (U. S. National Institutes of Health, Bethesda, MD, USA). The values used for the calculation are mean values calculated from 3 replicate tests.

The antibacterial activity was calculated by applying the formula reported in [45]:

$$\%\text{RIZD} = [(\text{IZD sample} - \text{IZD negative control})/\text{IZD gentamicin}] \times 100$$

where RIZD is the relative inhibition zone diameter (%) and IZD is the inhibition zone diameter (mm). As a negative control, the inhibition zones of 5% DMSO equal to 0 were taken. The inhibition zone diameter (IZD) was obtained by measuring the diameter of the transparent zone.

4. Conclusions

In this work, we have prepared five new Zn(II) complexes, $[Zn_4Cl_2(ClQ)_6]\cdot 2DMF$ (**1**), $[Zn_4Cl_2(ClQ)_6(H_2O)_2]\cdot 4DMF$ (**2**), $[Zn_2(ClQ)_3(HClQ)_3]I_3$ (**3**), $[Zn_2(dClQ)_2(H_2O)_6(SO_4)]$ (**4**) and $[Zn_2(dBrQ)_2(H_2O)_6(SO_4)]$ (**5**) (HClQ = 5-chloro-8-hydroxyquinoline, HdClQ = 5,7-dichloro-8-hydroxyquinoline and HdBrQ = 5,7-dibromo-8-hydroxyquinoline), which were characterized by IR spectroscopy, elemental analysis and single crystal X-ray structure analysis. Complexes **1** and **2** are tetranuclear molecular complexes with a complicated type of structure. Dinuclear complex **3** is an ionic compound and dinuclear complexes **4** and **5** have a similar molecular type of structure. The complexes **1** and **2** have symmetric structures in solution and ^{1}H NMR spectra display only one set of signals, corresponding to the appropriate units, and no fluxional phenomena were observed for these complexes. The broadened ^{1}H NMR signals of **3** arise from its low solubility and the presence of some particulate matter in the solution, but, on the other hand, the broadened ^{1}H NMR signals of **4** and **5** are due to the presence of a dynamic exchange process in solution. The time-dependent ^{1}H NMR spectra confirmed the stability of the studied complexes within 72 h. This study also provides information about the DNA binding mode and BSA binding potency of zinc complexes. According to the fairly low K_{sv} constant for EB displacement, we conclude that these complexes are only weak intercalators. We have also found that complex **3**, which shows considerable cytotoxic capacity, has the highest affinity to BSA of all the measured complexes. This outcome can be important for further potential pharmacokinetic measurements. The antiproliferative activity of the prepared complexes was studied using in vitro MTT assay against the HeLa, A549, MCF-7, MDA-MB-231, HCT116 and Caco-2 cancer cell lines and on Cos-7 non-cancerous cell line. The most sensitive to the tested complexes was Caco-2 cell line. Among the tested complexes, complex **3** showed the highest cytotoxicity against all cell lines. Complexes **3** and **4** showed better activity than cisplatin in almost all cases. Unfortunately, all complexes showed only poor selectivity to normal cells, except for complex **5**, which showed a certain level of selectivity. Only complex **5** showed antibacterial potential with a concentration of 33.6 μM.

Supplementary Materials: The following supporting information can be downloaded at: https://www.mdpi.com/article/10.3390/inorganics11020060/s1, Figure S1. ^{1}H NMR (600 MHz, DMSO-d_6) spectrum of complex **3**; Figure S2. ^{1}H (600 MHz, DMSO-d_6) and ^{13}C (150 MHz, DMSO-d_6) NMR spectrum of complex **5**; Figure S3. Time-dependent ^{1}H NMR (600 MHz, DMSO-d_6) spectra of complex **2**; Figure S4. Time-dependent ^{1}H NMR (600 MHz, DMSO-d_6) spectra of complex **3**; Figure S5. Time-dependent ^{1}H NMR (600 MHz, DMSO-d_6) spectra of complex **4**; Figure S6. Time-dependent ^{1}H NMR (600 MHz, DMSO-d_6) spectra of complex **5**; Figure S7. Part of the one-dimensional structure of **3** viewed along the *c* axis with π-π interactions (black dashed lines); Figure S8. UV-vis spectrum of complex **1** (6.14×10^{-6} M) with ctDNA. The arrow indicates changes in absorbance upon increasing DNA concentration. Inset: UV-vis absorption spectrum of **1**; Figure S9. UV-vis spectrum of complex **2** (6.14×10^{-6} M) with ctDNA. The arrow indicates changes in absorbance upon increasing DNA concentration. Inset: UV-vis absorption spectrum of **2**; Figure S10. Fluorescence spectrum of DNA-EB complex in the absence (black line) and presence of complex **1**. Inset: The corresponding Stern-Volmer plot for quenching process of EB by **1**; Figure S11. Fluorescence spectrum of DNA-EB complex in the absence (black line) and presence of complex **2**. Inset: The corresponding Stern-Volmer plot for quenching process of EB by **2**; Figure S12. Fluorescence quenching spectra of BSA in presence of complex **1**. Inset: The corresponding Stern-Volmer plot for **1** at 25 °C; Figure S13. Fluorescence quenching spectra of BSA in presence of complex **2**. Inset: The corresponding Stern-Volmer plot for **2** at 25 °C; Table S1. ^{13}C NMR (150 MHz, DMSO-d_6) chemical shifts δC [ppm] for complexes **1–5**. Table S2. Data describing different polyhedral distortions for **1–5**; Table S3. Cg···Cg distances and angles (Å, °) characterizing π-π interactions in **1** and **2**; Table S4. Cg···Cg distances and angles (Å, °) characterizing π-π interactions in **3**; Table S5. Cg···Cg distances and angles (Å, °) characterizing π-π interactions in **5**.

Author Contributions: Conceptualization, I.P. and M.H.; investigation, M.H., M.K., M.G., Ľ.T., M.V., D.S., S.S., E.S., M.L., V.K., J.K. and I.P.; resources, M.K., M.G., E.S., M.L. and I.P.; writing—original draft preparation, M.H., M.K., M.G., M.V., D.S., E.S. and I.P.; Writing—Review and Editing, M.H. and I.P.; visualization, M.H., M.V. and D.S.; supervision, I.P.; project administration, I.P.; funding acquisition, M.K., E.S., M.L. and I.P. All authors have read and agreed to the published version of the manuscript.

Funding: The financial support of Slovak grant agencies, VEGA 1/0126/23, 1/0037/22, 1/0653/19 and APVV-18-0016, as well as P.J. Šafárik University in Košice (VVGS-PF-2020-1425, VVGS-2021-1772 and VVGS-PF-2022-2134), are gratefully acknowledged. Moreover, this publication is the result of the project implementation: "Open scientific community for modern interdisciplinary research in medicine (OPENMED)", ITMS2014 +: 313011V455, supported by the Operational Programme Integrated Infrastructure, funded by the ERDF. CzechNanoLab project LM2018110 funded by MEYS CR is gratefully acknowledged for the financial support of the measurements at LNSM Research Infrastructure. We also thank the Research Infrastructure NanoEnviCz project, supported by the Ministry of Education, Youth and Sports of the Czech Republic, Project No. LM2018124 for instrumentation.

Data Availability Statement: CCDC 2219114–2219118 contain the supplementary crystallographic data for **1–5**. These data can be obtained free of charge via http://www.ccdc.cam.ac.uk/conts/retrieving.html, or from the Cambridge Crystallographic Data Centre, 12 Union Road, Cambridge CB2 1EZ, UK; fax: (+44) 1223-336-033; or e-mail: deposit@ccdc.cam.ac.uk.

Conflicts of Interest: The authors declare no conflict of interest. The funders had no role in the design of the study; in the collection, analyses, or interpretation of data; in the writing of the manuscript; or in the decision to publish the results.

References

1. Sung, H.; Ferlay, J.; Siegel, R.L.; Laversanne, M.; Soerjomataram, I.; Jemal, A.; Bray, F. Global Cancer Statistics 2020: GLOBOCAN Estimates of Incidence and Mortality Worldwide for 36 Cancers in 185 Countries. *CA Cancer J. Clin.* **2021**, *71*, 209–249. [CrossRef]
2. Peng, K.; Liang, B.B.; Liu, W.; Mao, Z.W. What blocks more anticancer platinum complexes from experiment to clinic: Major problems and potential strategies from drug design perspectives. *Coord. Chem. Rev.* **2021**, *449*, 214210. [CrossRef]
3. Dasari, S.; Tchounwou, B.P. Cisplatin in cancer therapy: Molecular mechanisms of action. *Eur. J. Pharmacol* **2014**, *740*, 364–378. [CrossRef]
4. Medici, S.; Peana, M.; Nurchi, V.M.; Lachowicz, J.I.; Crisponi, G.; Zoroddu, M.A. Noble metals in medicine: Latest advances. *Coord. Chem. Rev.* **2015**, *284*, 329–350. [CrossRef]
5. Afzal, O.; Kumar, S.; Haider, M.R.; Ali, M.R.; Kumar, R.; Jaggi, M.; Bawa, S. A review on anticancer potential of bioactive heterocycle quinoline. *Eur. J. Med. Chem.* **2015**, *97*, 871–910. [CrossRef]
6. Gupta, R.; Luxami, V.; Paul, K. Insights of 8-hydroxyquinolines: A novel target in medicinal chemistry. *Bioorgan. Chem.* **2021**, *108*, 104663. [CrossRef]
7. Colson, P.; Rolain, J.M.; Lagier, J.C.; Brouqui, P.; Raoult, D. Chloroquine and hydroxychloroquine as available weapons to fight COVID-19. *Int. J. Antimicrob. Agents* **2020**, *55*, 105932. [CrossRef]
8. Amolegbe, S.A.; Adewuyi, S.; Akinremi, C.A.; Adediji, J.F.; Lawal, A.; Atayese, A.O.; Obaleye, J.A. Iron(III) and copper(II) complexes bearing 8-quinolinol with amino-acids mixed ligands: Synthesis, characterization and antibacterial investigation. *Arab. J. Chem.* **2015**, *8*, 742–747. [CrossRef]
9. Scarim, C.B.; Lira de Farias, R.; Vieira de Godoy Netto, A.; Chin, C.M.; Leandro dos Santos, J.; Pavan, F.R. Recent advances in drug discovery against Mycobacterium tuberculosis: Metal-based complexes. *Eur. J. Med. Chem.* **2021**, *214*, 113166. [CrossRef]
10. Pippi, B.; Lopes, W.; Reginatto, P.; Silva, F.É.K.; Joaquim, A.R.; Alves, R.J.; Silveira, G.P.; Vainstein, M.H.; Andrade, S.F.; Fuentefria, A.M. New insights into the mechanism of antifungal action of 8-hydroxyquinolines. *Saudi Pharm. J.* **2019**, *27*, 41–48. [CrossRef]
11. Yadav, P.; Shah, K. Quinolines, a perpetual, multipurpose scaffold in medicinal chemistry. *Bioorg. Chem.* **2021**, *109*. [CrossRef] [PubMed]
12. Saadeh, H.A.; Sweidan, K.A.; Mubarak, M.S. Recent advances in the synthesis and biological activity of 8-hydroxyquinolines. *Molecules* **2020**, *25*, 4321. [CrossRef] [PubMed]
13. Zhiwei, R.; Songbai, X.; Qi, H. 5,7-Dihalo-8-quinolinol complex inhibits growth of ovarian cancer cells via the downregulation of expression of Wip1. *Trop. J. Pharm. Res.* **2020**, *19*, 1417–1422. [CrossRef]
14. Mrozek-Wilczkiewicz, A.; Kuczak, M.; Malarz, K.; Cieślik, W.; Spaczyńska, E.; Musiol, R. The synthesis and anticancer activity of 2-styrylquinoline derivatives. A p53 independent mechanism of action. *Eur. J. Med. Chem.* **2019**, *177*, 338–349. [CrossRef] [PubMed]
15. Cherdtrakulkiat, R.; Boonpangrak, S.; Sinthupoom, N.; Prachayasittikul, S.; Ruchirawat, S.; Prachayasittikul, V. Derivatives (halogen, nitro and amino) of 8-hydroxyquinoline with highly potent antimicrobial and antioxidant activities. *Biochem. Biophys. Rep.* **2016**, *6*, 135–141. [CrossRef]

16. Rbaa, M.; Haida, S.; Tuzun, B.; Hichar, A.; Hassane, A.E.; Kribii, A.; Lakhrissi, Y.; Hadda, T.B.; Zarrouk, A.; Lakhrissi, B.; et al. Synthesis, characterization and bioactivity of novel 8-hydroxyquinoline derivatives: Experimental, molecular docking, DFT and POM analyses. *J. Mol. Struct.* **2022**, *1258*, 132688. [CrossRef]
17. Budimir, A.; Humbert, N.; Elhabiri, M.; Osinska, I.; Biruš, M.; Albrecht-Gary, A.M. Hydroxyquinoline based binders: Promising ligands for chelatotherapy? *J. Inorg. Biochem.* **2011**, *105*, 490–496. [CrossRef]
18. Farkasová, V.; Drweesh, S.A.; Lüköová, A.; Sabolová, D.; Radojević, I.D.; Čomić, L.R.; Vasić, S.M.; Paulíková, H.; Fečko, S.; Balašková, T.; et al. Low-dimensional compounds containing bioactive ligands. Part VIII: DNA interaction, antimicrobial and antitumor activities of ionic 5,7-dihalo-8-quinolinolato palladium(II) complexes with K+ and Cs+ cations. *J. Inorg. Biochem.* **2017**, *167*, 80–88. [CrossRef]
19. Litecká, M.; Hreusová, M.; Kašpárková, J.; Gyepes, R.; Smolková, R.; Obuch, J.; David, T.; Potočňák, I. Low-dimensional compounds containing bioactive ligands. Part XIV: High selective antiproliferative activity of tris(5-chloro-8-quinolinolato)gallium(III) complex against human cancer cell lines. *Bioorgan. Med. Chem. Lett.* **2020**, *30*, 127206. [CrossRef]
20. Kuchárová, V.; Kuchár, J.; Lüköová, A.; Jendželovský, R.; Majerník, M.; Fedoročko, P.; Vilková, M.; Radojević, I.D.; Čomić, L.R.; Potočňák, I. Low-dimensional compounds containing bioactive ligands. Part XII: Synthesis, structures, spectra, in vitro antimicrobial and cytotoxic activities of zinc(II) complexes with halogen derivatives of quinolin-8-ol. *Polyhedron* **2019**, *170*, 447–457. [CrossRef]
21. Vranec, P.; Potočňák, I. Low-dimensional compounds containing bioactive ligands. Part III: Palladium(II) complexes with halogenated quinolin-8-ol derivatives. *J. Mol. Struct.* **2013**, *1041*, 219–226. [CrossRef]
22. Vranec, P.; Potočňák, I.; Sabolová, D.; Farkasová, V.; Ipóthová, Z.; Pisarčíková, J.; Paulíková, H. Low-dimensional compounds containing bioactive ligands. V: Synthesis and characterization of novel anticancer Pd(II) ionic compounds with quinolin-8-ol halogen derivatives. *J. Inorg. Biochem.* **2014**, *131*, 37–46. [CrossRef]
23. El-Dissouky, A.; Fahmy, A.; Amer, A. Complexing Ability of some r-Lactone Derivatives. Thermal, Magnetic and Spectral Studies on Cobalt(II), Nickel(I1) and Copper(I1) Complexes and their Base Adducts. *Inorganica Chim. Acta* **1987**, *133*, 311–316. [CrossRef]
24. Ananthanarayanan, V. Studies on the vibrational spectrum of the so4 Ion in Crystalline M2′M″(SO4)2· 6H2O (M′ =K or NH4 and M″ =Mg, Zn, Ni, or Co): Observations on the symmetry of the sulfate ion in crystals. *J. Chem. Phys.* **1968**, *48*, 573–581. [CrossRef]
25. Arjunan, V.; Mohan, S.; Ravindran, P.; Mythili, C.V. Vibrational spectroscopic investigations, ab initio and DFT studies on 7-bromo-5-chloro-8-hydroxyquinoline. *Spectrochim. Acta-Part A Mol. Biomol. Spectrosc.* **2009**, *72*, 783–788. [CrossRef] [PubMed]
26. Arici, K. Vibrational Spectra of 4-hydroxy-3-cyano-7-chloro-quinoline by Density Functional Theory and ab initio Hartree-Fock Calculations. *Int. J. Chem Technol.* **2017**, *1*, 24–29. Available online: http://dergipark.gov.tr/ijct (accessed on 26 October 2017). [CrossRef]
27. Zhang, H.R.; Liu, Y.C.; Chen, Z.F.; Guo, J.; Peng, Y.X.; Liang, H. Crystal Structures, Cytotoxicity, Cell Apoptosis Mechanism, and DNA Binding of Two 8-Hydroxylquinoline Zinc(II) Complexes. *Russ. J. Coord. Chem.* **2018**, *44*, 322–334. [CrossRef]
28. Addison, A.W.; Rao, T.N.; Reedijk, J.; Van Rijn, J.; Verschoor, G.C. Synthesis, structure, and spectroscopic properties of copper(II) compounds containing nitrogen–sulphur donor ligands; the crystal and molecular structure of aqua [1,7-bis(N-methylbenzimidazol-2′-yl)-2,6-dithiaheptane]copper(II) perchlorate. *J. Chem. Soc. Dalton Trans.* **1984**, *7*, 1349–1356. [CrossRef]
29. Llunell, M.; Casanova, D.; Cirera, J.; Alemany, P.; Alvarez, S. *SHAPE, Version 2.1*; Universitat de Barcelona: Barcelona, Spain, 2013.
30. Khakhlary, P.; Baruah, J.B. Studies on cluster, salt and molecular complex of zinc-quinolinate. *Chem. Sci. J.* **2015**, *127*, 215–223. [CrossRef]
31. Liu, Y.C.; Wei, J.H.; Chen, Z.F.; Liu, M.; Gu, Y.Q.; Huang, K.-B.; Li, Z.Q.; Liang, H. The antitumor activity of zinc(II) and copper(II) complexes with 5,7-dihalo-substituted-8-quinolinoline. *Eur. J. Med. Chem.* **2013**, *69*, 554–563. [CrossRef]
32. Zhang, H.R.; Liu, Y.C.; Meng, T.; Qin, Q.P.; Tang, S.F.; Chen, Z.F.; Zou, B.Q.; Liu, Y.N.; Liang, H. Cytotoxicity, DNA binding and cell apoptosis induction of a zinc(II) complex of HBrQ. *MedChemComm* **2015**, *6*, 2224–2231. [CrossRef]
33. Drweesh, E.A.; Kuchárová, V.; Volarevic, V.; Miloradovic, D.; Ilic, A.; Radojević, I.D.; Raković, I.R.; Smolková, R.; Vilková, M.; Sabolová, D.; et al. Low-dimensional compounds containing bioactive ligands. Part XVII: Synthesis, structural, spectral and biological properties of hybrid organic-inorganic complexes based on [PdCl4]2− with derivatives of 8-hydroxyquinolinium. *J. Inorg. Biochem.* **2022**, *228*, 111697. [CrossRef] [PubMed]
34. Palmeira-Mello, M.V.; Caballero, A.B.; Lopez-Espinar, A.; Guedes, G.P.; Caubet de Souza, A.M.T.A.; Lanznaster, M.; Gamez, P. DNA-interacting properties of two analogous square-planar cis-chlorido complexes: Copper versus palladium. *J. Biol. Inorg. Chem.* **2021**, *26*, 727–740. [CrossRef] [PubMed]
35. Mansouri-Torshizi, H.; Khosravi, F.; Ghahghaei, A.; Shahraki, S.; Zareian-Jahromi, S. Investigation on the interaction of newly designed potential antibacterial Zn(II) complexes with CT-DNA and HSA. *J. Biomol. Struct. Dyn.* **2018**, *36*, 2713–2737. [CrossRef]
36. Yallur, B.C.; Katrahalli, U.; Krishna, P.M.; Hadagali, M.D. BSA binding and antibacterial studies of newly synthesized 5,6-Dihydroimidazo [2,1-b]thiazole-2-carbaldehyde. *Spectrochim. Acta A Mol. Biomol.* **2019**, *222*, 117192. [CrossRef] [PubMed]
37. Butkus, J.M.; O'Riley, S.; Chohan, B.S.; Basu, S. Interaction of Small Zinc Complexes with Globular Proteins and Free Tryptophan. *Int. J. Spectrosc.* **2016**, *2016*, 1378680. [CrossRef]
38. Rigaku. *CrysAlisPRO, Version 1.0.43*; Rigaku Oxford Diffraction: Yarnton, UK, 2020.
39. Petříček, V.; Dušek, M.; Palatinus, L. Crystallographic Computing System JANA2006: General features. *Z. Krist.-Cryst. Mater. Online* **2014**, *229*, 345–352. [CrossRef]

40. Sheldrick, G.M. SHELXT—Integrated space-group and crystal-structure determination. *Acta Crystallogr. Sect. A Found. Adv.* **2015**, *71*, 3–8. [CrossRef] [PubMed]
41. Sheldrick, G.M. Crystal structure refinement with SHELXL. *Acta Crystallogr. Sect. C Struct. Chem.* **2015**, *71*, 3–8. [CrossRef] [PubMed]
42. Farrugia, L.J. WinGX and ORTEP for Windows: An update. *J. Appl. Crystallogr.* **2012**, *45*, 849–885. [CrossRef]
43. Spek, A.L. Structure validation in chemical crystallography. *Acta Crystallogr. Sect. D Biol. Crystallogr.* **2009**, *65*, 148–155. [CrossRef] [PubMed]
44. Brandenburg, K. *Diamond. Version 3.2K.*; Crystal Impact GbR: Bonn, Germany, 2014.
45. Rojas, J.J.; Ochoa, V.J.; Ocampo, S.A.; Muñoz, J.F. Screening for antimicrobial activity of ten medicinal plants used in Colombian folkloric medicine: A possible alternative in the treatment of non-nosocomial infections. *BMC Complement. Altern. Med.* **2006**, *6*, 2. [CrossRef] [PubMed]

Article

Reaching the Maximal Unquenched Orbital Angular Momentum *L* = 3 in Mononuclear Transition-Metal Complexes: Where, When and How?

Vladimir S. Mironov [1,2]

1 Shubnikov Institute of Crystallography of Federal Scientific Research Centre "Crystallography and Photonics" of Russian Academy of Sciences, Leninskiy Prospekt 59, Moscow 119333, Russia; mirsa@list.ru
2 Institute of Problems of Chemical Physics, IPCP RAS, Chernogolovka 142432, Russia

Abstract: The conditions for achieving the maximal unquenched orbital angular momentum $L = 3$ and the highest magnetic anisotropy in mononuclear 3d complexes with axial coordination symmetry are examined in terms of the ligand field theory. It is shown that, apart from the known linear two-coordinate $3d^7$ complex $Co^{II}(C(SiMe_2ONaph)_3)_2$ characterized by record magnetic anisotropy and single-molecule magnet (SMM) performance (with the largest known spin-reversal barrier $U_{eff} = 450\ cm^{-1}$), the maximal orbital angular momentum $L = 3$ can also be obtained in linear two-coordinate $3d^2$ complexes (V^{3+}, Cr^{4+}) and in trigonal-prismatic $3d^3$ (Cr^{3+}, Mn^{4+}) and $3d^8$ (Co^+, Ni^{2+}) complexes. A comparative assessment of the SMM performance of the $3d^2$, $3d^3$ and $3d^8$ complexes indicates that they are unlikely to compete with the record linear complex $Co^{II}(C(SiMe_2ONaph)_3)_2$, whose magnetic anisotropy is close to the physical limit for a 3d metal.

Keywords: magnetic anisotropy; single-molecule magnet; unquenched orbital momentum; spin-reversal barrier; mononuclear 3d complexes; spin-orbit coupling

Citation: Mironov, V.S. Reaching the Maximal Unquenched Orbital Angular Momentum *L* = 3 in Mononuclear Transition-Metal Complexes: Where, When and How? *Inorganics* **2022**, *10*, 227. https://doi.org/10.3390/inorganics10120227

Academic Editor: Peter Segl'a

Received: 28 October 2022
Accepted: 24 November 2022
Published: 27 November 2022

1. Introduction

Mononuclear paramagnetic metal complexes with high magnetic anisotropy are of great interest in recent years, largely due to the development of low-dimensional molecule-based magnetically bistable materials featuring slow magnetic relaxation and blocking of magnetization at low temperature—single-molecule magnets (0D, SMMs) [1–9] and single-chain magnets (1D, SCMs) [10–12]. The overall performance of SMMs and SCMs is quantified by two key parameters, the effective energy barrier to magnetization reversal (U_{eff}) and the blocking temperature (T_B), below which the magnetization persists in zero external fields [6–12]. The U_{eff} and T_B values depend strongly on the magnetic anisotropy of spin carriers incorporated in the high-spin molecule. The early research was mainly focused on polynuclear transition-metal complexes based on high-spin 3d-ions with enhanced single-ion magnetic anisotropy arising from the zero-field splitting (ZFS) of the ground spin state, such as Mn^{3+} and Ni^{2+} [1–3]. Then, after the discovery of slow magnetic relaxation in mononuclear lanthanide complexes (2003) [13] and later in mononuclear transition metal compounds (2010) [14,15], research interest has turned to SMMs involving single paramagnetic ion with high magnetic anisotropy due to the first-order unquenched orbital angular momentum, 4f-complexes [16–24] and special orbitally degenerate 3d-complexes with high axial symmetry [25,26]. These mononuclear 4f- and 3d-SMMs are known as single-ion magnets (SIMs).

At present, it is generally recognized that the unquenched orbital angular momentum of the magnetic metal ions is the most important factor governing large magnetic anisotropy and high SMM characteristics [8]. In this respect, lanthanide ions are especially attractive. In fact, all Ln^{3+} ions with open $4f^N$-shell exhibit large unquenched orbital angular momentum *L* (ranging from 3 to 6, except Gd^{3+} with $L = 0$) due to very weak ligand-field (LF) splitting energy and strong spin-orbit coupling (SOC) of 4f-electrons; the maximal orbital momentum $L = 6$ occurs in Nd^{3+} ($4f^3$, $^4I_{9/2}$), Pm^{3+} ($4f^4$, $^5I_4$), Ho^{3+} ($4f^{10}$, $^5I_8$) and Er^{3+} ($4f^{11}$, $^4I_{15/2}$) ions. Due to the strong SOC of 4f electrons, the orbital momentum L and spin S are coupled into the total angular momentum J; the ground state of free $Ln^{3+}(4f^N)$ ions corresponds to the $^{2S+1}L_J$ multiplet with $J = L - S$ ($N < 7$) and $J = L + S$ ($N > 7$). For Ln^{3+} ions in a ligands coordination environment, the ground J-multiplet undergoes LF splitting into $2J + 1$ sublevels (called Stark sublevels) creating a highly anisotropic ground state [8,20]. In particular, pure Ising-type uniaxial magnetic anisotropy occurs in monometallic lanthanide complexes with distinct axial symmetry, in which axial LF lifts the $(2J + 1)$-fold degeneracy of the ground multiplet into $\pm M_J$ microstates with a definite projection of the total angular momentum [16–24]. The energy splitting diagram of $\pm M_J$ states (E vs. M_J) depends on both the strength and the intrinsic structure of the axial LF, which is specified by three axial LF parameters B_{20}, B_{40}, B_{60} [27]. In the optimal case, the strong field stabilizes the doubly degenerate ground state with maximal M_J projection ($M_J = \pm J$) and with large energy separation from excited LF states thereby resulting in a double-well potential with the high-energy barrier U_{eff} inherent to high-performance SMMs. In particular, such LF-splitting pattern is observed in square antiprismatic [13], pentagonal-bipyramidal [28,29] and quasi-linear metallocene lanthanide complexes [30,31], many of which exhibit exceptionally high U_{eff} and T_B values, including a record-setting dysprosium metallocene SMM with $U_{eff} = 1541$ cm^{-1} and $T_B = 80$ K [31]. Recent trends in high-performance 4f SIM are reviewed in ref. [32].

Later on, since the report on the first mononuclear trigonal pyramidal Fe^{II} complexes with SMM behavior (2010) [14,15], extensive efforts have been taken to the development of 3d SIMs using diverse approaches to increase magnetic anisotropy and energy barrier; numerous 3d SIMs were reported in the past decade [8,25,26]. However, in contrast to f-block element complexes, in transition metal complexes the orbital angular momentum is commonly quenched ($L = 0$) by a strong LF to produce the spin-only ground state with low second-order magnetic anisotropy [33]. In the majority of high-spin 3d complexes, magnetic anisotropy is typically due to zero-field splitting (ZFS) of the ground spin multiplet $2S + 1$, which is generally a weak effect resulting from second-order SOC [34]. Hence, the barriers U_{eff} of 3d SIMs are mostly low, within a few tens of cm^{-1} [8,25,26,33]. The most efficient strategy toward high-performance 3d SIMs takes advantage of unquenched orbital angular momentum ($L \neq 0$) giving rise to the first-order spin-orbit splitting of the ground spin stated with $L \neq 0$, which provides considerably stronger magnetic anisotropy and high energy barriers [16–24]. Unquenched first-order orbital angular momentum can only appear in orbitally degenerate transition metal complexes with high symmetry, such as octahedral Co^{II} and $[Fe^{III}(CN)_6]^3$ complexes (O_h) [33,34], trigonal-pyramidal (C_{3v}), [14,15] trigonal-bipyramidal (D_{3h}), [35] trigonal-prismatic (D_{3h}) [36,37], pentagonal-bipyramidal (D_{5h}) [38] and linear ($D_{\infty h}$) two-coordinate complexes [39–43].

However, apart from unquenched orbital angular momentum, another important condition for the highest SMM performance of mononuclear 3d complexes is the axial limit of the magnetic anisotropy, which has a purely Ising-type nature, with zero transverse components. This case occurs in the LF of axial symmetry, which is produced by tuning the geometric axiality of the ligand environment of the 3d-ion. Depending on the specific type of the coordination polyhedron and electronic configuration of the 3d ion, the axial LF may stabilize the ground orbital doublet with the projection of the orbital angular momentum $M_L = 0, \pm 1, \pm 2$, or ± 3. The latter is further split by the first-order SOC into energy levels $\pm M_J$ with definite projection M_J of the total angular momentum J along the magnetic axis, which ranges from $L + S$ to $|L - S|$. In this regime, spin energy levels are represented by pure $|\pm M_J>$ spin wave functions, in which quantum tunneling of magnetization (QTM) is suppressed both in the ground and exited spin states. In these systems the barrier U_{eff} is controlled by the first-order SOC energy $\zeta_{3d}\boldsymbol{LS}$, which is proportional to the value of the ground-state orbital angular momentum M_L; approximately, U_{eff} is specified by $\zeta_{3d}M_L/2S$, which is the energy separation between the ground $\pm M_J$ and first exited

$\pm(M_J - 1)$ states [42]. This leads to a pure Ising magnetic anisotropy and double-well potential with two lowest $M_J = \pm J$ states similar to those in 4f SIMs [16–24]. Accordingly, the highest barriers U_{eff} are observed in 3d SIMs with large orbital momentum M_L, namely, in linear $3d^7$ complexes $[Fe^I(C(SiMe_3)_3)_2]^-$ (226 cm^{-1}) [40] and $[(sIPr)Co^{II}NDmp]$ (413 cm^{-1}) [41] and also in trigonal-prismatic Co^{II} complex (152 cm^{-1}) [36] featuring $L = 2$, $S = 3/2$ and $M_J = \pm 7/2$ in the ground state. The record barrier belongs to linear two-coordinate cobalt complex, $Co^{II}(C(SiMe_2ONaph)_3)_2$ ($U_{eff} = 450$ cm^{-1}) [42] and for the monocoordinated cobalt(II) adatom on a MgO surface ($U_{eff} = 468$ cm^{-1}) [43] which have $L = 3$, $S = 3/2$ and $M_J = \pm 9/2$ in the ground state.

Considering that a large unquenched orbital angular momentum is extremely important for attaining the highest SMM performance in 3d-SIMs, the relevant question is in which mononuclear 3d complexes, besides the already known record-breaking linear two-coordinate $3d^7$ complexes [42,43], the maximum orbital angular momentum $L = 3$ (represented by the ground orbital doublet $M_L = \pm 3$) is feasible. In this paper, general conditions for the appearance of the maximum orbital angular momentum $L = 3$ in mononuclear 3d complexes are analyzed in terms of the LF theory. We show that, apart from the linear two-coordinate $3d^7$ complex $Co^{II}(C(SiMe_2ONaph)_3)_2$ [42], the ground-state orbital doublet $M_L = \pm 3$ can occur in linear $3d^2$ complexes and in trigonal-prismatic $3d^3$ and $3d^8$ complexes. Specific conditions for the LF splitting pattern of 3d orbitals leading to $L = 3$ are established.

2. Results

Below, the basic conditions for the realization of the maximal orbital angular momentum $M_L = \pm 3$ in mononuclear 3d complexes with the axial LF are analyzed. First of all, it is clear that the ground orbital doublet $M_L = \pm 3$ can only occur in 3d ions, which in the free-ion state have a ${}^{2S+1}F$ ground atomic term featuring the maximum orbital angular momentum $L = 3$ for 3d ions, i.e., $3d^2$, $3d^3$, $3d^7$ and $3d^8$ ions. In transition–metal complexes with the LF of axial symmetry, the lowest ${}^{2S+1}F$ term splits into four orbital components ${}^{2S+1}F(M_L)$ with a projection of orbital momentum $M_L = 0, \pm 1, \pm 2, \pm 3$ (Figure 1). The objective of the present study is to establish conditions for the $M_L = \pm 3$ orbital doublet to be the ground state.

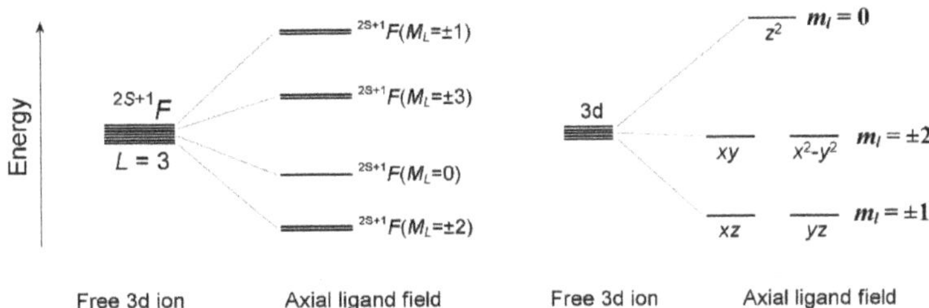

Figure 1. Energy splitting of the ground-state atomic term ${}^{2S+1}F(L = 3)$ and 3d orbitals in a ligand field of axial symmetry. The orbital composition of the ${}^{2S+1}F(M_L = \pm 3)$ states is shown below in the insets to the corresponding figures.

In an axial LF, five 3d orbitals are split into three groups of orbitals with a definite projection of the one-electron orbital angular momentum m_l on the anisotropy axis z, $3d_{z2}$ ($m_l = 0$), ($3d_{xz}$, $3d_{yz}$) ($m_l = \pm 1$) and $3d_{xy}$, $3d_{x2-y2}$ ($m_l = \pm 2$) (Figure 1). A necessary (but not sufficient) condition for the $M_L = \pm 3$ ground state is the presence of one unpaired electron in each of the two doubly degenerate 3d orbitals with magnetic numbers $m_l = \pm 1$, and $m_l = \pm 2$, i.e., $(xz, yz)^1(xy, x2-y2)^1$ or $(xz, yz)^3(xy, x2-y2)^3$. For each of the 3d ions selected above, the LF splitting pattern of 3d-orbitals leading to the ${}^{2S+1}F(M_L = \pm 3)$ ground state is examined in terms of the conventional LF theory, which takes into account the interelectron repulsion (quantified in terms of the Racah parameters B and C) and SOC, $\zeta_{3d}\Sigma_i \mathbf{l}_i \mathbf{s}_i$. LF calculations are performed with the full basis set of $3d^N$ configurations involving 45 ($3d^2$, $3d^8$) and 120 ($3d^3$, $3d^7$) $|LM_LSM_S>$ microstates. Details of LF calculations are available in the Supplementary Materials.

2.1. $3d^2$ Complexes (V^{3+}, Cr^{4+})

The necessary condition for the $^3F(M_L = \pm 3)$ orbital doublet to be the ground state is to have the LF splitting pattern $E(z^2) > E(xz, yz) > E(xy, x2 - y2)$. Such an orbital energy scheme can occur in complexes with linear or quasi-linear two-coordinate complexes [44] similar to the quasi-linear Fe^{II} [39] and Co^{II} complexes [42]. For a quantitative evaluation, LF calculations for the energy levels of the $3d^2$ configuration were performed using the Racah parameters $B = 600$, $C = 2900$ cm^{-1} and SOC constant $\zeta = 150$ cm^{-1}, which are typical for V^{3+} ion [45]. The calculations were performed with fixed orbital energies $E(z^2) = 10{,}000$ cm^{-1}, $E(xy, x2 - y2) = 0$ and variable orbital energy $E(xz, yz) = \Delta$ (see inset in Figure 2a). The orbital doublet $^3F(M_L = \pm 3)$ (shown in the solid red line in Figure 2a) is found to be the ground state in the energy range $0 \le \Delta \le 5400$ cm^{-1}; at larger energies, the ground state is the orbital singlet $^3F(M_L = 0)$, whose wave function $Det||xy\uparrow, x2 - y2\uparrow||$ corresponds to two singly occupied xy and $x2 - y2$ orbitals with parallel electron spins (Figure 2a). From the point of view of coordination geometry, the only suitable case for the $^3F(M_L = \pm 3)$ ground state is a linear two-coordination of $3d^2$ ion. Numerous 3d complexes of this type were reported in the literature, as outlined in a review article [44].

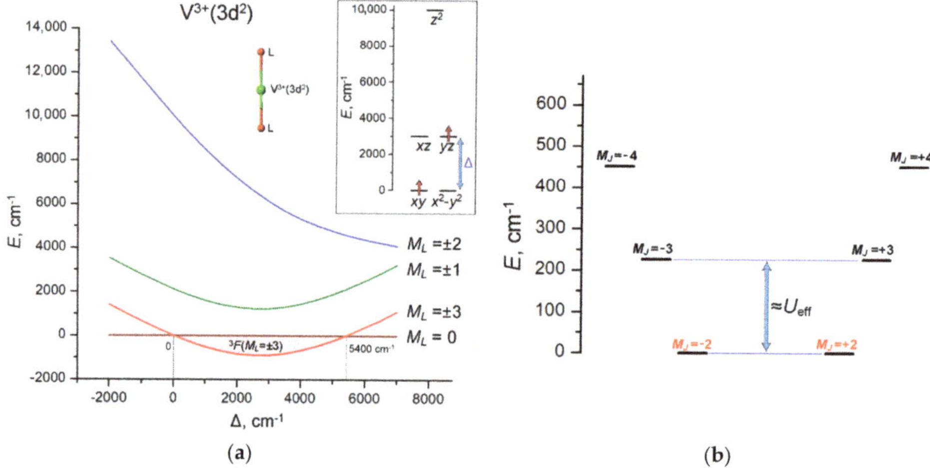

Figure 2. (**a**) Variation of the energies of the $^3F(M_L)$ LF states of $V^{3+}(3d^2)$ ion in linear coordination with increasing energy separation $\Delta = E(xz, yz) - E(xy, x2 - y2)$ between the lowest and first excited sets of 3d orbitals. The range of the stabilization of the $^3F(M_L = \pm 3)$ ground state ($0 \le \Delta \le 5400$ cm^{-1}) is marked with light pink. Inset: the LF splitting pattern of 3d orbitals in a linear two-coordination, $E(xy, x2 - y2) = 0$, $E(z^2) = 10{,}000$ cm^{-1} (fixed) and $E(xz, yz) = \Delta$ (variable); (**b**) E vs. M_J diagram calculated at $\Delta = 3000$ cm^{-1}. The spin-reversal barrier $U_{eff} \approx 200$ cm^{-1} is approximately estimated by the energy separation between the ground $M_J = \pm 2$ and the first excited $M_J = \pm 3$ states.

The calculated E vs. M_J diagram indicates that such a $3d^2$ complex may be a SMM with a barrier U_{eff} of about 200 cm^{-1}, as estimated from the energy gap between ground $M_J = \pm 2$ and first excited $M_J = \pm 3$ states (Figure 2b). Remarkably, the E vs. M_J diagram has an inverse order of $\pm M_J$ levels compared to the usual double-well spin energy profile (with the maximum M_J at the bottom and the minimum M_J at the top) due to the antiparallel coupling of the orbital momentum $L = 3$ and spin $S = 1$ of $V^{3+}(3d^2)$ ion. However, the actual SMM performance of the linear two-coordinate complex of V^{3+} is expected to be much lower due to the non-Kramers nature of the ground doublet $M_J = \pm 2$, which is subject to splitting due to possible deviations from the axiality of the molecular structure, leading to fast under-barrier quantum tunneling of magnetization (QTM) that shortcuts the energy barrier. Hence, in terms of SMM performance, the situation with the linear two-coordinate $3d^2$ complex is much less favorable compared to the record cobalt $3d^7$ complex $Co^{II}(C(SiMe_2ONaph)_3)_2$ [42].

2.2. $3d^3$ Complexes (Cr^{3+}, Mn^{4+})

The ground state of a free $3d^3$ ion is a ^4F atomic term, which is split by the axial LF into four orbital components, $^4F(M_L = \pm 3)$, $^4F(M_L = \pm 2)$, $^4F(M_L = \pm 1)$ and $^4F(M_L = 0)$. The $^4F(M_L = \pm 3)$ state originates from the $(xy, x2 - y2)^1(xz, yz)^1(z^2)^1$ electronic configuration. LF analysis indicates that this configuration can only occur at the $E(xy, x2 - y2) > E(xz, yz) > E(z^2)$ LF splitting pattern of 3d orbitals. Comparative LF calculations with atomic parameters $B = 650$, $C = 3000$ cm^{-1} and $\zeta = 250$ cm^{-1} for Cr^{3+} ion reveal that the orbital doublet $^4F(M_L = \pm 3)$ is the ground state when the energy difference $\Delta = E(xy, x2 - y2) - E(xz, yz)$ between the energy levels of the $(xy, x2 - y2)$ and (xz, yz) orbitals is within the range $0 \leq \Delta \leq 5850$ cm^{-1}. Beyond this range, the ground state turns to the $^4F(M_L = 0)$ orbital singlet, which is represented by a mixture of the $(xy)^1(x^2 - y^2)^1(z^2)^1$ and $(xz)^1(yz)^1(z^2)^1$ electronic configurations (Figure 3).

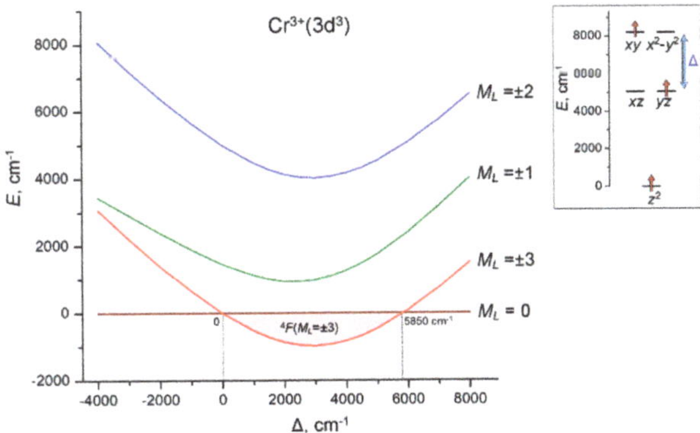

Figure 3. Dependence of the energies of the $^4F(M_L)$ LF states of $Cr^{3+}(3d^3)$ ion on the energy separation $\Delta = E(xy, x2 - y2) - E(xz, yz)$ between the $(xy, x2 - y2)$ and (xz, yz) sets of 3d orbitals. The range of the stabilization of the $^4F(M_L = \pm 3)$ ground state ($0 \leq \Delta \leq 5850$ cm^{-1}) is marked with light pink. Inset: the LF splitting pattern of 3d orbitals, $E(z^2) = 0$, $E(xz, yz) = 5000$ cm^{-1} (fixed) and $E(xy, x2 - y2) = 5000$ cm^{-1} $+ \Delta$ (variable). LF calculations were done with atomic parameters $B = 650$, $C = 3000$ and $\zeta = 250$ cm^{-1}.

Evidently, the linear two-coordinate geometry of $3d^3$ complexes cannot provide the required orbital pattern $E(xy, x2 - y2) > E(xz, yz) > E(z^2)$ since linear coordination is characterized by another order of orbital energies, $E(z^2) > E(xz, yz) > E(xy, x2 - y2)$, which can result in $M_L = \pm 1$ or $M_L = \pm 2$ ground state. Therefore, unlike the record linear two-coordinate $Co^{2+}(3d^7)$ complexes with the $M_L = \pm 3$ ground state, linear $3d^3$ complexes have a maximum achievable orbital angular momentum of only $M_L = \pm 2$.

Among a variety of axial symmetry coordination geometries, the necessary orbital splitting pattern $E(xy, x2 - y2) > E(xz, yz) > E(z^2)$ can be obtained in the trigonal-prismatic coordination of the $3d^3$ ion. However, in trigonal-prismatic complexes, the energy order of 3d orbital strongly depends on the polar angle θ of the trigonal prism. In order to evaluate the conditions for stabilization of the $M_L = \pm 3$ ground state, LF calculations are performed in combination with the angular overlap model (AOM) [46–51] using the AOM parameters $e_\sigma = 9000$ cm^{-1} and $e_\sigma/e_\pi = 4$ for Cr^{3+} ions. These calculations show that the ground state

$^4F(M_L = \pm 3)$ with the maximal orbital momentum $M_L = \pm 3$ stabilizes in a compressed trigonal prism with a polar angle θ ranging between 61° and 66.5° (Figure 4a).

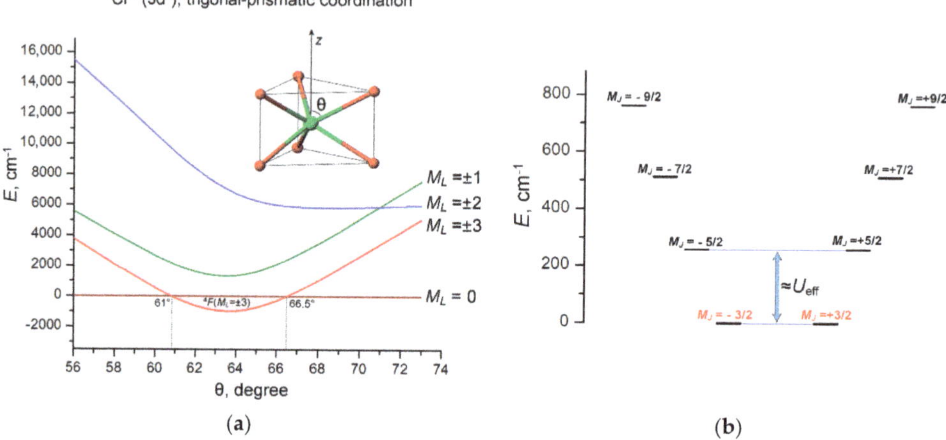

Figure 4. (**a**) Variation of the energies of the $^4F(M_L)$ LF states of $Cr^{3+}(3d^3)$ ion with increasing polar angle θ in a trigonal-prismatic coordination polyhedron. The range of the stabilization of the $^4F(M_L = \pm 3)$ ground state (61° ≤ θ ≤ 66.5°) is marked with light pink; (**b**) E vs. M_J diagram calculated at θ = 64°. The spin-reversal barrier $U_{eff} \approx 200 = 250$ cm^{-1} is approximately estimated by the energy separation between the ground $M_J = \pm 3/2$ and the first excited $M_J = \pm 5/2$ states.

SOC splits the ground orbital doublet $M_L = \pm 3$ into four Kramer doublets with the projection of the total angular momentum $M_J = \pm 3/2, \pm 5/2, \pm 7/2$ and $\pm 9/2$, listed in order of increasing energy; their energy diagram with the total splitting energy of 762 cm^{-1} is shown in Figure 4b. These results indicate that such a complex should have a SMM behavior with a barrier of about 250 cm^{-1}, corresponding to the energy separation between the ground $M_J = \pm 3/2$ and first excited $M_J = \pm 5/2$ states. However, as opposed to linear two-coordinate Co^{II} complexes [42,43], the trigonal-prismatic 3d^3 complexes with doubly degenerate ground state $M_L = \pm 3$ is Jahn–Teller active, subject to distortions that remove orbital degeneracy and thereby reduce the magnetic anisotropy of the complex. Nevertheless, LF/AOM calculations indicate that with moderate departures from the regular D_{3h} geometry, the overall energy pattern of the J_M states (Figure 4b) does not change much, so the trigonal-prismatic 3d^3 complex behaves as an SMM with an estimated barrier of about 150–200 cm^{-1}.

2.3. 3d^7 Complexes (Fe^+,Co^{2+})

Linear two-coordinate 3d^7 cobalt complex $Co^{II}(C(SiMe_2ONaph)_3)_2$ [42] is currently the only known individual mononuclear 3d complex featuring the maximal orbital angular momentum $L = 3$. From the point of view of the LF theory, in the 3d^7 configuration the orbital LF splitting pattern $E(z^2) > E(xz, yz) > E(xy, x2 - y2)$ is a necessary condition for the stabilization of the $M_L = \pm 3$ ground state. Note that the order of the 3d orbitals in the 3d^7 configuration is opposite to that in the 3d^3 configuration because of the electron–hole symmetry. As in the case of 3d^2 and 3d^3 complexes, the additional condition is the restriction on the energy separation between the $m_l = \pm 1$ and $m_l = \pm 2$ orbitals, $\Delta = E(xz, yz) - E(xy, x2 - y2)$, which is $0 \leq \Delta \leq 6750$ cm^{-1} (Figure 5). Outside this range, the ground state is the orbital singlet $^4F(M_L = 0)$. Formally, an extra condition is a limitation on the total LF splitting energy $\Delta_{LF} = E(z^2) - E(xy, x2 - y2) < 22{,}000$ cm^{-1}, which is, however, too large to occur in a linear two-coordinate 3d^7 complex exhibiting much lower total LF splitting energy, around 6000 cm^{-1} [42].

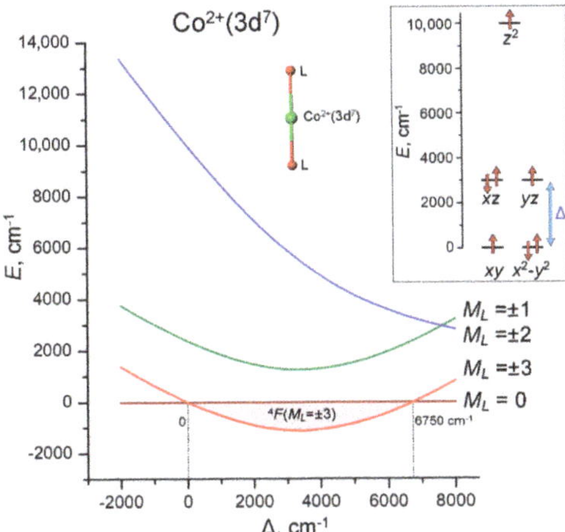

Figure 5. Energies of the $^4F(M_L)$ LF states of $Co^{2+}(3d^7)$ ion in a linear coordination as a function of the energy gap $\Delta = E(xz, yz) - E(xy, x2 - y2)$. The range of the stabilization of the $^4F(M_L = \pm3)$ ground state ($0 \leq \Delta \leq 6750$ cm^{-1}) is marked with light pink. Inset: the LF splitting pattern of 3d orbitals, $E(xy, x2 - y2) = 0$, $E(z^2) = 10{,}000$ cm^{-1} (fixed) and $E(xz, yz) = \Delta$ (variable). LF calculations are performed with atomic parameters $B = 750$, $C = 3800$ and $\zeta = 480$ cm^{-1}.

In light of these results, it is interesting to consider the difference in the nature of the ground state of the isoelectronic linear two-coordinate $3d^7$ complexes $[Fe^I(C(SiMe_3)_3)_2]^-$ ($M_L = \pm2$) [40] and $Co^{II}(C(SiMe_2ONaph)_3)_2$ ($M_L = \pm3$) [42]. The underlying reason is the different order of the actual 3d orbitals, which is $E(xz, yz) > E(xy, x2 - y2) > E(z^2)$ (ca. 5200, 3200 and 0 cm^{-1}, respectively) in the Fe^I complex [40] and $E(z^2) > E(xz, yz) > E(xy, x2 - y2)$ ((ca. 5700, 3000 and 0 cm^{-1}) in Co^{II} complex [42]. A significant decrease in the energy of the $3d_z{}^2$ orbital in $Fe^I(C(SiMe_3)_3)_2]^-$ is caused by a very strong $4s$–$3d_z{}^2$ mixing [40]. The combination of the ground orbital state $M_L = \pm3$, the parallel coupling of the orbital momentum $L = 3$ and spin $S = 3/2$ and a sufficiently strong SOC leads to the most favorable conditions for obtaining a record spin-reversal barrier $U_{eff} = 450$ cm^{-1} of $Co^{II}(C(SiMe_2ONaph)_3)_2$ [42].

2.4. $3d^8$ Complexes (Ni^{2+}, Co^{1+})

As in the case of $3d^2$ complexes with $L = 3$, the necessary condition for the occurrence of the $M_L = \pm3$ ground state in a $3d^8$ complex is the presence of one unpaired electron on the doubly degenerate 3d orbitals with magnetic quantum numbers $m_l = \pm1$ and $m_l = \pm2$, i.e., $(x2 - y2, xy)^3(xz, yz)^3(z^2)^2$. For this, the order of the 3d orbitals should be the same as in $3d^3$ complexes, i.e., $E(xy, x2 - y2) > E(xz, yz) > E(z^2)$. Therefore, this LF splitting pattern can occur in trigonal-prismatic complexes (see Figure 4a), but not in linear two-coordinate complexes (Figure 5). As in the case of other above 3d complexes, there is an additional condition that constrains the energy separation between $m_l = \pm1$ and $m_l = \pm2$ orbitals, $\Delta = E(xy, x2 - y2) - E(xz, yz)$, which is $0 \leq \Delta \leq 7200$ cm^{-1}.(Figure 6). LF/AOM calculations for a trigonal-prismatic $3d^8$ complex with variable polar angle θ indicate that the $M_L = \pm3$ ground state stabilizes at $61^\circ \leq \theta \leq 74.2^\circ$, corresponding to a compressed trigonal prism (Figure 7a). Several trigonal-prismatic cage complexes of Ni^{II} were recently reported in [52]. In these complexes, the Ni^{2+} ion is encapsulated in a frame-type clathrochelate cage, that forces robust trigonal-prismatic coordination. However, in these complexes, the polar angle θ is about 52°, which is well below the critical angle $\theta > 61^\circ$ (Figure 7a). Accordingly, the ground state of these complexes was found to be orbital singlet $^3F(M_L = 0)$ with a second-order magnetic anisotropy due to ZFS [48], which agrees well with the results of LF/AOM calculations indicating the $M_L = 0$ ground state at $\theta \approx 52^\circ$ (Figure 7a). One more example of a trigonal-prismatic $3d^8$ complex is given by similar clathrochelate complexes $Co^I(GmCl_2)_3(BPh)_2$ reported in ref. [53]. Unfortunately, these complexes also do not fall into the range of existence of the ground state $M_L = \pm 3$ due to violation of the $\Delta > 0$ criterion (which is actually $\Delta \approx -4500$ cm^{-1}), resulting in the ground state being the orbital singlet $^3F(M_L = 0)$.

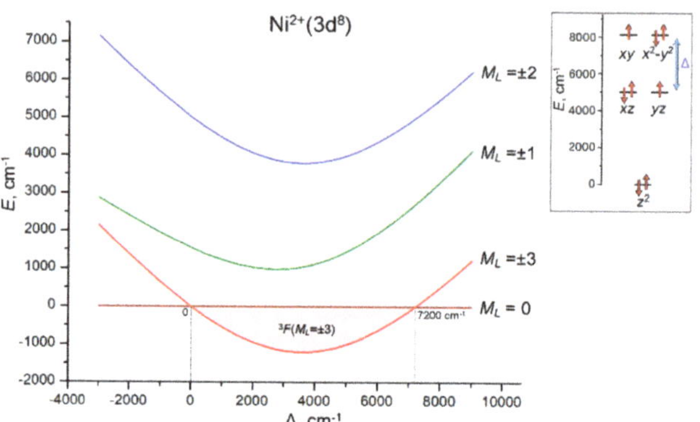

Figure 6. Variation of the energies of the $^3F(M_L)$ LF states of $Ni^{2+}(3d^8)$ ion with increasing energy gap $\Delta = E(xy, x2 - y2) - E(xz, yz)$. The range of the stabilization of the $^3F(M_L = \pm 3)$ ground state ($0 \leq \Delta \leq 7200$ cm^{-1}) is marked with light pink. Inset: the LF splitting pattern of 3d orbitals, $E(z^2) = 0$, $E(xz, yz) = 5000$ cm^{-1} (fixed) and $E(xy, x2 - y2) = 5000$ cm^{-1} + Δ (variable). LF calculations are performed with atomic parameters $B = 800$, $C = 3200$ and $\zeta = 600$ cm^{-1}.

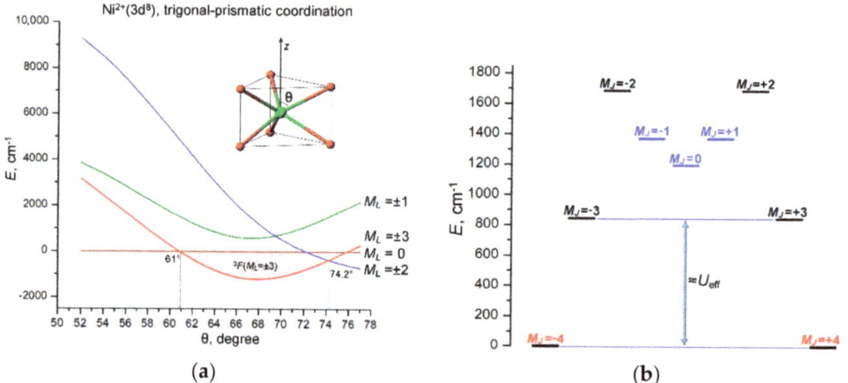

Figure 7. (**a**) Variation of the energies of the $^3F(M_L)$ LF states of $Ni^{2+}(3d^8)$ ion with increasing polar angle θ in trigonal-prismatic coordination, as obtained from LF/AOM calculations. The range of the stabilization of the $^3F(M_L = \pm 3)$ ground state ($61^\circ \leq \theta \leq 74.2^\circ$) is marked with light pink; (**b**) E vs. M_J diagram calculated at $\theta = 64^\circ$. The spin-reversal barrier $U_{eff} \approx 850$ cm^{-1} is approximately estimated by the energy separation between the ground $M_J = \pm 3/2$ and the first excited $M_J = \pm 5/2$ states; the $M_J = 0$ and $M_J = \pm 1$ states (marked in blue) originate from ^{2S+1}L atomic terms other than ^3F. LF/AOM calculations were performed with $e_\sigma = 5000$ cm^{-1}, $e_\sigma/e_\pi = 4$ AOM parameters and atomic parameters $B = 800$, $C = 3200$ and $\zeta = 600$ cm^{-1}.

The calculated E vs. M_J diagram suggests potentially high SMM characteristics of the compressed ($\theta > 61^\circ$) trigonal-prismatic Ni^{2+} complex with a perfect D_{3h} symmetry. In particular, the spin-reversal barrier U_{eff}, assigned to the energy separation between the ground state $J_M = \pm 4$ and the first excited state $J_M = \pm 3$, is as high as 857 cm^{-1} (Figure 7b), which is significantly larger than in the record cobalt complex $Co^{II}(C(SiMe_2ONaph)_3)_2$

(450 cm^{-1}) [42]. However, it should be considered that, similarly to the trigonal-prismatic $3d^3$ complex, this complex is Jahn–Teller active and is subject to significant distortions that tend to reduce magnetic anisotropy. In fact, LF/AOM calculations for distorted complexes show that even subtle departures from the regular trigonal-prismatic geometry (D_{3h}) of the complex cause noticeable splitting of the ground non-Kramers doublet $J_M = \pm 4$ (tens of cm^{-1}), which should lead to complete suppression of the SMM behavior due to fast QTM processes in the split ground state $J_M = \pm 4$ (Figure 7b).

3. Materials and Methods

Ligand Field and Angular Overlap Model Calculations

Ligand-field calculations for many-electron $3d^N$ ions ($3d^2$, $3d^3$, $3d^7$ and $3d^8$) were performed in terms of the conventional LF theory involving the one-electron LF operator (specified by the set of energies of 3d orbitals in an axial LF, $E(xy, x2 - y2)$, $E(xz, yz)$ and $E(z^2)$), interelectron Coulomb repulsion (treated in terms of the Racah parameters B and C) and the spin-orbit coupling $\zeta_{3d}\Sigma_i \boldsymbol{l}_i \boldsymbol{s}_i$. Angular-overlap model calculations [46–48] for trigonal-prismatic complexes were carried out with the AOM parameters $e_\sigma = 5000$ cm^{-1} (for Ni^{2+}) and 9000 cm^{-1} (Cr^{3+}) at a fixed ratio of $e_\sigma/e_\pi = 4$. Details of LF and AOM calculations are available in the Supplementary Materials.

4. Discussion and Conclusions

This study has established that the maximum orbital angular momentum $L = 3$ observed in the linear two-coordinate $3d^7$ complex $Co^{II}(C(SiMe_2ONaph)_3)_2$ [42] (as well as in the monocoordinated Co adatom on a MgO surface [43]) is not a unique single phenomenon, but under specific conditions, it can also be obtained in some complexes of $3d^2$, $3d^3$ and $3d^8$ transition metal ions. The basic point is that in the free-ion state these ions have the ground-state atomic term ^{2S+1}F with maximal orbital angular momentum $L = 3$, which, when split in the ligand field of axial symmetry, can produce the ground orbital doublet $^{2S+1}F(M_L = \pm 3)$ with a maximum projection $M_L = \pm 3$ of the orbital momentum $L = 3$. For each of these electronic configurations, the LF splitting pattern of 3d orbitals leading to the ground orbital doublet $^{2S+1}F(M_L = \pm 3)$ has been established. Then, with the help of LF and AOM calculations, it was shown that the maximal orbital angular moment $L = 3$ is realized in the linear two-coordinate geometry for the $3d^2$ and $3d^7$ ions and in the compressed trigonal-prismatic coordination (with polar angle $\theta > 61°$) for the $3d^3$ and $3d^8$ ions. In this regard, it is worth noting that the linear two-coordinate geometry, which intuitively appears to be the most preferable for obtaining maximum magnetic anisotropy, in the case of $3d^3$ and $3d^8$ ions yields only an orbital momentum $L = 2$, not $L = 3$.

These results can be used to assess the prospects for improving the SMM performance of single-ion 3d SMMs. The calculated E vs. M_J spin energy diagrams show that the SMM performance of the linear $3d^2$ complex and the trigonal-prismatic $3d^3$ complex is significantly lower than that of the linear cobalt complex due to weaker spin-orbit coupling and antiparallel coupling of L and S. On the other hand, although the calculated spin energy diagram for the trigonal-prismatic $Ni^{2+}(3d^8)$ complex formally indicates a higher barrier ($U_{eff} \approx 850$ cm^{-1}) than for the record linear cobalt complex (450 cm^{-1}) [42], in reality, this complex is unlikely to be a good SMM because of the Jahn–Teller distortions resulting in the significant splitting of the ground non-Kramer doublet $M_J = \pm 4$ (Figure 7b), that causes fast QTM processes bypassing the barrier. Thus, despite the possible presence of maximum orbital momentum $L = 3$ in the aforementioned $3d^2$, $3d^3$ and $3d^8$ complexes, they cannot compete as SMMs with linear two-coordinate $3d^7$ cobalt complex $Co^{II}(C(SiMe_2ONaph)_3)_2$ [42].

In summary, this study has shown that the maximal orbital angular momentum $L = 3$ can occur only in $3d^2$, $3d^3$, $3d^7$ and $3d^8$ complexes with axial symmetry of the ligand field. However, the SMM performance of $3d^2$, $3d^3$ and $3d^8$ complexes appeared significantly lower than that of the linear two-coordinate $3d^7$ complex $Co^{II}(C(SiMe_2ONaph)_3)_2$ with $L = 3$. Thus, these results confirm the earlier conclusion [42] that linear two-coordinate

$3d^7$ complexes provide the maximal SMM performance of mononuclear 3d complexes (U_{eff} = 450 cm^{-1}), which is apparently near the physical limit for a single 3d metal ion.

Supplementary Materials: The following supporting information can be downloaded at: https://www.mdpi.com/article/10.3390/inorganics10120227/s1, details of LF and AOM calculations.

Author Contributions: Conceptualization, validation, methodology, writing—original draft preparation, writing—review and editing, supervision, V.S.M. All authors have read and agreed to the published version of the manuscript.

Funding: This work was supported by the Russian Science Foundation, project No. 18-13-00264.

Data Availability Statement: The data presented in this study are available in this article and in the Supplementary Materials.

Acknowledgments: V.S.M. acknowledges support by the Ministry of Science and Higher Education within the State assignment FSRC 'Crystallography and Photonics' RAS in part of the development of computational approaches for analysis of extreme magnetic anisotropy in transition-metal complexes.

Conflicts of Interest: The author declares no conflict of interest.

References

1. Sessoli, R.; Gatteschi, D.; Caneschi, A.; Novak, M.A. Magnetic bistability in a metal-ion cluster. *Nature* **1993**, *365*, 141–143. [CrossRef]
2. Gatteschi, D.; Sessoli, R. Quantum tunneling of magnetization and related phenomena in molecular materials. *Angew. Chem. Int. Ed.* **2003**, *42*, 268–297. [CrossRef] [PubMed]
3. Gatteschi, D.; Sessoli, R.; Villain, J. *Molecular Nanomagnets*; Oxford University Press: Oxford, UK, 2006. [CrossRef]
4. Leuenberger, M.N.; Loss, D. Quantum computing in molecular magnets. *Nature* **2001**, *410*, 789–793. [CrossRef] [PubMed]
5. Bogani, L.; Wernsdorfer, W. Molecular spintronics using single-molecule magnets. *Nat. Mater.* **2008**, *7*, 179. [CrossRef] [PubMed]
6. Bartolomé, J.; Luis, F.; Fernández, J.F. (Eds.) *Molecular Magnets: Physics and Applications*; NanoScience and Technology; Springer: Berlin/Heidelberg, Germany, 2014; pp. 1–395, ISBN 1010:9783642406089.
7. McInnes, E.J.L.; Winpenny, R.E.P. *4.14—Molecular Magnets. Comprehensive Inorganic Chemistry II*; Elsevier: Edinburgh, UK, 2013; Volume 4, pp. 371–396. [CrossRef]
8. Feng, M.; Tong, M.-L. Single Ion Magnets from 3d to 5f: Developments and Strategies. *Chem.—Eur. J.* **2018**, *24*, 7574–7594. [CrossRef]
9. Coronado, E. Molecular magnetism: From chemical design to spin control in molecules, materials and devices. *Nat. Rev. Mat.* **2020**, *5*, 87–104. [CrossRef]
10. Caneschi, A.; Gatteschi, D.; Lalioti, N.; Sangregorio, C.; Sessoli, R.; Venturi, G.; Vindigni, A.; Rettori, A.; Pini, M.G.; Novak, M.A. Cobalt(II)-Nitronyl Nitroxide Chains as Molecular Magnetic Nanowires. *Angew. Chem. Int. Ed.* **2001**, *40*, 1760–1763. [CrossRef]
11. Coulon, C.; Miyasaka, H.; Clérac, R. Single-Chain Magnets: Theoretical Approach and Experimental Systems. In *Single-Molecule Magnets and Related Phenomena. Structure and Bonding*; Winpenny, R., Ed.; Springer: Berlin/Heidelberg, Germany, 2006; Volume 122, pp. 163–206. [CrossRef]
12. Sun, H.-L.; Wang, Z.-M.; Gao, S. Strategies towards Single-Chain Magnets. *Coord. Chem. Rev.* **2010**, *254*, 1081–1100. [CrossRef]
13. Ishikawa, N.; Sugita, M.; Ishikawa, T.; Koshihara, S.-Y.; Kaizu, Y. Lanthanide Double-Decker Complexes Functioning as Magnets at the Single-Molecular Level. *J. Am. Chem. Soc.* **2003**, *125*, 8694–8695. [CrossRef]
14. Freedman, D.E.; Harman, W.H.; Harris, T.D.; Long, G.J.; Chang, C.J.; Long, J.R. Slow magnetic relaxation in a high-spin iron(II) complex. *J. Am. Chem. Soc.* **2010**, *132*, 1224–1225. [CrossRef]
15. Harman, W.H.; Harris, T.D.; Freedman, D.E.; Fong, H.; Chang, A.; Rinehart, J.D.; Ozarowski, A.; Sougrati, M.T.; Grandjean, F.; Long, G.J.; et al. Slow magnetic relaxation in a family of trigonal pyramidal iron(II) pyrrolide complexes. *J. Am. Chem. Soc.* **2010**, *132*, 18115–18126. [CrossRef]
16. Rinehart, J.D.; Long, J.R. Exploiting Single-Ion Anisotropy in the Design of f-Element Single-Molecule Magnets. *Chem. Sci.* **2011**, *2*, 2078–2085. [CrossRef]
17. Woodruff, D.N.; Winpenny, R.E.P.; Layfield, R.A. Lanthanide Single-Molecule Magnets. *Chem. Rev.* **2013**, *113*, 5110–5148. [CrossRef]
18. Liddle, S.T.; van Slageren, J. Improving f-Element Single Molecule Magnets. *Chem. Soc. Rev.* **2015**, *44*, 6655–6669. [CrossRef]
19. Zhang, P.; Zhang, L.; Tang, J. Lanthanide single molecule magnets: Progress and perspective. *Dalton Trans.* **2015**, *44*, 3923–3929. [CrossRef]
20. Layfield, R.; Murugesu, M. *Lanthanides and Actinides in Molecular Magnetism*; John Wiley & Sons: New York, NY, USA, 2015; pp. 1–368. ISBN 978-3-527-33526-8.
21. Lu, J.; Guo, M.; Tang, J. Recent Developments in Lanthanide Single-Molecule Magnets. *Chem.—Asian J.* **2017**, *12*, 2772–2779. [CrossRef]

22. Liu, J.-L.; Chen, Y.-C.; Tong, M.-L. Symmetry Strategies for High Performance Lanthanide-Based Single-Molecule Magnets. *Chem. Soc. Rev.* **2018**, *47*, 2431–2453. [CrossRef]
23. Feltham, H.L.C.; Brooker, S. Review of Purely 4f and Mixed-Metal nd-4f Single-Molecule Magnets Containing only One Lanthanide Ion. *Coord. Chem. Rev.* **2014**, *276*, 1–33. [CrossRef]
24. Layfield, R.A. Organometallic Single-Molecule Magnets. *Organometallics* **2014**, *33*, 1084–1099. [CrossRef]
25. Craig, G.A.; Murrie, M. 3d single-ion magnets. *Chem. Soc. Rev.* **2015**, *44*, 2135–2147. [CrossRef]
26. Frost, J.M.; Harriman, K.L.M.; Murugesu, M. The rise of 3-d single-ion magnets in molecular magnetism: Towards materials from molecules? *Chem. Sci.* **2016**, *7*, 2470–2491. [CrossRef] [PubMed]
27. Wyborne, B.G. *Spectroscopic Properties of Rare Earths*; John Wiley & Sons, Inc.: Hoboken, NJ, USA, 1965; pp. 1–241, ISBN 978-0470965078.
28. Liu, J.; Chen, Y.-C.; Liu, J.-L.; Vieru, V.; Ungur, L.; Jia, J.-H.; Chibotaru, L.F.; Lan, Y.; Wernsdorfer, W.; Gao, S.; et al. A stable pentagonal bipyramidal Dy(III) single-ion magnet with a record magnetization reversal barrier over 1000 K. *J. Am. Chem. Soc.* **2016**, *138*, 5441–5450. [CrossRef] [PubMed]
29. Ding, Y.-S.; Chilton, N.F.; Winpenny, R.E.P.; Zheng, Y.-Z. On approaching the limit of molecular magnetic anisotropy: A near-perfect pentagonal bipyramidal dysprosium(III) single-molecule magnet. *Angew. Chem. Int. Ed.* **2016**, *55*, 16071–16074. [CrossRef] [PubMed]
30. Goodwin, C.A.P.; Ortu, F.; Reta, D.; Chilton, N.F.; Mills, D.P. Molecular magnetic hysteresis at 60 kelvin in dysprosocenium. *Nature* **2017**, *548*, 439–442. [CrossRef] [PubMed]
31. Guo, F.-S.; Day, B.M.; Chen, Y.-C.; Tong, M.-L.; Mansikkamäki, A.; Layfield, R.A. Magnetic Hysteresis up to 80 Kelvin in a Dysprosium Metallocene Single-Molecule Magnet. *Science* **2018**, *362*, 1400–1403. [CrossRef]
32. Zhu, Z.; Tang, J. Lanthanide single-molecule magnets with high anisotropy barrier: Where to from here? *Natl. Sci. Rev.* **2022**, nwac194. [CrossRef]
33. Bar, A.K.; Pichon, C.; Sutter, J.-P. Magnetic Anisotropy in Two- to Eight-Coordinated Transition–Metal Complexes: Recent Developments in Molecular Magnetism. *Coord. Chem. Rev.* **2016**, *308*, 346–380. [CrossRef]
34. Boca, R. Zero-field splitting in metal complexes. *Coord. Chem. Rev.* **2004**, *248*, 757–815. [CrossRef]
35. Marriott, K.E.R.; Bhaskaran, L.; Wilson, C.; Medarde, M.; Ochsenbein, S.T.; Hill, S.; Murrie, M. Pushing the limits of magneticanisotropy in trigonal bipyramidal Ni(II). *Chem. Sci.* **2015**, *6*, 6823–6828. [CrossRef]
36. Novikov, V.V.; Pavlov, A.A.; Nelyubina, Y.V.; Boulon, M.-E.; Varzatskii, O.; Voloshin, Y.Z.; Winpenny, R.E.P. A Trigonal Prismatic Mononuclear Cobalt(II) Complex Showing Single-Molecule Magnet Behavior. *J. Am. Chem. Soc.* **2015**, *137*, 9792–9795. [CrossRef]
37. Yao, B.; Singh, M.K.; Deng, Y.-F.; Wang, Y.-N.; Dunbar, K.R.; Zhang, Y.-Z. Trigonal Prismatic Cobalt(II) Single-Ion Magnets: Manipulating the Magnetic Relaxation Through Symmetry Control. *Inorg. Chem.* **2020**, *59*, 8505–8513. [CrossRef]
38. Sutter, J.P.; Béreau, V.; Jubault, V.; Bretosh, K.; Pichon, C.; Duhayon, C. Magnetic anisotropy of transition metal and lanthanide ions in pentagonal bipyramidal geometry. *Chem. Soc. Rev.* **2022**, *51*, 3280–3313. [CrossRef]
39. Zadrozny, J.M.; Atanasov, M.; Bryan, A.M.; Lin, C.-Y.; Rekken, B.D.; Power, P.P.; Neese, F.; Long, J.R. Slow magnetization dynamics in a series of two-coordinate iron(II) complexes. *Chem. Sci.* **2013**, *4*, 125–138. [CrossRef]
40. Zadrozny, J.M.; Xiao, D.J.; Atanasov, M.; Long, G.J.; Grandjean, F.; Neese, F.; Long, J.R. Magnetic blocking in a linear iron(I) complex. *Nat. Chem.* **2013**, *5*, 577–581. [CrossRef]
41. Yao, X.-N.; Du, J.-Z.; Zhang, Y.-Q.; Leng, X.-B.; Yang, M.-W.; Jiang, S.-D.; Wang, Z.-X.; Ouyang, Z.-W.; Deng, L.; Wang, B.-W.; et al. Two-Coordinate Co(II) Imido Complexes as Outstanding Single-Molecule Magnets. *J. Am. Chem. Soc.* **2017**, *139*, 373–380. [CrossRef]
42. Bunting, P.C.; Atanasov, M.; Damgaard-Møller, E.; Perfetti, M.; Crassee, I.; Orlita, M.; Overgaard, J.; van Slageren, J.; Neese, F.; Long, J.R. A linear cobalt(II) complex with maximal orbital angular momentum from a non-Aufbau ground state. *Science* **2018**, *362*, eaat7319. [CrossRef]
43. Rau, G.; Baumann, S.; Rusponi, S.; Donati, F.; Stepanow, S.; Gragnaniello, L.; Dreiser, J.; Piamonteze, C.; Nolting, F.; Gangopadhyay, S.; et al. Reaching the magnetic anisotropy limit of a 3d metal atom. *Science* **2014**, *344*, 988–992. [CrossRef] [PubMed]
44. Powell, P.P. Stable Two-Coordinate, Open-Shell (d^1–d^9) Transition Metal Complexes. *Chem. Rev.* **2012**, *112*, 3482–3507. [CrossRef]
45. Reber, C.; Güdel, H.U. Near-infrared luminescence spectroscopy of Al_2O_3:V^{3+} and YP_3O_9:V^{3+}. *Chem. Phys. Lett.* **1989**, *154*, 425–431. [CrossRef]
46. Schaeffer, C.E.; Jorgensen, C.K. The angular overlap model, an attempt to revive the ligand field approaches. *Mol. Phys.* **1965**, *9*, 401–412. [CrossRef]
47. Schäffer, C.E. A perturbation representation of weak covalent bonding. *Struct. Bonding.* **1968**, *5*, 68–95. [CrossRef]
48. Jorgensen, C.K. The Nephelauxetic Series. *Prog. Inorg. Chem.* **1962**, *4*, 73–124. [CrossRef]
49. Mironov, V.S.; Bazhenova, T.A.; Manakin, Y.V.; Lyssenko, K.A.; Talantsev, A.D.; Yagubskii, E.B. A new Mo(IV) complex with the pentadentate (N_3O_2) Schiff-base ligand: The first non-cyanide pentagonal–bipyramidal paramagnetic 4d complex. *Dalton Trans.* **2017**, *46*, 14083–14087. [CrossRef] [PubMed]
50. Manakin, Y.V.; Mironov, V.S.; Bazhenova, T.A.; Lyssenko, K.A.; Gilmutdinov, I.F.; Bikbaev, K.S.; Masitov, A.A.; Yagubskii, E.B. $(Et_4N)[Mo^{III}(DAPBH)Cl_2]$, the first pentagonal-bipyramidal Mo(III) complex with a N_3O_2-type Schiff-base ligand: Manifestation of unquenched orbital momentum and Ising-type magnetic anisotropy. *Chem. Commun.* **2018**, *54*, 10084–10087. [CrossRef]

51. Bazhenova, T.A.; Zorina, L.V.; Simonov, S.V.; Mironov, V.S.; Maximova, O.V.; Spillecke, L.; Koo, C.; Klingeler, R.; Manakin, Y.V.; Vasiliev, A.N.; et al. The first pentagonal-bipyramidal vanadium(iii) complexes with a Schiff-base N_3O_2 pentadentate ligand: Synthesis, structure and magnetic properties. *Dalton Trans.* **2020**, *49*, 15287–15298. [CrossRef]
52. Campanella, A.J.; Ozvat, T.M.; Zadrozny, J.M. Ligand design of zero-field splitting in trigonal prismatic Ni(ii) cage complexes. *Dalton Trans.* **2022**, *51*, 3341–3348. [CrossRef]
53. Nehrkorn, J.; Veber, S.L.; Zhukas, L.A.; Novikov, V.V.; Nelyubina, Y.V.; Voloshin, Y.Z.; Holldack, K.; Stoll, S.; Schnegg, A. Determination of Large Zero-Field Splitting in High-Spin Co(I) Clathrochelates. *Inorg. Chem.* **2018**, *57*, 15330–15340. [CrossRef]

Article

The Role of the Bridge in Single-Ion Magnet Behaviour: Reinvestigation of Cobalt(II) Succinate and Fumarate Coordination Polymers with Nicotinamide

Marek Brezovan [1], Jana Juráková [2], Ján Moncol [1], Ľubor Dlháň [1], Maria Korabik [3], Ivan Šalitroš [1,2,4], Ján Pavlik [1,*] and Peter Segľa [1]

1. Faculty of Chemical and Food Technology, Slovak University of Technology, Radlinského 9, 81237 Bratislava, Slovakia
2. Central European Institute of Technology, Brno University of Technology, Purkyňova 123, 61200 Brno, Czech Republic
3. Faculty of Chemistry, University of Wrocław, F. Joliot-Curie 14, 50-383 Wrocław, Poland
4. Faculty of Science, Palacký University, 17. Listopadu 12, 77146 Olomouc, Czech Republic

* Correspondence: jan.pavlik@stuba.sk

Abstract: Two previously synthesized cobalt(II) coordination polymers; $\{[Co(\mu_2\text{-suc})(nia)_2(H_2O)_2]\cdot 2H_2O\}n$ (suc = succinate(2−), nia = nicotinamide) and $[Co(\mu_2\text{-fum})(nia)_2(H_2O)_2]n$ (fum = fumarate(2−)) were prepared and thoroughly characterized. Both complexes form 1D coordination chains by bonding of $Co(nia)_2(H_2O)_2$ units through succinate or fumarate ligands while these chains are further linked through hydrogen bonds to 3D supramolecular networks. The intermolecular interactions of both complexes are quantified using Hirshfeld surface analysis and their infrared spectra, electronic spectra and static magnetic properties are confronted with DFT and state-of-the-art ab-initio calculations. Dynamic magnetic measurements show that both complexes exhibit single-ion magnet behaviour induced by a magnetic field. Since they possess very similar chemical structure, differing only in the rigidity of the bridge between the magnetic centres, this chemical feature is put into context with changes in their magnetic relaxation.

Keywords: single-ion magnet; cobalt(II) coordination polymer; ab-initio calculations

Citation: Brezovan, M.; Juráková, J.; Moncol, J.; Dlháň, Ľ.; Korabik, M.; Šalitroš, I.; Pavlik, J.; Segľa, P. The Role of the Bridge in Single-Ion Magnet Behaviour: Reinvestigation of Cobalt(II) Succinate and Fumarate Coordination Polymers with Nicotinamide. *Inorganics* **2022**, *10*, 128. https://doi.org/10.3390/inorganics10090128

Received: 1 August 2022
Accepted: 26 August 2022
Published: 30 August 2022

1. Introduction

Coordination polymers have been studied for a long time in terms of architecture, topology, and their potential applications in catalysis [1–3], gas adsorption [4–6], chemical adsorption [7], luminescence [8,9], and design of molecular magnetic materials [10–12]. The proper choice of relevant ligands and metal centre is the key to form fascinating and useful coordination polymers [13–23].

An extremely appealing class of molecular magnetic materials is formed by *single-molecule magnets* (SMMs) [24]. Very simply speaking, in these materials, the magnetic dipole moments tend to keep their orientation with respect to the molecular or polymeric frame. In reality, however, any imposed orientation disappears after some time, mostly due to their interference with thermal bath of molecular surroundings. In current SMMs it happens typically within milliseconds, albeit the limit of second has already been breached [25]. This measurable process is called *slow relaxation of magnetization* and its time constant is used as a characteristic of SMM systems. In special cases of 1D polymers, the term *single-chain magnets* (SCM) is used [26]. In both SMMs and SCMs, the slow relaxation of magnetization exists as a collective property of the magnetic centres. If, on the other hand, magnetically isolated centres are capable of slow relaxation of magnetization, a *single-ion magnet* (SIM) is encountered [27]. Counterintuitively, the chemical and magnetic structure of a system need not coincide, one can find, e.g., a chain of SIMs [28]. Although the best performing SIMs are based on the lanthanide and actinide central atoms [29], among few suitable 3*d*-elements, cobalt (II) occupies a prominent position forming SIMs with two-, three- four-, five-, six-, seven- and eight-coordinate Co(II) centres and various geometries of coordination environment [30]. The vast majority of them, however, are six-coordinate, which is thanks to the high unquenched angular momentum they bear [31]. The role of the angular momentum and symmetry of coordination environment plays important role for the mechanism of magnetic relaxation and there is a lot of attention being paid to the fine tuning of the relaxation process by design of the coordination environment. For example, the effect of linearity of coordinated pseudohalides was systematically addressed by Herchel et al. [32] or Wang et al. [33], and distortion of octahedral coordination environment was studied by groups of Pardo [34], Gao [35], Song [36,37], Herchel [38,39], or Colacio [40], concluding sometimes that the closest environment of Co(II) ion is decisive for SIM behaviour [41]. Nevertheless, there occur some clues that the focus on the central atom itself cannot uncover the full story behind slow relaxation of magnetization in SIMs. As discussed e.g., by Ren et al. [42] or Boča et al. [43] and very explicitly stated by Dunbar et al. [44]: "... *the deciding factor for SMM behavior is not the degree of distortion which, a priori, would be expected to be the case, but rather the interactions between neighboring molecules in the solid state*". Nonetheless, the stiffness of molecular and supramolecular structures is somehow overlooked in this context and the research in this sake is almost exclusively theoretical [29]. In one of very scarce works, Marinho et al. synthesized by carefully controlled conditions four variants of Co(II) coordination polymer which differed in their folding. The conformation of bridges manifested itself in the relaxation of the chain-arranged SIMs [45]. However, to the best of our knowledge no study yet addressed the question of the effect of rigidity of bridges connecting SIM centres.

Keeping this in mind, herein we attempted to compare the effect of bridge saturation upon the SIM behaviour in two analogous Co(II) coordination polymers. The succinate polymer $\{[Co(\mu_2\text{-suc})(nia)_2(H_2O)_2]\cdot 2H_2O\}_n$ (suc = succinate(2−), nia = nicotinamide, complex **I**) was synthesized in 2009 by Demir et al. and characterized by single-crystal X-ray crystallography, IR spectroscopy, photoluminescence and TG-DTA [46]. The fumarate polymer $[Co(\mu_2\text{-fum})(nia)_2(H_2O)_2]_n$ (fum = fumarate(2−), complex **II**) was prepared in 2018 by Kansız et al. and characterized by FT-IR, X-ray crystallography and DFT calculations [47].

We demonstrated an alternative synthetic route for complex **I** employing *N*-(hydroxym ethyl)nicotinamde (hmnia) instead of nicotinamide. Crystal structures of both complexes were refined using aspheric atomic scattering factors by Hirshfeld Atomic Refinement and analysed using Hirshfeld surface analysis. Infrared and electronic spectra were interpreted and confronted with the prediction from DFT. The magnetic properties of complexes were measured and thoroughly interpreted with assistance of state-of-the-art quantum chemistry method CASSCF-NEVPT2-SOI. Finally, magnetic relaxation was discussed for these systems in context of mutual variances in their chemical structure.

2. Results

2.1. Syntheses and Characterization

Direct preparation of polymeric cobalt(II) carboxylates from nicotinamide—nia is represented in Scheme 1. Complex $\{[Co(\mu_2\text{-suc})(nia)_2(H_2O)_2]\cdot 2H_2O\}_n$ (**I**) was also prepared from *N*-(hydroxymethyl)nicotinamide—hmnia (Scheme 2). Such a tendency of Co(II) salts to degradation of hmnia was not yet described in literature despite the fact that only very few Ni(II), Co(II) and Cu(II) complexes with hmnia are known [48–50]. A similar reaction occurs in the Chłopicki's preparation of the pyridinium salts of nicotinamide [51]. In Chłopicki's procedure the nitrogen atom of the amide group of nicotinamide is protected with methanal and after subsequent alkylation at the pyridine nitrogen atom, the methanal is released in water media at 37 °C. Cobalt (II) dichloride serves as Lewis acid, coordinating the pyridine nitrogen atom (instead of quarternization like it was in the patent) thus enabling easier methanal release (Scheme 2). In contrast with Chłopicki's procedure, however, in our case the reaction mixture was refluxed for two hours.

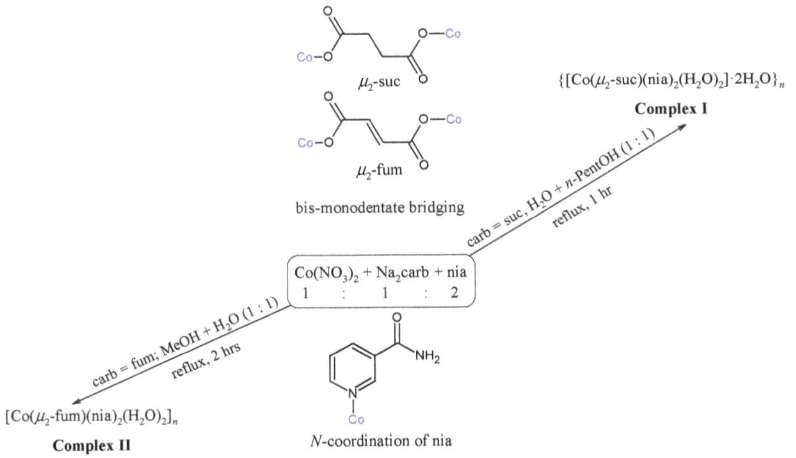

Scheme 1. Synthesis of polymeric cobalt(II)dicarboxylate complexes (**I** and **II**) from nicotinamide.

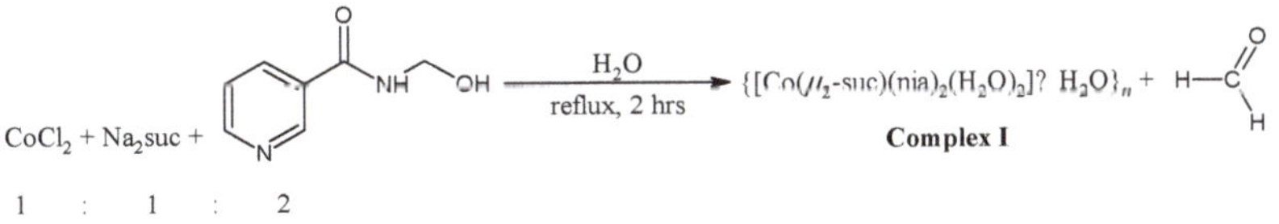

Scheme 2. Synthesis of polymeric cobalt(II)succinate complex (**I**) from N-(hydroxymethyl) nicotinamide.

The preparation of complexes **I** and **II** was carried out by short refluxing from a water + n-pentanol (complex **I**, prepared from nia, Scheme 1), water + methanol (complex **I**, prepared from hmnia, Scheme 2) and water solution (complex **II**). The molar ratio of the reactants was 1:1:2 for cobalt(II) salts, sodium salts of dicarboxylic acids and nicotinamide or *N*-(hydroxymethyl)nicotinamide, respectively. To synthesize the discussed complexes, various molar ratios of cobalt(II) chloride hexahydrate or cobalt(II) nitrate hexahydrate to sodium carboxylate (succinate or fumarate)—Na$_2$carb (where carb is suc or fum anion) and corresponding nicotinamide or *N*-(hydroxymethyl)nicotinamide were tested, specifically 1:1:1; 1:2:1; 1:1:2 and 1:2:2. Well-defined crystalline products were obtained only for the case of 1:1:2 (Schemes 1 and 2). The syntheses of complexes **I** and **II** was very well reproducible in terms of the product quality and yield. Prepared complexes are non-hygroscopic and stable in air and soluble in hot water.

2.2. Description of the Structures

The crystal structure of both compounds **I** and **II** was previously determined using standard single-crystal X-ray diffraction at room temperature, and their models were refined using the standard IAM [46,47]. In this study, the structural parameters of both compounds are significantly more accurate as they were obtained by refining the structure model (all H atoms being refined isotropically and independently) using aspheric atomic scattering factors and the HAR method. Selected bond distances are given in Table S2 (see Supplementary Materials).

The cobalt atoms of both complexes lie in the centre of symmetry and are octahedrally coordinated by two oxygen atoms (O1) of the carboxyl groups of the succinate (**I**) [Co1–O1 = 2.0931(9) Å] or fumarate (**II**) [Co1–O1 = 2.0683(8) Å] anionic ligands, two nitrogen atoms (N1) of the pyridine rings of the nicotinamide ligands [Co1–N1 = 2.1672(12) Å for **I** and 2.1689(10) Å for **II**], and a pair of oxygen atoms (O1W) of the coordinated water molecules [Co1–O1W = 2.1090(10) Å for **I** and 2.0924(9) Å for **II**] in trans positions (Figure 1). Both substances form 1D coordination chains formed by bonding of $Co(nia)_2(H_2O)_2$ units bridged through succinate (**I**) or fumarate (**II**) ligands. The closest Co···Co distances are 9.465 Å in **I** and 9.740 in **II**.

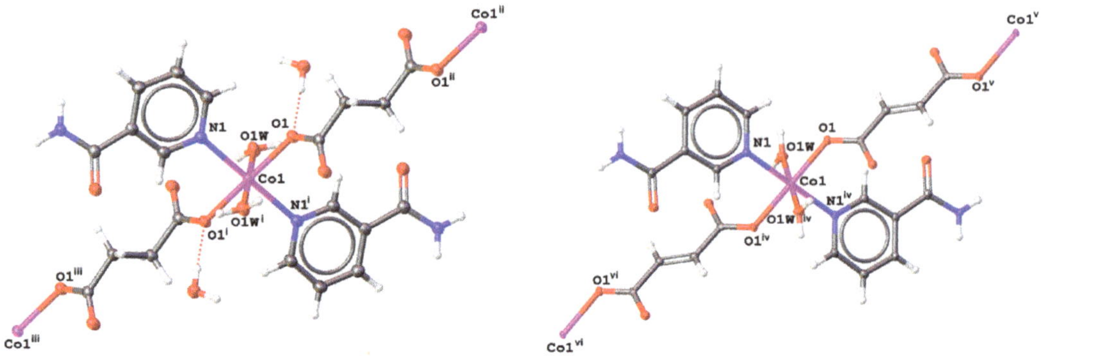

(**a**) (**b**)

Figure 1. Comparison of perspective view of the molecular structure of the complexes: (**a**) **I** and (**b**) **II**. Thermal ellipsoids are drawn at the 50% probability level. Symmetry codes: (i) $1-x, 1-y, 2-z$; (ii) $-x, 2-y, 2-z$; (iii) $1+x, -1+y, +z$; (iv) $1-x, 1-y, -z$; (v) $-x, -y, -z$; (vi) $1+x, 1+y, +z$.

Several ideal six-coordinate geometries were compared with **I** and **II** through SHAPE structural analysis, proposed by S. Alvarez et al. [52–55] (Table S3, see Supplementary Materials). The obtained symmetry measure parameters undoubtedly prove the octahedral shape of coordination polyhedra of reported compounds ($S(O_h) \approx 0.2$ for **I** and 0.3 for **II**) and the next lowest values for the trigonal prism shape ($S(D_{3h}) \approx 16.1$ for both **I** and **II**) suggest that the Bailar twist does not occur in the reported complexes. Furthermore, the distortion parameter Σ [56,57], calculated from the twelve *cis* angles of the hexacoordinated polyhedron (Table S3, see Supplementary Materials) acquiring zero values when ideal octahedral symmetry is present, can express the degree of angular distortion of coordination polyhedra of reported compounds. Both reported complexes indicate only moderate angular distortion, although more pronounced in the complex **II** (**Σ** = 43°), since its value of distortion parameter is notably higher comparing the one observed in **I** (**Σ** = 30°).

The 1D coordination chains of both complexes are linked through hydrogen bonds to 3D supramolecular hydrogen-bonding networks. Parameters of hydrogen bonds are given in Table S4 (see Supplementary Materials). Figure S1 (see Supplementary Materials) shows the O–H···O hydrogen bonds between the two 1D coordination chains of compound **I** (top) and compound **II** (bottom). Two coordination chains of **I** are joined via O–H···O hydrogen bonds between coordinated water molecules (O1W) and oxygen atoms (O2) of the carboxyl groups of succinate anions of adjacent coordination chains [O1W–H1WB···O2, with O···O distance of 2.857(1) Å (Table S4, see Supplementary Materials)]. The oxygen atoms (O1W) of coordinated water molecules are linked by O–H···;O hydrogen bonds to the oxygen atoms of uncoordinated water molecules (O2W) [O1W–H1WA···O2W, with O···O distance of 2.758(1) Å]. Uncoordinated water molecules (O2W) are further involved as donor atoms in other O–H···O hydrogen bonds where the oxygen atoms (O1) of carboxyl groups [O2W–H2WA···O1, with O···O distance of 2.794(1) Å], and oxygen atoms (O3) of carboxamide groups of nicotinamide ligands [O2W–H2WB···O3, with O···O distance of 2.834(1) Å] serve as acceptor atoms of H-bonds. On the other hand, the O–H···O hydrogen bonds in crystal structure of **II** are observed only between the oxygen atoms (O1W) of the coordinated water molecules and the oxygen atoms (O2) of the carboxyl groups of fumarate anions [O1W–H1WB···O2, with O···O distance of 2.758(1) Å], and between the oxygen atoms (O1W) of the coordinated water molecules and the oxygen atoms (O3) of carboxamide groups of nicotinamide ligands [O1W–H1WA··· O3, with O··· O distance of 2.822(1) Å].

The N–H··· O and C–H··· O hydrogen bonds of both complexes are shown in the Figures S2 and S3 (see Supplementary Materials). Crystal structures of both complexes show formation of supramolecular rings $R_2^2(8)$ [58] by linking two carboxamide groups via a pair of N–H··· O hydrogen bonds [N2–H2A··· O3, with N··· O distance of 2.943(1) Å for **I** and 2.972(1) Å for **II**, (Table S4 and Figure S2, see Supplementary Materials)]. The crystal structure of **II** also shows π-π stacking interactions of the pyridine rings [N1/C2–C5] of nicotinamide ligands (distances between two planes of 3.52 Å, centroid-centroid distance of 3.95 Å, shift distance of 1.77 Å [59]). In Figure S3 (see Supplementary Materials) the supramolecular rings $R_2^1(7)$ [58] observed in the crystal structure of both complexes is displayed. These supramolecular rings form nicotinamide molecules and oxygen atoms (O2) of carboxylate groups of succinate (**I**) or fumarate (**II**) ligands through N–H··· O hydrogen bonds between nitrogen atom (N2) of carboxamide group of nicotinamide ligand and carboxylate oxygen atom (O2) [N2–H2B··· O2, with N··· O distance of 3.046(1) Å for **I** and 2.901(1) Å for **II**, (Table S4 and Figure S3, see Supplementary Materials)], and C3–H3··· O2 hydrogen bonds between carbon atom (C3) of pyridine ring of nicotinamide ligand and carboxylate oxygen atom (O2) [C3–H3··· O3, with C··· O distance of 3.357(1) Å for **I** and 3.232(1) Å for **II**].

2.2.1. Hirshfeld Surface Analysis

The intermolecular interactions of both complexes have been quantified using Hirshfeld surface analysis. Figures S4 and S5 (see Supplementary Materials) show transparent 3D Hirshfeld surface mapped over d_{norm} shape index for complex **I** and complex **II**, respectively. Deep red spots on these surfaces indicate close-contact interactions majority of which is due to intermolecular hydrogen bonds. In the case of **II** π-π stacking interactions between pyridine rings of nicotinamide ligands are also visible (Figure S5, see Supplementary Materials). As shown in the 2D fingerprint plots (Figures S6 and S7, see Supplementary Materials), H··· H interactions cover 37.3–40.6% range of the total Hirshfeld surface, H··· O/O··· H interactions span between 31.1–35.3%, H··· C/C··· H interactions cover between 9.1 and 11.1% and C··· C interactions between 9.0 and 10.6%.

2.2.2. X-ray Powder Diffraction

The Le Bail analysis of both samples shows that they are of good crystalline quality without any significant amount of foreign impurity (Figures S8 and S9, see Supplementary Materials). A small residual intensity in difference plot can be assigned to the effect of real structure of powder sample. For example, artifacts of "first derivative" shape on difference plot originate from peak asymmetry of strong diffraction lines at low angle region. It can be concluded that both powder samples of **I** and **II** possess the same crystal structure as the structure of the corresponding single crystals.

2.3. *Spectral Characterization*

Infrared spectra of complexes **I** and **II** comprise bands confirming the presence of all characteristic functional groups. Some characteristic bands in the IR mid region of the nicotinamide (nia), the sodium salts ($Na_2suc \cdot 6H_2O$ and Na_2fum) and cobalt(II) complex **I** (prepared from nia and hmnia) and complex **II** are given in Table S5 (see Supplementary Materials).

IR spectra of model molecules **1** and **2** (Figures S10–S12, see Supplementary Materials) were calculated at the B3LYP/def2-TZVP level of theory after successful geometry optimization. Comparison of experimental and theoretical IR spectra is displayed in Figures S10–S12 and assigned bands of both complexes are collected in Table S5 (see Supplementary Materials).

In the IR spectra of **I**, **II** and $Na_2suc \cdot 6H_2O$ the broad bands at ca. 3500 cm^{-1} were assigned to ν(O–H) groups from coordinated and uncoordinated water molecules [60].

In theoretical spectra of **1** and **2** the O–H stretching modes were recorded at 3653 and 3628 cm^{-1} (Table S5, see Supplementary Materials).

In the experimental IR spectra of **I**, **II** and nia, the bands assigned to the symmetric and antisymmetric stretching vibrations of NH_2 groups from nicotinamide are observed in region about 3400–3140 cm^{-1} (Table S5, see Supplementary Materials). In calculated spectra only symmetric $\nu_s(NH_2)$ (and none antisymmetric) stretching vibration modes were obtained at 3293 cm^{-1} for both model molecules.

The C–H aromatic stretching modes are observed in experimental spectra at the same wavenumber 3062 cm^{-1}, while calculation gives 3012 cm^{-1} (complex **I**) and 3018 cm^{-1} (complex **II**). Aliphatic stretching vibrations are observed only for complex **I**, the antisymmetric $\nu_{as}(CH_2)$ at 2988 cm^{-1} and symmetric $\nu_s(CH_2)$ at 2955 cm^{-1} while the calculation predicts 2957 cm^{-1} and 2939 cm^{-1}, respectively (Table S6, see Supplementary Materials).

Most characteristic bands for dicarboxylate complexes are due to symmetric $\nu_s(COO^-)$ and antisymmetric $\nu_{as}(COO^-)$ stretching vibrations of carboxylate groups [61]. While the former are observed at ca. 1370 cm^{-1}, the latter occur at ca. 1560 cm^{-1} (Table S5, see Supplementary Materials). The difference (Δ) between the wavenumber of antisymmetric and symmetric vibration of carboxylate group gives information on the carboxylate bonding mode of complexes. When confronted with disodium succinate (141 cm^{-1}) and disodium fumarate (188 cm^{-1}) which possess ionic carboxylic groups, complexes **I** and **II** show higher values of Δ (180 cm^{-1} and 201 cm^{-1}, respectively) which are typical for monodentate *O*-coordination of carboxylate groups [61]. Bis(monodentate) bridging coordination mode of both carboxylate group in all complexes agrees with the structure of complexes as was determined by X-ray analysis.

In the recorded experimental IR spectra for the nia as well as for the complexes under study very strong or strong bands at about 1670, 1620 and 1390 cm^{-1} were assigned to Amide I (mainly $\nu(C{=}O)$), Amide II (mainly $\delta(NH_2)$) and Amide III (mainly $\nu(C–N)$ stretching) (S4). The vibration mode of Amide II and Amide III was calculated at about 1610 and 1370 cm^{-1}, respectively (Table S6, see Supplementary Materials). Relatively close positions of the bands assigned to Amide I, Amide II and Amide III for complexes under study to positions of bands in corresponding free molecules of nicotinamide are typical for non-coordinated amide groups [60].

In the UV-Vis absorption spectra, bands at about 220 and 265 nm can be observed, which are assigned to ligand (suc, fum and nia) internal transitions (Figures S13–S15, see Supplementary Materials). The shoulder at 350 nm is assigned to ligand-to-metal charge transfer transition [62] (LMCT, OαCo, NαCo). A band in the visible region between 470 and 480 nm with shoulder between 490 and 506 nm can be assigned to the $^4T_{1g}(F) \rightarrow {}^4T_{1g}(P)$ transition. Obtained electronic spectra are consistent with a strict-octahedral or a tetragonally-distorted-octahedral structure of coordination environment [63].

The electronic spectra of model molecules **1** and **2** were calculated at the same level of theory like the IR spectra. Only one dominant absorption was found for both cases, centred at 354 nm for **1** and 357 nm for **2** (Figure S16, Table S7, see Supplementary Materials). Inspection of the natural transition orbitals (NTOs) [64] associated with this electronic transition suggests that it corresponds to metal-to-ligand charge transfer in both cases (Figure S17, see Supplementary Materials) In both cases, a corresponding peak can be found in the experimental spectra, overlaid by strong intraligand absorption peak.

2.4. Static Magnetic Properties

Magnetic behaviour of complexes **I** and **II** is displayed in Figure 2. In general, the decrease of magnetic moment with decreasing temperature is mostly due to local magnetic anisotropy and/or magnetic coupling interaction. Usually, the simple isotropic form of the interaction is applicable for octahedral Co(II) complexes [65]. A DFT assessment on model systems **11** and **22** (Figure S18, Table S8, see Supplementary Materials) indicated however its negligible contribution and studied systems cannot be considered single-chain magnets. The local magnetism of Co(II) complexes with octahedral coordination environment is governed by both, spin magnetic momentum and angular magnetic momentum of the ground state [66]. Nevertheless, in some cases the angular momentum can be omitted, and it can be effectively described by simple spin Hamiltonian. As a rule of thumb this approximation can be applied if two lowest Kramers' doublets of the ground term $^4T_{1g}$ are separated from excited doublets by more than 1000 cm^{-1} [67].

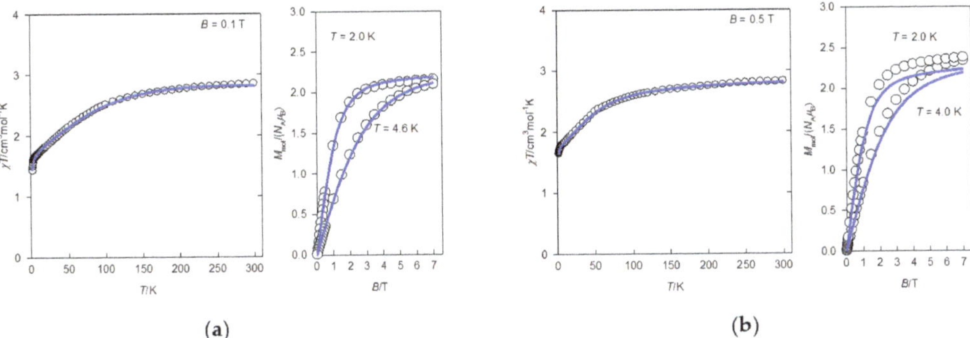

Figure 2. Product of magnetic susceptibility and temperature as a function of temperature (left) and magnetization as a function of magnetic field (right) per one magnetic centre for: (**a**) **I** and (**b**) **II**. Circles: experiment, solid line: optimum parameter fit.

To obtain this kind of insight, the magnetic energetic levels of mononuclear model systems **1** and **2** (Figure 3) were inspected using the method SA-CAS[7,5]SCF-NEVPT2-SOI. The resulting energy values of relevant Kramers' doublets are presented in Table S9 (see Supplementary Materials) revealing that the spin Hamiltonian is not justified in these systems (separation of Γ_c and Γ_b by 420.1 cm^{-1} and 272.9 cm^{-1} for **1** and **2**, respectively).

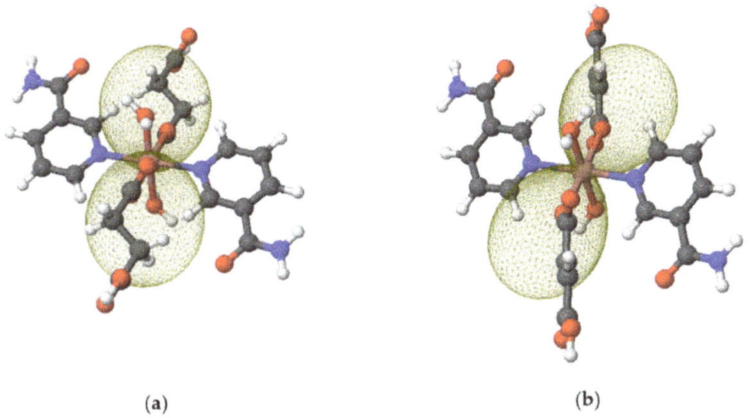

Figure 3. A cut-off from respective polymeric systems **I** and **II** forming a neutral mononuclear model molecule: (**a**) **1**; (**b**) **2** (right). The yellow transparent surface shows calculated directional dependence of molecular magnetization.

Therefore, more appropriate Griffith-Figgis Hamiltonian was employed instead [19,38,66], which in atomic units adopts the form of Equation (1).

$$\hat{H} = \sigma\lambda \vec{\hat{L}} \cdot \vec{\hat{S}} + \sigma^2 \Delta_{ax}\left(\hat{L}_z^2 - \hat{L}^2/3\right) + \sigma^2 \Delta_{rh}\left(\hat{L}_x^2 - \hat{L}_y^2\right) + \left(\sigma \vec{\hat{L}} + g\vec{\hat{S}}\right) \cdot \vec{B} \quad (1)$$

Here $\vec{\hat{L}}$ and $\vec{\hat{S}}$ are vector operators of angular momentum and spin, respectively, and $\vec{B}$ is vector of magnetic field. Parameters of the model are the constant of spin-orbit interaction λ, combined parameter σ (accounting for covalence of chemical bonds between central ion and ligands along with configuration interaction between ground and excited terms of matching symmetry), parameter of crystal field of axial symmetry Δ_{ax}, parameter of crystal field of rhombic symmetry Δ_{rh} and gyromagnetic factor g which was fixed equal to 2.00. In this Hamiltonian the convention with σ^2 absorbed in to Δ_{ax} and Δ_{rh} can be often encountered [66]. To improve the agreement between the model and experiment the effect of molecular field was included in the fitting procedure (quantified by parameter zj). The optimum values of parameters Δ_{ax}, Δ_{rh}, σ and λ obtained from fitting of experimental curves are collected in Table 1 and the match of experimental and optimum fit curves is displayed in Figure 2. The product of error residuals $R(\chi T) \times R(M)$ gains value considerably lower than 0.05 indicating very good accordance between model and experiment. The values resulting from ab-initio calculation are collected in Table 2. Comparing the calculated and fitted values, a satisfactory agreement can be concluded for the constant of spin-orbit interaction and parameter of crystal field of axial symmetry, for the parameter of rhombic splitting; however, there is a discrepancy of one order of magnitude in the case of couple **I**/**1**. Since the values obtained as optimum fit of experimental data (Table 1) are model-dependent and simultaneous fit for **II** is not perfect, the values obtained by ab-initio calculation (Table 2) can be considered somewhat more reliable. Finally, the visual assessment of directional dependence of molecular magnetization derived from all SA-CAS[7,5]SCF-NEVPT2-SOI states [38,67] shows, that magnetic anisotropy is of axial-like type (which is in accordance with negative value of parameter Δ_{ax}) and that the preferred orientation of molecular magnetization is towards the connecting bridge (Figure 3).

Table 1. Magnetic parameters extracted from optimum fit for complexes **I** and **II**.

Complex	λ/cm^{-1}	σ	Δ_{ax}/cm^{-1}	Δ_{rh}/cm^{-1}	zj/cm^{-1}	$R(\chi T) \times R(M)$
I	−177.4	−1.14	−424.7	−11.6	−0.06	0.0006
II	−161.5	−1.39	−395.1	−92.1	0.007	0.0094

Table 2. Calculated magnetic parameters for model molecules **1** and **2**.

	λ/cm^{-1}	Δ_{ax}/cm^{-1}	Δ_{rh}/cm^{-1}
1	−175.3	−522.7	−151.2
2	−175.2	−425.3	−128.8

2.5. Dynamic Magnetic Properties

To examine the presence of the slow relaxation of magnetization (SRM), which is the proof of SIM behaviour, the temperature and frequency dependence of the alternating-current (AC) susceptibility was measured at low temperatures for both complexes (see Supplementary Materials, Tables S10 and S11 for a detailed experimental description of AC susceptibility measurements and data analysis). The DC field scan for a limited number of frequencies over four orders of magnitude shows that out-of-phase component of AC susceptibility is silent at B_{DC} = 0 T (Figure S19, see Supplementary Materials). This indicates very fast SRM, probably due to the quantum tunnelling of magnetization induced by hyperfine interactions with the nuclear spin and/or dipolar interactions between the spin centres in the lattice. In order to determine the optimal B_{DC} to suppress the quantum tunnelling effect, AC susceptibility measurements under various B_{DC} fields were applied at 2 K (Figure S19). Upon increasing DC field up to B_{DC} = 0.8 T, the out-of-phase component varies, but differently for individual frequencies. This confirms that compounds **I** and **II** show field-induced SRM and the subsequent temperature and frequency dependent measurements were carried out by choosing B_{DC} fields at which the out-of-phase components

χ'' reach the maximal response (B_{DC} = 0.1 T for **I** and B_{DC} = 0.2 T for **II**). Furthermore, in order to detect changes in mechanisms of the SRM upon the various static magnetic fields, the temperature variable AC susceptibility measurements have been recorded also at B_{DC} = 0.02 T for complex **I** as well as at B_{DC} = 0.05 T, B_{DC} = 0.1 T and B_{DC} = 0.15 T for complex **II** (Figures S20–S25).

At 2 K and 0.1 T, the out-of-phase component χ'' of **I** does not yet show maximum, which indicates that the SRM acquires relaxation times τ longer than 0.16 s (Figure 4a). However, the reduction of static magnetic field to 0.02 T has already caused the appearance of maxima even at the lowest temperatures of measurement (14 Hz, τ = 73 ms at 1.8 K; Figure S20, Table S12). At both fields, the increase of temperature resulted in the obvious shift of the maxima in the χ'' vs. f dependencies towards higher frequencies and proves that **I** is field-induced SIM. On the contrary, the out-of-phase component χ'' of complex **II** shows maxima around 63 Hz at 1.8 K which indicates much faster relaxation of magnetisation ($\tau \approx 2.54$ ms) in comparison to complex **I**. Also here, the increase of temperature shifts those maxima towards higher frequencies (shorter relaxation times). All herein reported AC susceptibility measurements recorded at various static magnetic field B_{DC} were satisfactorily fitted by one-set Debye model for a single relaxation channel relaxation of magnetisation (Equations (S1) and (S2), see Supplementary Materials). This analysis resulted in the set of four parameters—adiabatic χ_S and isothermal χ_T susceptibilities, the distribution parameters α_i and relaxation times τ_i (see Supplementary Materials, Tables S12 and S13 for **I**, Tables S14–S17 for **II**).

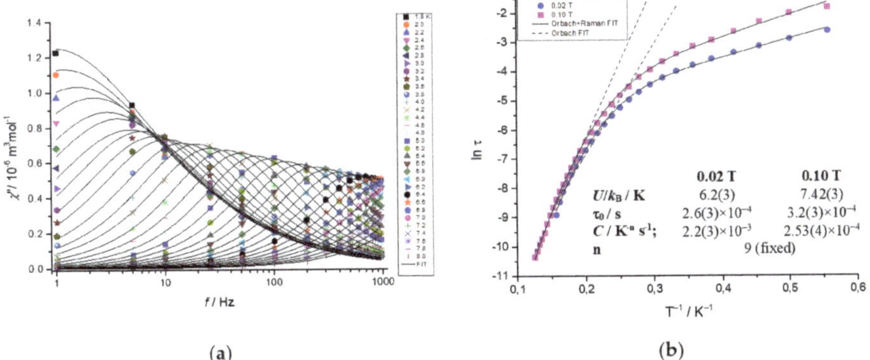

Figure 4. (**a**) Frequency dependent out-of-phase χ'' components of AC susceptibility for compound **I** recorded at the applied static magnetic field B_{DC} = 0.1 T (Solid lines present fits using the one-component Debye's model, Equations (S1) and (S2); see Supplementary Materials); (**b**) The $\ln\tau$ vs. $1/T$ dependency obtained from AC susceptibility measurements at two static magnetic fields.

The obtained thermal dependency of relaxation time, presented in the form ln τ vs. $1/T$, was analysed according to extended relaxation Equation (2) [68,69]

$$\frac{1}{\tau} = \frac{1}{\tau_0}\exp(-\frac{U}{kT}) + CT^{\mathrm{n}} + AB^{\mathrm{m}}T, \quad (2)$$

where the terms correspond to thermally activated Orbach, Raman and direct processes, respectively (Tables S18–S22, see Supplementary Materials). At first, the high temperature regions of ln τ vs. $1/T$ dependencies were fitted to the Arrhenius-like plot, which considers the Orbach relaxation process only (Figures 4b and 5b, dashed lines). Such simple analysis resulted in the preliminary evaluation of the effective energy barrier of spin reversal U, which in the case of **I** showed a decrease upon the increase of the B_{DC} from 0.02 T (U = 52 K) to 0.1 T (U = 38 K). On the other hand, complex **II** shows field-independent values that are twice as small (U = 26 K) in the range 0.1–0.2 T, while the linear fit of the low field measurement at 0.05 T has not yielded reliable results (Table S18).

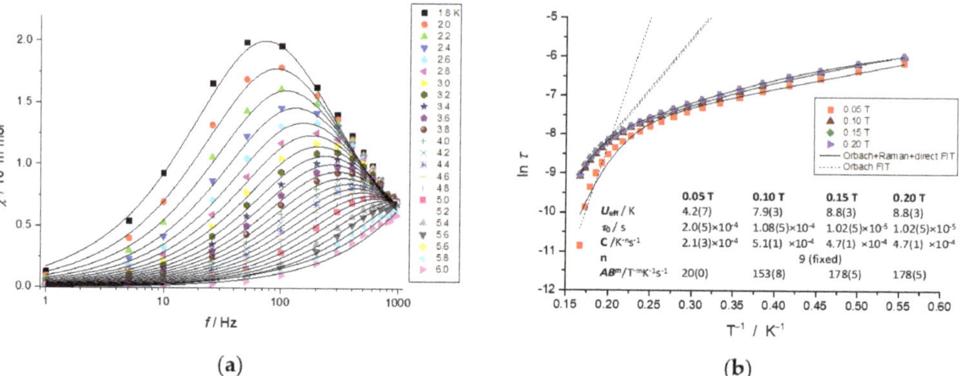

(a) (b)

Figure 5. (**a**) Frequency dependent out-of-phase χ'' components of AC susceptibility for compound **II** recorded at the applied static magnetic field B_{DC} = 0.05 T (Solid lines present fits using the one-component Debye's model, Equations (S1) and (S2); see Supplementary Materials); (**b**) The $\ln\tau$ vs. $1/T$ dependency obtained from AC susceptibility measurements at four static magnetic fields.

In the more complex analysis of ln τ vs. $1/T$ dependencies, combinations of two or all three relaxation processes from Equation (2) have been used for the fitting procedure (Tables S18–S22, see Supplementary Materials). The most reliable results were obtained for the combination of Orbach and Raman processes in the case of complex **I** and for the combination of all three relaxation processes in the case of complex **II** (Figures 4b and 5b) [68,69].

Although the contribution of Orbach process to the overall slow relaxation magnetisation in hexacoordinated Co(II) complexes is quite rare [70], some recent studies suggest that this two-phonon thermally activated relaxation can be operative also in this type of SIMs [70,71]. Obtained values of energy barriers U are relatively small for the class of hexacoordinated Co(II) single-ion magnets [67,71,72]. The values of Raman exponent were fixed to the recommended value n = 9 for Kramers' systems, pre-exponential constant C are rather field independent and pre-exponential factor of direct process A (if m = 4 for Kramers' system [68,69]) has an decreasing trend when the static magnetic field is increased.

To put the observed changes in magnetic relaxation in context with the change of geometry of coordination environment, the difference of symmetry measure parameter S(O_h) from 0.198 (**I**) to 0.306 (**II**) is associated with decrease of relaxation time limit τ_0 from 3.2×10^{-4} s to 1.0×10^{-5} s at the field of 0.1 T. For comparison, a recent work by Liu et al. [73] reported the effect of subtle geometry changes in isolated octahedral complexes with the same coordination environment CoN_2O_4 but an easy-plane magnetic anisotropy. In their work, the difference of symmetry measure parameter from S(O_h) = 0.025 (Co1) and S(O_h) = 0.067 (Co2) to S(O_h) = 0.117 induced a small increase of relaxation time limit τ_0 from 1.67×10^{-8} s to 1.88×10^{-8} s at the field of 0.2 T.

3. Materials and Methods

3.1. Chemical Reagents

The chemicals were of reagent grade (Sigma-Aldrich, Darmstadt, Germany) and used without further purification. The organic reagents were purchased from Sigma-Aldrich (Darmstadt, Germany) and TCI Chemicals (Tokyo, Japan); their purity was checked by IR spectra.

3.2. Syntheses of the Complexes

3.2.1. Synthesis of {[Co(μ_2-suc)(nia)$_2$(H_2O)$_2$]·2H_2O}$_n$ (Complex **I**)

(a) From nicotinamide (nia)

Crystals of {[Co(μ_2-suc)(nia)$_2$(H_2O)$_2$]·2H_2O}$_n$ (**I**) were obtained by dissolving cobalt(II) nitrate hexahydrate (2 mmol) with equimolar quantity of sodium succinate hexahydrate in 20 cm^3 of water. Nicotinamide (4 mmol) was dissolved in 20 cm^3 of *n*-pentanol. This solution was slowly added to an aqueous solution of cobalt(II) nitrate and $Na_2suc·6H_2O$ (Scheme 1). The resulting solution was refluxed for 1 h and during the reflux the precipitate was formed. Then the solution was slowly cooled down. The solution was filtered and two layers were created—a layer of pink water solution, which was covered by *n*-Pentanol. These two layers were left to slowly diffuse at ambient temperature. After several weeks purple crystals of complex **I** were filtered off.

Yield: 54% based on Co. Elemental analysis for $C_{16}H_{24}CoN_4O_{10}$ (M_W = 491.32) found % (expected %): C 38.7 (38.3); N 11.0 (10.8); H 4.9 (4.7); Co 12.11 (12.30). IR (ATR, cm^{-1}): 3424 sh, 3367 s, 3202 s, 3060 w, 2988 w, 2955 w, 1661 vs, 1559 vs, br, 1395 s, 1367 vs, 1151 m, 1138 m, 780 m, 652 vs, 516 vs. Electronic spectra (nujol mulls, nm): 223, 265, 330 sh, 480, 495 sh.

(b) From *N*-(hydroxylmethyl)nicotinamide (hmnia)

Complex {[Co(μ_2-suc)(nia)$_2$(H_2O)$_2$]·2H_2O}$_n$, which has the same composition as complex **I**, was prepared by reaction of cobalt(II) chloride hexahydrate (2 mmol) with disodium succinate hexahydrate (2 mmol) and *N*-(hydroxymethyl)nicotinamide—hmnia (4 mmol) in 50 cm^3 of water (Scheme 2). The resulting solution was stirred under reflux. After 20 min of reflux the colour of the solution changed to purple and after another 20 min the colour changed to pink and a light precipitate was formed. After 2 h of reflux, the solution was slowly cooled down. The precipitate was filtered off and pink solution was left to evaporate at ambient temperature. After few days, purple crystals of complex {[Co(μ_2-suc)(nia)$_2$(H_2O)$_2$]·2H_2O}$_n$ were collected.

Yield: 48% based on Co. Elemental analysis for $C_{16}H_{24}CoN_4O_{10}$ (M_W = 491.32) found % (expected %): C 38.8 (38.3); N 11.3 (10.8); H 5.1 (4.7); Co 12.2 (12.3). IR (ATR, cm^{-1}): 3430 sh, 3360 sh, 3200 vs, 3061 w, 2989 w, 2959 w, 1656 s, 1597 s, 1546 vs, br, 1395 s, 1366 s, 1150 m, 1133 m, 779 m, 652 vs, 515 vs. Electronic spectra (nujol mull, nm): 208, 267, 325 sh, 480, 506 sh.

3.2.2. Synthesis of [Co(μ_2-fum)(nia)$_2$(H_2O)$_2$]$_n$ (Complex **II**)

Pink crystals of [Co(μ_2-fum)(nia)$_2$(H_2O)$_2$]$_n$ (**II**) were acquired by dissolving cobalt(II) nitrate hexahydrate (2 mmol), disodium fumarate (2 mmol) and nicotinamide (4 mmol) in 50 cm^3 mixture of water and methanol (1:1). Solution was refluxed for 2 h (Scheme 1). After 20 min of reflux the precipitate was formed. After 2 h, the solution was slowly cooled down and a pink precipitate was filtered off. The resulting pink solution was left to evaporate at ambient temperature. Pink crystals of complex [Co(μ_2-fum)(nia)$_2$(H_2O)$_2$]$_n$ were separated after few weeks.

Yield: 61% based on Co. Elemental analysis for $C_{16}H_{18}CoN_4O_8$ (M_W = 453.27) found % (expected %): C 41.8 (42.4); N 12.1 (12.4); H 4.2 (4.0); Co 12.2 (12.4). IR (ATR, cm^{-1}): 3500 m, 3312 s, 3193 s, br, 3060 w, 1687 s, 1622 m, 1595 sh, 1573 vs, 1557 vs, 1393 s, 1368 vs, 1153 m, 1099 m, 754 m, 653 vs, 638 vs, 504 vs. Electronic spectra (nujol mull, nm): 229, 265, 320 sh, 472, 506 sh.

3.3. Analysis and Physical Measurements

Analytical grade (Mikrochem, Pezinok, Slovakia; Acros Organics, Geel, Belgium and TCI Chemical, Tokyo, Japan) chemicals and solvents were used without further purification. Cobalt was determined by electrolysis after mineralization of the complexes; carbon, hydrogen and nitrogen were determined by microanalytical methods (Thermo Electron Flash EA 1112). Electronic spectra (9000–50,000 cm^{-1}) of the powdered samples in nujol mull were recorded at room temperature on Specord 240 spectrophotometer (Carl Zeiss, Jena, Germany). Infrared spectra in the region of 400–4000 cm^{-1} were recorded on a Nicolet 5700 FT-IR spectrometer (Thermo Scientific, Waltham, MA, USA). Spectra of the solid samples were obtained by ATR technique at room temperature. Magnetism of all complexes was measured using a SQUID magnetometer (MPMS-XL7, Quantum Design, San Deigo, CA, USA). The temperature dependence of magnetization was recorded at a constant magnetic field $B = 0.1$ (complex **I**) T or $B = 0.5$ T (complex **II**), corrected for diamagnetic contribution, and displayed as the product of temperature and molar susceptibility (in the cgs-emu unit system). The dependence of magnetization on the magnetic field was measured at two constant temperatures: $T = 2.0$ K and $T = 4.6$ K (complex **I**) or $T = 2.0$ K and $T = 4.0$ K (complex **II**).

3.4. Computational Details

Fitting of the DC magnetic susceptibility and magnetization of both compounds was performed with the program PHI 3.1.3 [74]. Fitting of AC magnetic susceptibility was realized with the help of a home-made program. Calculations of magnetic exchange coupling parameter was performed within ORCA 5.0.2 [75] using model molecules **11** and **22** (Figure S18, see Supplementary Materials), all other calculations were carried out within the program ORCA 4.2.0 [76] with the model molecules **1** and **2** (Figures 3 and S17, see Supplementary Materials). Magnetic coupling was assessed with exchange-correlation density functional approximations B3LYP [77–79], PBE0 [80] and TPSSh [81]. The resolution of identity and chain-of-spheres approximations for Coulomb and exchange integrals (RIJCOSX) [82] were set on. For all atoms the Ahlrichs' basis set def2-TZVP [83,84] was used with an auxiliary basis set def2/J [85]. Prior to this calculation, the positions of all hydrogen atoms were optimized on the model molecules **11** and **22** using the method PBEh-3c [86] and all other atoms were kept in their positions as obtained from the X-ray analysis. The energy levels of crystal-field terms in mononuclear model molecules **1** and **2** were obtained using the state averaged complete active space self-consistent field method (SA-CAS[7,5]SCF) [87] complemented by strongly-contracted *N*-electron valence perturbation theory of second-order (NEVPT2) [88–90] and spin-orbit interaction [91,92]. All 10 spin quartet states and 40 spin doublet reference states were taken into account. The resolution of identity approximation for Coulomb and exchange integrals (RI-JK) [82] were set on. For all atoms the basis set def2-TZVP was used, this time with an automatically generated auxiliary basis set [93]. In all calculations the increased integration grid was set (level 5 in ORCA convention). The positions of all hydrogen atoms were optimized on the model molecules **1** and **2** using the same approach as for **11** and **22**. The molecular magnetization isosurface was visualized by a home-made program using the approach described in [38]. The infrared spectra were calculated at model molecules **1** and **2** with the abovementioned basis set and hybrid exchange-correlation density functional approximation B3LYP [77–79]. No negative vibration frequencies were obtained. The electronic spectra were calculated for molecules **1** and **2** using the time-dependent DFT method with the same setting of basis and exchange-correlation functional like it was for the IR spectra, asking for 15 roots.

3.5. Crystal Structure Determination

Data collections and cell refinement were carried out using four-circle diffractometer STOE StadiVari using Pilatus3R 300K HPD detector, and microfocused X-ray source Xenocs Genix3D Cu HF (CuK_α radiation) at 100 K. The diffraction intensities were corrected for Lorentz and polarization factors. The absorption corrections were made by LANA [94]. The structures were solved with program SHELXT [95], and refined by the full-matrix least squares procedure of Independent Atom Model (IAM) [96] with SHELXL-2018/3 [97]. The Hirshfeld Atom Refinement (HAR) was carried out using IAM model as a starting point. The wave function was calculated using ORCA 4.2.0 software [76] with basis set def2-TZVPP [83,84] and hybrid exchange-correlation functional PBE0 [80]. The least-squares refinements of HAR model were then carried out with olex2.refine [98], while keeping the same constrains and restrains as for the SHELXL refinement. The NoSpherA2 implementation [99] of HAR makes used for tailor-made aspherical atomic factors calculated on-the-fly from a Hirshfeld-partitioned electron density. For the HAR approach, all H atoms were refined isotropically and independently. All calculations and structure drawings were done in the OLEX2 package [100]. Final crystal data and HAR's refinement parameters are given in Table S1 (see Supplementary Materials).

Hirshfeld Surface Analysis

The software Crystal Explorer [101] was used to calculate Hirshfeld surface [102,103] and associated fingerprint plots [104,105]. The Hirshfeld surfaces were obtained using CIF files from HAR model.

3.6. Powder X-ray Analysis

PXRD data of **I** were collected within the 2Θ range 3°–60° on a Brag-Brentano focusing powder diffractometer PHILIPS, model 1730/10. The instrument was equipped with X-ray tube providing Co α radiation, wavelength (0.179021 nm). The experimental conditions were as follow: exciting voltage: 40 kV, anode current: 35 mA, step size: 0.02°, time on step 2.4 s.

In the case of **II**, PXRD were collected within the 2Θ range 5°–60° on a Brag-Brentano automated focusing powder diffractometer EMPYREAN. The instrument was equipped with X-ray tube providing Cu Kα radiation. wavelength (0.15405980 nm) and PIXcel3D-Medipix3 1 × 1 detector. The exciting voltage was 45 kV, and the anode current was 40 mA, continuous mode was used.

The simulated powder patterns of **I** and **II** were obtained from single-crystal data employing the Le Bail analysis in the computer program Jana 2006 [106].

4. Conclusions

In conclusion, we have thoroughly reinvestigated two previously described analogous cobalt (II) coordination polymers, where the isolated metal centres are bridged by succinate or fumarate anion. The molecular structures of both compounds were determined with much higher accuracy then before, showing the shortest intermetallic distances in **I** and **II** about 9.5 Å and 9.7 Å, respectively, which imply isolated magnetic environments for both systems.

Static magnetic studies and ab-initio calculations further supported that the Co(II) centres can be considered magnetically isolated and they both show easy axis magnetic anisotropy pointing towards the connecting bridge.

The AC magnetic study showed that the fumarate analogue **II** relaxes comparatively faster at the field 0.1 T than the succinate complex **I**. Although the slow relaxation of magnetization is very susceptible to even small changes in local anisotropy of coordination environment, we can suppose that the enhanced rigidity of the bridge is a non-negligible factor for conservation of molecular magnetization. Indeed, as discussed in a few recent works [107–109], magnetic relaxation is faster if the material possesses low-lying avoided-crossing points between acoustic and optical phonons, or, in simpler words, if the collective thermal vibrations spread easily to vibrations around the magnetic centre. In this sense we can state that in the studied couple of SIMs, the fumarate bridge could act as better "transmitter" of the vibrational perturbations onto the magnetic centre than the succinate bridge. We can thus conjugate that the less stiff bridges are more appropriate components for targeted design of single-molecule magnets.

Supplementary Materials: The following supporting information can be downloaded at: https://www.mdpi.com/article/10.3390/inorganics10090128/s1, Figures S1–S7: Information on the molecular and supramolecular structure of complexes; Figures S8 and S9: PXRD spectra; Figures S10–S12: Experimental and theoretical IR spectra; Figures S13–S16: Experimental and theoretical electron spectra; Figure S17: NTOs of electron transitions in mononuclear model systems; Figure S18: Binuclear model molecules; Figure S19: Out-of-phase susceptibility component at 2 K;

Figures S20–S25: AC susceptibility data at various fields; Tables S1–S4: Crystallographic data and parameters; Tables S5–S6: Characteristic band in IR spectra; Table S7: Calculated bands in electron spectra; Table S8: Calculated magnetic coupling interaction; Table S9: Calculated energies of Kramers' doublets; Tables S10 and S11: Conditions of AC magnetic experiments; Tables S12–S17: Parameters of the extended one-set Debye model, Tables S18–S22: Relaxation parameters using various combinations of mechanisms.

Author Contributions: Conceptualization, J.P. and P.S.; syntheses, M.B.; crystallographic measurements and interpretation, J.M.; spectroscopic measurements and interpretation, M.B. and P.S.; magnetism measurements, L'.D. and M.K.; magnetism interpretation, J.J., I.Š. and J.P.; computational studies. J.P.; writing—original draft preparation, M.B., J.M., I.Š., J.P. and P.S.; writing—review and editing, J.P.; supervision, J.P. and P.S. All authors have read and agreed to the published version of the manuscript.

Funding: Grant Agencies (Slovakia: APVV-18-0197, APVV-18-0016, APVV-19-0087, VEGA 1/0029/22, KEGA 018-STU-4; Czech Republic: (GA Č R 22-23760S) are acknowledged for the financial support. This article was written thanks to the generous support under the Operational Program Integrated Infrastructure for the project: "Strategic research in the field of SMART monitoring, treatment and preventive protection against coronavirus (SARS-CoV-2)", Project no. 313011ASS8, co-financed by the European Regional Development Fund.

Institutional Review Board Statement: Not applicable.

Informed Consent Statement: Not applicable.

Data Availability Statement: Not applicable.

Acknowledgments: I.Š. acknowledges the financial support from institutional sources of the Department of Inorganic Chemistry, Palacký University Olomouc, Czech Republic. J.P. is grateful to the HPC centre at the Slovak University of Technology in Bratislava, which is a part of the Slovak Infrastructure of High-Performance Computing (SIVVP project, ITMS code 26230120002, funded by European regional development funds, ERDF), for computational time and resources made available. All authors are thankful to Vladimír Jorík (Slovak University of Technology in Bratislava) for PXRD measurements and analyses.

Conflicts of Interest: The authors declare no conflict of interest.

References

1. Wang, C.-C.; Li, J.-R.; Lv, X.-L.; Zhang, Y.-Q.; Guo, G. Photocatalytic organic pollutants degradation in metal-organic frameworks. *Energy Environ. Sci.* **2014**, *7*, 2831–2867. [CrossRef]
2. Lee, J.; Farha, O.K.; Roberts, J.; Scheidt, K.A.; Nguyen, S.T.; Hupp, J.T. Metal-organic framework materials as catalysts. *Chem. Soc. Rev.* **2009**, *38*, 1450–1459. [CrossRef] [PubMed]
3. Reinsch, H.; van der Veen, M.A.; Gil, B.; Marszalek, B.; Verbiest, T.; de Vos, D.; Stock, N. Sctructures, Sorption Characteristics, and Nonlinear Optical Properties of a New Series of Highly Stable Aluminum MOFs. *Chem. Mater.* **2013**, *25*, 17–26. [CrossRef]
4. Khan, S.; Masum, A.A.; Giri, P.; Islam, M.; Harms, K.; Chattopadhyay, S. Chirality-Induced Variation In Interaction of Two Similar Copper(II) Coordination Polymers with Calf Thymus DNA: Exploration of Their Antimicrobial Activity and Cytotoxicity. *Chem. Select.* **2018**, *3*, 7112–7122. [CrossRef]
5. Li, J.-R.; Kruppler, R.J.; Zhou, H.-C. Selective gas adsorption and separation in metal-organic frameworks. *Chem. Soc. Rev.* **2009**, *38*, 1477–1504. [CrossRef] [PubMed]
6. Li, L.; Wang, X.; Liang, J.; Huang, Y.; Li, H.; Lin, Z.; Cao, R. Water-Stable Anionic Metal-Organic Framework for Highly Selective Separation of Methane from Natural Gas and Pyrolysis Gas. *ACS Appl. Mater. Interfaces* **2016**, *8*, 9777–9781. [CrossRef]
7. Wilmer, C.E.; Farha, O.K.; Yildimir, T.; Eryazici, I.; Krungleviciute, V.; Sarjeant, A.A.; Snurr, R.Q.; Hupp, J.T. Gram-scale, high-yield synthesis of a robust metal-organic framework for storing methane and other gases. *Energy Environ. Sci.* **2013**, *6*, 1158–1163. [CrossRef]
8. Beauvais, L.G.; Shores, M.P.; Long, J.R. Cyano-Bridged Re_6Q_8 (Q = S, Se) Cluster-Cobalt(II) Framework Materials: Versatile Solid Chemical Sensors. *J. Am. Chem. Soc.* **2000**, *122*, 2763–2772. [CrossRef]
9. Jianghua, H.; Jihong, Y.; Yuetao, Z.; Qinhe, P.; Ruren, X. Synthesis, Structure, and Luminescent Property of a Heterometallic Metal−Organic Framework Constructed from Rod-Shaped Secondary Building Blocks. *Inorg. Chem.* **2005**, *44*, 9279–9289.
10. Kahn, O. Chemistry and Physics of Supramolecular Magnetic Materials. *Acc. Chem. Res.* **2000**, *33*, 647–657. [CrossRef]
11. Marinho, M.V.; Marques, L.F.; Diniz, R.; Stumpf, H.O.; do Canto Visentin, L.; Yoshida, M.I.; Machado, F.C.; Llorent, F.; Julve, M. Solvothermal synthesis, crystal structure and magnetic properties of homometallic Co(II) and Cu(II) chains with double di(4-pyridyl)sulfide as bridges. *Polyhedron* **2012**, *45*, 1–8. [CrossRef]

12. Zheng, Y.Z.; Evangelisti, M.; Winpenny, R.E.P. Co-Gd phosphonate complexes as magnetic refrigerants. *Chem. Sci.* **2011**, *2*, 99–102. [CrossRef]
13. James, S.L. Metal-organic frameworks. *Chem. Soc. Rev.* **2003**, *32*, 276–288.
14. Natarajan, R.; Sevitha, G.; Dominiak, P.; Wozniak, K.; Moorthy, J.N. Corundum, Diamond, and PtS Metal–Organic Frameworks with a Difference: Self-Assembly of a Unique Pair of 3-Connecting D_{2d}-Symmetric 3,3′,5,5′-Tetrakis(4-pyridyl)bimesityl. *Angew. Chem. Int. Ed.* **2005**, *44*, 2115–2119.
15. Xue, Y.-H.; Liu, Y.; Xu, D.-J. *catena*-Poly[[diaquabis(1*H*-benzimidazole-κN^3)cobalt(II)]-μ-succinato-κ*O*]. *Acta Cryst.* **2003**, *E59*, m944–m946. [CrossRef]
16. Zhang, J.; Kang, Y.; Zhang, R.-B.; Li, Z.-J.; Cheng, J.-K.; Yao, Y.-G. A twisting chiral 'dense' 7^5.9 net, incorporating a helical sub-structure. *CrystEngComm* **2005**, *7*, 177–178.
17. Duangthonyou, T.; Jirakulpattana, S.; Phakawatchai, C.; Kurmoo, M.; Siripaisarnpipat, S. Comparison of crystal structures and magnetic properties of two Co(II) complexes containing different dicarboxylic acid ligands. *Polyhedron* **2010**, *29*, 1156. [CrossRef]
18. Bora, S.J.; Das, B.K. Synthesis and properties of a few 1-D cobaltous fumarates. *J. Solid State Chem.* **2012**, *192*, 93–101.
19. Brezovan, M.; Kuchtanin, V.; Moncol', J.; Pavlik, J.; Dlháň, L'.; Segl'a, P. Four novel cobalt(II) succinate coordination polymers with *N*-heterocyclic ligands: Crystal structures, spectral properties, magnetism and computational study. *Chem. Pap.* **2020**, *74*, 3741–3753.
20. Beatty, A.M. Hydrogen bonded networks of coordination complexes. *CrystEngComm* **2001**, *3*, 243–255.
21. Beatty, A.M. Open-framework coordination complexes from hydrogen-bonded networks: Toward host/guest complexes. *Coord. Chem. Rev.* **2003**, *246*, 131–143. [CrossRef]
22. Moncol', J.; Kuchtanin, V.; Polakovičová, P.; Mrozinski, J.; Kalinska, B.; Koman, M.; Padělková, Z.; Segl'a, P.; Melník, M. Study of copper(II) thiophenecarboxylate complexes with nicotinamide. *Polyhedron* **2012**, *45*, 94–102. [CrossRef]
23. Puchoňová, M.; Maroszová, J.; Mazúr, M.; Valigura, D.; Moncol, J. Structures with different supramolecular interactions and spectral properties of monomeric, dimeric and polymeric benzoatocopper(II) complexes. *Polyhedron* **2021**, *197*, 115050. [CrossRef]
24. Ganzhorn, M.; Wernsdorfer, W. *Molecular Magnets*; Bartolome, J., Luis, F., Fernandez, J.F., Eds.; Springer: Berlin/Heidelberg, Germany, 2014; pp. 319–364.
25. Boča, R.; Rajnák, C.; Moncol, J.; Titiš, J.; Valigura, D. Breaking the Magic Border of One Second for Slow Magnetic Relaxation of Cobalt-Based Single Ion Magnets. *Inorg. Chem.* **2018**, *57*, 14314–14321. [CrossRef] [PubMed]
26. Clérac, R.; Miyasaka, H.; Yamashita, M.; Coulon, C. Evidence for Single-Chain Magnetic Behavior in a Mn(III)-N(II) Chain Designed with High Spin Magnetic Units: A Route to High Temperature Metastable Magnets. *J. Am. Chem. Soc.* **2002**, *124*, 12837–12844. [CrossRef]
27. Frost, J.M.; Harriman, K.L.M.; Murugesu, M. The Rise of 3-d Single-Ion Magnets in Molecular Magnetism: Towards Materials from Molecules. *Chem. Sci.* **2016**, *7*, 2470–2491. [CrossRef]
28. Jiménez, J.-R.; Xy, B.; El Said, H.; Li, Y.; von Bardeleben, J.; Chamoreau, L.-M.; Lescouëzec, R.; Shova, S.; Visinescu, S.; Alexandru, M.-G.; et al. Field-induced single ion magnet behaviour of discrete and one-dimensional complexes containing [bis(1-methylimidazol-2-yl)ketone]-cobalt(II) building units. *Dalton Trans.* **2021**, *50*, 16353–16363. [CrossRef]
29. Chilton, N.E. Molecular Magnetism. *Annu. Rev. Mater. Res.* **2022**, *52*, 79–101. [CrossRef]
30. Sarkar, A.; Dey, S.; Rajaraman, G. Role of Coordination Number and Geometry in Controlling the Magnetic Anisotropy in Fe^{II}, Co^{II} and Ni^{II} Single-Ion Magnets. *Chem. Eur. J.* **2020**, *26*, 14036–14058. [CrossRef]
31. Perlepe, P.S.; Maniaki, D.; Pilichos, E.; Katsoulakou, E.; Perlepes, S.P. Smart Ligands for Efficient 3d-, 4d- and 5d-Metal Single-Molecule Magnets and Single-Ion Magnets. *Inorganics* **2020**, *8*, 39. [CrossRef]
32. Váhovská, L.; Vitushkina, S.; Potočňák, I.; Trávníček, Z.; Herchel, R. Effect of linear and non-linear pseudohalides on the structural and magnetic properties of Co(II) hexacoordinate single-molecule magnets. *Dalton Trans.* **2018**, *47*, 1498–1512. [CrossRef] [PubMed]
33. Cui, H.-H.; Zhang, Y.-Q.; Chen, X.-T.; Wang, Z.; Xue, Z.-L. Magnetic anisotropy and slow magnetic relaxation processes of cobalt(II)-pseudohalide complexes. *Dalton Trans.* **2019**, *48*, 10743–10752. [CrossRef]
34. Vallejo, J.; Castro, I.; Ruiz-García, R.; Cano, J.; Julve, M.; Lloret, F.; De Munno, G.; Wernsdorfer, W.; Pardo, E. Field-Induced Slow Magnetic Relaxation in a Six-Coordinate Mononuclear Cobalt(II) Complex with a Positive Anisotropy. *J. Am. Chem. Soc.* **2012**, *134*, 15704–15707. [CrossRef]
35. Ding, Z.-Y.; Meng, Y.-S.; Xiao, Y.; Zhang, Y.-Q.; Zhu, Y.-Y.; Gao, S. Probing the influence of molecular symmetry on magnetic anisotropy of octahedral cobalt(II) complexes. *Inorg. Chem. Front.* **2017**, *4*, 1909–1916. [CrossRef]
36. Zhang, J.; Li, J.; Yang, L.; Yuan, C.; Zhang, Y.-Q.; Song, Y. Magnetic Anisotropy from Trigonal Prismatic to Trigonal Antiprismatic Co(II) Complexes: Experimental Observation and Theoretical Prediction. *Inorg. Chem.* **2018**, *57*, 3903–3912. [CrossRef] [PubMed]
37. Chen, L.; Zhou, J.; Cui, H.-H.; Yuan, A.-H.; Wang, Z.; Zhang, Y.-Q.; Ouyang, Z.-W.; Song, Y. Slow magnetic relaxation influenced by change of symmetry ideal Ci to D3d in cobalt(II)-based single-ion magnets. *Dalton Trans.* **2018**, *47*, 2506–2510. [CrossRef] [PubMed]
38. Nemec, I.; Herchel, R.; Trávníček, Z. Two polymorphic Co(II) field-induced single-ion magnets with enormous angular distortion from the ideal octahedron. *Dalton Trans.* **2018**, *47*, 1614–1623. [CrossRef]
39. Nemec, I.; Fellner, O.I.; Indruchová, B.; Herchel, R. Trigonally Distorted Hexacoordinate Co(II) Single-Ion Magnets. *Materials* **2022**, *15*, 1064. [CrossRef]

40. Landart-Gereka, A.; Quesada-Moreno, M.M.; Díaz-Ortega, I.; Nojiri, H.; Ozerov, M.; Krzystek, J.; Palacios, M.A.; Colacio, E. Large easy-axis magnetic anisotropy in a series of trigonal prismatic mononuclear cobalt(II) complexes with zero-field hidden single-molecule magnet behaviour: The important role of the distortion of the coordination sphere and intermolecular interactions in the slow relaxation. *Inorg. Chem. Front.* **2022**, *9*, 2810–2831.
41. Roy, S.; Oyarzabal, I.; Vallejo, J.; Cano, J.; Colacio, E.; Bauza, A.; Frontera, A.; Kirillov, A.M.; Drew, M.G.B.; Das, S. Two Polymorphic Forms of a Six-Coordinate Mononuclear Cobalt(II) Complex with Easy-Plane Anisotropy: Structural Features, Theoretical Calculations, and Field-Induced Slow Relaxation of the Magnetization. *Inorg. Chem.* **2016**, *55*, 8502–8513. [CrossRef]
42. Peng, G.; Qian, Y.-F.; Wang, Z.-W.; Chen, Y.; Yadav, T.; Fink, K.; Ren, X.-M. Tuning the Coordination Geometry and Magnetic Relaxation of Co(II) Single-Ion Magnets by Varying the Ligand Substitutions. *Cryst. Growth Des.* **2021**, *21*, 1035–1044. [CrossRef]
43. Rajnák, C.; Titiš, J.; Moncol, J.; Boča, R. Effect of the distant substituent on the slow magnetic relaxation of the mononuclear Co(II) complex with pincer-type ligands. *Dalton Trans.* **2020**, *49*, 4206–4210. [CrossRef] [PubMed]
44. Woods, T.J.; Ballesteros-Rivas, M.F.; Gómez-Coca, S.; Ruiz, E.; Dunbar, K.R. Relaxation Dynamics of Identical Trigonal Bipyramidal Cobalt Molecules with Different Local Symmetries and Packing Arrangements: Magnetostructural Correlations and ab inito Calculations. *J. Am. Chem. Soc.* **2016**, *138*, 16407–16416. [CrossRef] [PubMed]
45. de Campos, N.R.; Simosono, C.A.; Landre Rosa, I.M.; da Silva, R.M.R.; Doriguetto, A.C.; do Pim, W.D.; Gomes Simões, T.R.; Valdo, A.K.S.M.; Martins, F.T.; Sarmiento, C.V.; et al. Building-up host–guest helicate motifs and chains: A magneto-structural study of new field-induced cobalt-based single-ion magnets. *Dalton Trans.* **2021**, *50*, 10707–10728. [CrossRef]
46. Demir, S.; Yilmaz, V.T.; Yilmaz, F.; Buyukgungor, O. One-Dimensional Cobalt(II) and Zinc(II) Succinato Coordination Polymers with Nicotinamide: Synthesis, Structural, Spectroscopic, Fluorescent and Thermal Properties. *J. Inorg. Organomet. Polym.* **2009**, *19*, 342–347. [CrossRef]
47. Kansız, S.; Dege, N. Synthesis, crystallographic structure, DFT calculations and Hirshfeld surface analysis of a fumarate bridged Co(II) coordination polymer. *J. Mol. Struct.* **2018**, *1173*, 42–51. [CrossRef]
48. Masárová, P.; Mazúr, M.; Segl'a, P.; Moncol, J. Supramolecular assembled networks in crystal structure built up of copper(II) dipicolinates with pyrazine- and pyridinecarboxamides connected through hydrogen bonds. *Polyhedron* **2018**, *149*, 25–33. [CrossRef]
49. Shamuratov, E.B.; Batsanov, A.S.; Struchkov, Y.T.; Yunuskhodzhaev, A.N.; Dusmatov, A.F.; Sabirov, V.K. Structure of Co(II) and Cu(II) complexes with *N*-(hydroxymethyl)nicotinamide. *Russ. J. Coord. Chem.* **1990**, *16*, 1526–1528.
50. Shamuratov, E.B.; Batsanov, A.S.; Struchkov, Y.T.; Yunuskhodzhaev, A.N. Crystal structures of Ni(II) and Co(II) chloride and nitrate complexes with *N*-(hydroxymethyl)nicotinamide. *Russ. J. Coord. Chem.* **1991**, *17*, 1668–1675.
51. Chłopicki, S.; Ku, K.; Kalvins, I.; Adrianov, V.; Loza, E. Quaternary Nicotinamide Derivatives as Precursors for Release of MNA. European Patent Organization. Patent EP3505512 A1, 3 July 2019.
52. Alvarez, S.; Avnir, D.; Llunell, M.; Pinsky, M. Continuous symmetry maps and shape classification. The case of six-coordinated metal compounds. *New J. Chem.* **2002**, *26*, 996–1009. [CrossRef]
53. Alvarez, S. Relationships between Temperature, Magnetic Moment, and Continuous Symmetry Measures in Spin Crossover Complexes. *J. Am. Chem. Soc.* **2003**, *125*, 6795–6802. [CrossRef] [PubMed]
54. Alvarez, S. Distortion Pathways of Transition Metal Coordination Polyhedra Induced by Chelating Topology. *Chem. Rev.* **2015**, *115*, 13447–13483. [CrossRef]
55. Kershaw Cook, L.J.; Mohammed, R.; Sherborne, G.; Roberts, T.D.; Alvarez, S.; Halcrow, M.A. Spin state behavior of iron(II)/dipyrazolylpyridine complexes. New insights from crystallographic and solution measurements. *Coord. Chem. Rev.* **2015**, *289–290*, 2–12. [CrossRef]
56. Guionneau, P.; Marchivie, M.; Bravic, G.; Létard, J.-F.; Chasseau, D. Structural Aspects of Spin Crossover. Example of the $[Fe^{II}L_n(NCS)_2]$ Complexes. *Top. Curr. Chem.* **2004**, *234*, 97–128.
57. Halcrow, M.A. Structure:function relationships in molecular spin-crossover complexes. *Chem. Soc. Rev.* **2011**, *40*, 4119–4142. [CrossRef]
58. Bernstein, J.; Davis, R.E.; Shimoni, L.; Chang, N.-L. Patterns in Hydrogen Bonding: Functionality and Graph Set Analysis in Crystals. *Angew. Chem. Int. Ed.* **1995**, *34*, 1555–1573. [CrossRef]
59. Janiak, C. A critical account on π–π stacking in metal complexes with aromatic nitrogen-containing ligands. *J. Chem. Soc. Dalton Trans.* **2000**, *21*, 3885–3896. [CrossRef]
60. Dziewulska-Kułaczkowska, A.; Mazur, L.; Ferenc, W. Thermal, spectroscopic and structural studies of zinc(II) complex with nicotinamide. *J. Therm. Anal. Calorim.* **2009**, *96*, 255–260. [CrossRef]
61. Nakamoto, K. *Infrared and Raman Spectra of Inorganic and Coordination Compounds, Part B*; Wiley-VCH: Weinheim, Germany, 1997; pp. 384–387; ISBN 9780471743392.
62. Bando, Y.; Nagakura, S. The electronic structure and electronic spectrum of dichlorodipyridinecobalt(II). Charge-transfer band due to interaction between halide ion and cobalt(II) cation. *Inorg. Chem.* **1968**, *7*, 893–897. [CrossRef]
63. Lever, A.B.P. *Inorganic Electronic Spectroscopy*; Elsevier: Amsterdam, The Netherlands, 1984; ISBN 0-444-42389-3.
64. Martin, R.L. Natural Transition Orbitals. *J. Chem. Phys.* **2003**, *118*, 4775. [CrossRef]
65. Palii, A.; Tsukerblatt, B.; Klokishner, S.; Dunbar, K.R.; Clemente-Juan, J.M.; Coronado, E. Beyond the Spin Model: Exchange Coupling in Molecular Magnets with Unquenched Orbital Angular Momenta. *Chem. Soc. Rev.* **2011**, *40*, 3130–3156. [CrossRef] [PubMed]

66. Boča, R. Magnetic Parameters and Magnetic Functions in Mononuclear Complexes Beyond the Spin-Hamiltonian Formalism. In *Magnetic Functions Beyond the Spin-Hamiltonian. Structure and Bonding*; Springer: Berlin/Heidelberg, Germany, 2006; Volume 117. [CrossRef]
67. Juráková, J.; Dubnická Midlíková, J.; Hrubý, J.; Kliuikov, A.; Santana, V.T.; Pavlik, J.; Moncol', J.; Čižmár, E.; Orlita, M.; Mohelský, I.; et al. Pentacoordinate cobalt(II) single ion magnets with pendant alkyl chains: Shall we go for chloride or bromide? *Inorg. Chem. Front.* **2022**, *9*, 1179–1194. [CrossRef]
68. Singh, A.; Shrivastava, K. Optical-acoustic two-phonon relaxation in spin systems. *Phys. Status Solidi B* **1979**, *95*, 273–277. [CrossRef]
69. Shrivastava, K.N. Theory of Spin–Lattice Relaxation. *Phys. Status Solidi B* **1983**, *117*, 437–458. [CrossRef]
70. Juráková, J.; Šalitroš, I. Co(II) single-ion magnets: Synthesis, structure and magnetic properties. *Monatsh. Chem.* 2022; *in press*. [CrossRef] [PubMed]
71. Zahradníková, E.; Herchel, R.; Šalitroš, I.; Císařová, I.; Drahoš, B. Late first-row transition metal complexes of a 17-membered piperazine-based macrocyclic ligand: Structures and magnetism. *Dalton Trans.* **2020**, *419*, 9057–9069. [CrossRef] [PubMed]
72. Brachňaková, B.; Matejová, S.; Moncol', J.; Herchel, R.; Pavlik, J.; Moreno-Pineda, E.; Ruben, M.; Šalitroš, I. Stereochemistry of coordination polyhedral vs. single ion magnetism in penta- and hexacoordinated Co(II) complexes with tridentate rigid ligands. *Dalton Trans.* **2020**, *49*, 1249–1264. [CrossRef] [PubMed]
73. Wu, Y.; Tian, D.; Ferrando-Soria, J.; Cano, J.; Yin, L.; Ouyang, Z.; Wang, Z.; Luo, S.; Liu, X.; Pardo, E. Modulation of the magnetic anisotropy of octahedral cobalt(II) single-ion magnets by a fine-tuning of axial coordination microenvironment. *Inorg. Chem. Front.* **2019**, *6*, 848–856. [CrossRef]
74. Chilton, N.F.; Anderson, R.P.; Turner, L.D.; Soncini, A.; Murray, K.S. PHI: A powerful new program for the analysis of anisotropic monomeric and exchange-coupled polynuclear *d*- and *f*-block complexes. *J. Comput. Chem.* **2013**, *34*, 1164–1175. [CrossRef]
75. Neese, F. Software update: The ORCA program system, version 5.0. *WIREs Comput Mol Sci.* **2022**, *12*, e1606. [CrossRef]
76. Neese, F. Software update: The ORCA program system, version 4.0. *WIREs Comput. Mol. Sci.* **2018**, *8*, e1327. [CrossRef]
77. Vosko, S.H.; Wilk, L.; Nusair, M. Accurate spin-dependent electron liquid correlation energies for local spin density calculations: A critical analysis. *Can. J. Phys.* **1980**, *58*, 1200–1211. [CrossRef]
78. Lee, C.; Yang, W.; Parr, R.G. Development of the Cole-Salvetti correlation-energy formula into a functional of the electron density. *Phys. Rev. B.* **1988**, *37*, 785–789. [CrossRef]
79. Becke, A.D. Density-functional thermochemistry. III. The role of exact exchange. *J. Chem. Phys.* **1993**, *98*, 5648–5651. [CrossRef]
80. Adamo, C.; Barone, V. Toward reliable density functional methods without adjustable parameters: The PBE0 model. *J. Chem. Phys.* **1999**, *110*, 6158–6169. [CrossRef]
81. Tao, J.M.; Perdew, J.P.; Staroverov, V.N.; Scuseria, G.E. Climbing the density functional ladder: Nonempirical meta-generalized gradient approximation designed for molecules and solids. *Phys. Rev. Lett.* **2003**, *91*, 146401. [CrossRef]
82. Izsák, R.; Neese, F. An overlap fitted chain of spheres exchange method. *J. Chem. Phys.* **2011**, *135*, 144105. [CrossRef]
83. Weigend, F.; Ahlrichs, R. Balanced basis sets of split valence, triple zeta valence and quadruple zeta valence quality for H to Rn: Design and assessment of accuracy. *Phys. Chem. Chem. Phys.* **2005**, *7*, 3297–3305. [CrossRef]
84. Weigend, F.; Kattanneck, M.; Ahlrichs, R. Approximated electron repulsion integrals: Cholesky decomposition versus resolution of the identity methods. *J. Chem. Phys.* **2009**, *130*, 164106. [CrossRef]
85. Weigend, F. Accurate Coulomb-fitting basis sets for H-Rn. *Phys. Chem. Chem. Phys.* **2006**, *8*, 1057–1065. [CrossRef]
86. Grimme, S.; Brandenburg, J.G.; Bannwarth, C.; Hanse, A. Consistent structures and interactions by density functional theory with small atomic orbital basis sets. *J. Chem. Phys.* **2015**, *143*, 054107. [CrossRef] [PubMed]
87. Malmqvist, P.A.; Roos, B.O. The CASSCF state interaction method. *Chem. Phys. Lett.* **1989**, *155*, 189–194. [CrossRef]
88. Angeli, C.; Cimiraglia, R.; Malrieu, J.-P. *N*-electron valence state perturbation theory: A fast implementation of the strongly contracted variant. *Chem. Phys. Lett.* **2001**, *350*, 297–305. [CrossRef]
89. Angeli, C.; Cimiraglia, R.; Evangelisti, S.; Leininger, T.; Malrieu, J.-P. Introduction of *n*-electron valence states for multireference perturbation theory. *J. Chem. Phys.* **2001**, *114*, 10252–10264. [CrossRef]
90. Angeli, C.; Cimiraglia, R.; Malrieu, J.-P. *n*-electron valence state perturbation theory: A spinless formulation and an efficient implementation of the strongly contracted and of the partially contracted variants. *J. Chem. Phys.* **2002**, *117*, 9138–9153. [CrossRef]
91. Ganyushin, D.; Neese, F. A fully variational spin-orbit coupled complete active space self-consistent field approach: Application to electron paramagnetic resonance g-tensors. *J. Chem. Phys.* **2013**, *138*, 104113. [CrossRef]
92. Neese, F. Efficient and accurate approximations to the molecular spin-orbit coupling operator and their use in molecular g-tensor calculations. *J. Chem. Phys.* **2005**, *122*, 034107. [CrossRef] [PubMed]
93. Stoychev, G.L.; Auer, A.A.; Neese, F. Automatic Generation of Auxiliary Basis Sets. *J. Chem. Theory Comput.* **2017**, *13*, 554–562. [CrossRef] [PubMed]
94. Kožíšková, J.; Hahn, F.; Richter, J.; Kožíšek, J. Comparison of different absorption corrections on the model structure of tetrakis(μ_2-acetato)-diaqua-di-copper(II). *Acta Chim. Slovaca* **2016**, *9*, 136–140. [CrossRef]
95. Sheldrick, G.M. *SHELXT*—Integrated space-group and crystal-structure determination. *Acta Cryst.* **2015**, *A71*, 3–8. [CrossRef]
96. Coppens, P.; Hansen, N.K. Testing aspherical atom refinements on small-molecule data sets. *Acta Crystallogr.* **1978**, *A34*, 909–921.
97. Sheldrick, G.M. Crystal structure refinement with *SHELXL*. *Acta Cryst.* **2015**, *C71*, 3–8.

98. Bourhis, L.J.; Dolomanov, O.V.; Gildea, R.J.; Howard, J.A.K.; Puschmann, H. The anatomy of a comprehensive constrained, restrained refinement program for the modern computing environment—*Olex2* dissected. *Acta Cryst. A* **2015**, *A71*, 59–75. [CrossRef] [PubMed]
99. Kleemiss, F.; Dolomanov, O.V.; Bodensteiner, M.; Peyerimhoff, N.; Midgley, L.; Bourhis, L.J.; Genoni, A.; Malaspina, L.A.; Jayatilaka, D.; Spencer, J.L.; et al. Accurate crystal structures and chemical properties from NoSpherA2. *Chem. Sci.* **2021**, *12*, 1675–1692. [CrossRef] [PubMed]
100. Dolomanov, O.V.; Bourhis, L.J.; Gildea, R.J.; Howard, J.A.K.; Puschmann, H. *OLEX2*: A complete structure solution, refinement and analysis program. *J. Appl. Cryst.* **2009**, *42*, 339–341. [CrossRef]
101. Spackman, P.R.; Turner, M.J.; McKinnon, J.J.; Wolff, S.K.; Grimwood, D.J.; Jayalitaka, D.; Spackman, M.A. *CrystalExplorer*: A program for Hirshfeld surface analysis, visualization and quantitative analysis of molecular crystals. *J. Appl. Cryst.* **2021**, *54*, 1006–1011. [CrossRef]
102. Hirshfeld, F.L. Bonded-atom fragments for describing molecular charge densities. *Theor. Chim. Acta* **1977**, *44*, 129–138. [CrossRef]
103. Spackman, M.A.; Jayalitaka, D. Hirshfeld surface analysis. *CrystEngComm* **2009**, *11*, 19–32. [CrossRef]
104. Spackman, M.A.; McKinnon, J.J. Fingerprinting intermolecular interactions in molecular crystals. *CrystEngComm* **2002**, *4*, 378–392. [CrossRef]
105. Parkin, A.; Barr, G.; Dong, W.; Gilmore, C.J.; Jayalitaka, D.; McKinnon, J.J.; Spackman, M.A.; Wilson, C.C. Comparing entire crystal structures: Structural genetic fingerprinting. *CrystEngComm* **2007**, *9*, 648–652. [CrossRef]
106. Petříček, V.; Dušek, M.; Palatinus, L.J. *The Crystallographic Computing System*; Institute of Physics: Praha, Czech Republic, 2006.
107. Lunghi, A.; Totti, F.; Sanvito, S.; Sessoli, R. Intra-molecular origin of the spin-phonon coupling in slow-relaxing molecular magnets. *Chem. Sci.* **2017**, *8*, 6051–6059. [CrossRef] [PubMed]
108. Lunghi, A.; Sanvito, S. How do phonons relax molecular spins? *Sci. Adv.* **2019**, *5*, eaax7163. [CrossRef] [PubMed]
109. Garlatti, E.; Tesi, L.; Lunghi, A.; Atzori, M.; Voneshen, D.J.; Santini, P.; Sanvito, S.; Guidi, T.; Sessoli, R.; Carretta, S. Unveiling phonons in a molecular qubit with four-dimensional inelastic neutron scattering and density functional theory. *Nat. Commun.* **2020**, *11*, 1751. [CrossRef] [PubMed]

inorganics

MDPI

Article

Energy Levels in Pentacoordinate d^5 to d^9 Complexes

Ján Titiš, Cyril Rajnák and Roman Boča *

Department of Chemistry, Faculty of Natural Sciences, University of SS Cyril and Methodius, 917 01 Trnava, Slovakia
* Correspondence: roman.boca@ucm.sk

Abstract: Energy levels of pentacoordinate d^5 to d^9 complexes were evaluated according to the generalized crystal field theory at three levels of sophistication for two limiting cases of pentacoordination: trigonal bipyramid and tetragonal pyramid. The electronic crystal field terms involve the interelectron repulsion and the crystal field potential; crystal field multiplets account for the spin–orbit interaction; and magnetic energy levels involve the orbital– and spin–Zeeman interactions with the magnetic field. The crystal field terms are labelled according to the irreducible representations of point groups D_{3h} and C_{4v} using Mulliken notation. The crystal field multiplets are labelled with the Bethe notations for the respective double groups D'_3 and C'_4. The magnetic functions, such as the temperature dependence of the effective magnetic moment and the field dependence of the magnetization, are evaluated by employing the apparatus of statistical thermodynamics as derivatives of the field-dependent partition function. When appropriate, the formalism of the spin Hamiltonian is applied, giving rise to a set of magnetic parameters, such as the zero-field splitting *D* and *E*, magnetogyric ratio tensor, and temperature-independent paramagnetism. The data calculated using GCFT were compared with the *ab initio* calculations at the CASSCF+NEVPT2 level and those involving the spin–orbit interaction.

Keywords: electronic terms; spin–orbit multiplets; zero-field splitting; pentacoordinate complexes

Citation: Titiš, J.; Rajnák, C.; Boča, R. Energy Levels in Pentacoordinate d^5 to d^9 Complexes. *Inorganics* **2022**, *10*, 116. https://doi.org/10.3390/inorganics10080116

Academic Editor: Wolfgang Linert

Received: 15 July 2022
Accepted: 5 August 2022
Published: 12 August 2022

1. Introduction

A correct interpretation of electronic spectra for transition metal complexes (d-d transitions), magnetometric data (magnetic susceptibility and magnetization), and spectra of electron spin resonance requires appropriate theoretical support. A traditional approach is represented by the crystal field theory, which is well elaborated for octahedral complexes (O_h symmetry), even with tetragonal (trigonal) distortion (D_{4h}, D_{3d}) [1–5]. Analogously, tetrahedral patterns (T_d) and their distortion daughters to prolate and/or oblate bispheoids (D_{2d}) are also known. However, one is rather helpless when dealing with pentacoordination in its limiting cases represented by trigonal bipyramids (D_{3h}) and tetragonal pyramids (C_{4v}) and especially for intermediate geometries on the Berry rotation path (C_{2v}).

A target of the present work is to elucidate a comprehensive view of the crystal field terms and crystal field multiplets in the case of pentacoordinate d^5 to d^9 complexes. Whereas multielectron crystal field terms are labelled according to Mulliken notation (A, B, E, T), the involvement of the spin–orbit interaction requires a passage from common symmetry point groups to double groups; therefore, crystal field multiplets are labelled according to Bethe notation (Γ_1 to Γ_8).

Geometries belonging to point groups O_h, T_d, D_{4h}, or D_{2d} are mostly omitted hereafter; numerical computer-assisted treatment is necessary when ligands occupy arbitrary positions. This approach is slightly more complicated, involving algebra of complex numbers due to the occurrence of complex spherical harmonic functions fixing the ligand positions. The treatment used below is termed the Generalized Crystal Field Theory, as outlined elsewhere [6].

The eigenvalues of the model Hamiltonian refer to the energy levels at a given approximation. The eigenvectors bear all information about the symmetry of the wave function; therefore, they can be utilized to assign irreducible representations (IRs) either of the crystal field terms $|d^n : (LS); G : \Gamma\gamma a\rangle$ or the crystal field multiplets $|d^n : (J); G' : \Gamma'\gamma' a'\rangle$. The irreducible representation within point group G is Γ, γ (its component when IR is degenerate), and a (the branching (repetition) number). The same holds true for double group G'.

2. Results

The generalized crystal field theory (GCFT) applied below is fully described elsewhere, along with the closed formulae for the matrix elements of the involved operators in the basis set of the electronic atomic terms $|\Psi\rangle = |d^n : \nu LSM_L M_S\rangle$, where the apparatus of the irreducible tensor operators has been utilized [6,7]. (Here, the seniority number (ν) for the terms can distinguish between terms possessing the same set $\{LS\}$; the quantum numbers adopt their usual meaning [8]). These matrix elements refer to five operators:

(a) Interelectronic repulsion $\langle\Psi'|\hat{V}_{ee}|\Psi\rangle$ parametrized by the Racah parameters B_M and C_M;
(b) Crystal field potential $\langle\Psi'|\hat{V}_{cf}|\Psi\rangle$ depending upon crystal field poles (strengths) $F_k(R_L)$ of the k-th order (k = 4, 2) for each ligand (L) and its position;
(c) Spin–orbit interaction $\langle\Psi'|\hat{V}_{so}|\Psi\rangle$ depending upon the spin–orbit coupling constant (ξ_M);
(d) The orbital Zeeman term $\langle\Psi'|\hat{V}_{lB}(B)|\Psi\rangle$, which eventually involves the orbital reduction factors; and
(e) The spin Zeeman term $\langle\Psi'|\hat{V}_{sB}(B)|\Psi\rangle$, which contains the spin-only magnetogyric (g_e) factor.

The position of ligands (L) is arbitrary and fixed by the polar coordinates $\{\vartheta_L, \varphi_L\}$. The model Hamiltonian involves three important cases:

1. Diagonalization of (a) + (b) yields the energies of crystal field terms $|d^n : (LS); G : \Gamma\gamma a\rangle$, which span the IRs of point group G;
2. Diagonalization of (a) + (b) + (c) produces energies of the crystal field multiplets in the zero magnetic field $|d^n : (J); G' : \Gamma'\gamma' a'\rangle$, which span the IRs of double group G';
3. Diagonalization of (a) + (b) + (c) + (d) + (e) gives the magnetic energy levels in the applied magnetic field.

The energy levels of crystal field multiplets for the half-integral spin (S = 1/2, 3/2, 5/2) appear as Kramers doublets and remain doubly degenerate in the absence of a magnetic field. This is the case of high-spin Fe(III), Mn(II), Co(II), and Cu(II) complexes.

Traditional crystal field theory operates with a set of collective parameters, such as $10Dq = \Delta$, Ds, Dt, etc., and is useful for cases certain symmetry, such as O_h, T_d, and D_{4h}, all of which are derived from the crystal field poles (F_4(L) and eventually F_2(L)), e.g.,

- For O_h/D_{4h}: $10Dq = (10/6)F_4(xy)$;
- $Dt = (2/21)[F_4(xy) - F_4(z)]$;
- $Ds = (2/7)\,[F_2(xy) - F_2(z)]$; and
- For T_d: $10Dq = (20/27)F_4(xy)$.

The crystal field poles originate in the partitioning of the matrix elements of the crystal field potential into radial (R) and angular (A) parts in the polar coordinates $(R_K, \vartheta_K, \varphi_K)$.

$$\langle\Psi'(R,A)|\hat{V}_{cf}(R,A)|\Psi(R,A)\rangle = \langle\Psi'(R)|\hat{V}_{cf}(R)|\Psi(R)\rangle\cdot\langle\Psi'(A)|\hat{V}_{cf}(A)|\Psi(A)\rangle \quad (1)$$

The integration of the angular part yields some values (manually calculated for some cases, such as O_h symmetry). This part contains the spherical harmonic functions $Y_{k,q}(\vartheta_K, \varphi_K)$ for the positions of ligand K, and in general, it is a complex number. The radial part contains the metal–ligand distance (R_K) and defines the crystal field poles:

$$F_k(R_K) = \int_0^\infty R^*_{nl}(r)\frac{r^k_<}{r^{k+1}_>}R_{nl}(r)r^2\mathrm{d}r \approx \langle r^k\rangle / R_K^{k+1} \quad (2)$$

($k = 0, 2, 4$), where the integration runs over the electronic coordinates. The matrix elements of the crystal field operators can be expressed as:

$$\begin{aligned} &\langle l^n vLSM_LM_S|\hat{V}^{cf}|l^n v'L'S'M'_LM'_S\rangle \\ &= \delta_{S,S'}\delta_{M_S,M'_S}\sum_{k=0,2,4}^{2l}\sum_{q=-k}^{+k}\left[\left\langle l\left\|\mathbf{C}^k\right\|l\right\rangle\left(\tfrac{4\pi}{2k+1}\right)^{1/2}\sum_{K=1}^{N} z_K F_k(R_K)\cdot Y^*_{k,q}(\vartheta_K,\varphi_K)\right] \\ &\cdot\left[\left\langle l^n vLS\left\|\mathbf{U}^k\right\|l^n v'L'S'\right\rangle(-1)^{L-M_L}\cdot\begin{pmatrix} L & k & L' \\ -M_L & q & M'_L\end{pmatrix}\right] \end{aligned} \tag{3}$$

where for the reduced matrix elements $\left\langle l^n vLS\left\|\mathbf{U}^k\right\|l^n v'L'S'\right\rangle$, $\left\langle l\left\|\mathbf{C}^k\right\|l\right\rangle$ and the 3j symbols, closed formulae exist [6,7] and can be evaluated with a desktop computer.

In practice, the crystal field poles are not subject to evaluation; they are taken as parameters of the theory and depend on the quality of the ligand (halide, amine, phosphine, cyanide, carbonyl, etc.), as well as the quality and oxidation state of the central atom. For practical applications, the spectroscopic series is used according to the Δ-value [4]. The values of Δ can be deduced from the transitions observed in the electronic d-d spectra. Moreover, the Δ value can be estimated based on the empirically determined increments f_L for the ligands and g_M for the central atoms

$$\Delta = f_L \cdot g_M \tag{4}$$

However, the same ligand can produce different crystal field strengths depending on the actual metal–ligand distance (cf. Equation (2)). For instance, the -NCS^- group can be attached at distance R(Ni–N) = 2.2 or 2.0 Å. In the second case, it produces a much stronger crystal field.

For the hexacoordinate complexes, value of $F_4 = 5000$ cm^{-1} refers to $\Delta(O_h) = 8300$ cm^{-1}, which is a weak crystal field (appropriate for the halido ligand). Then, $F_4 = 15{,}000$ cm^{-1} is equivalent to $\Delta(O_h) = 25{,}000$ cm^{-1}, which refers to the strong crystal field (appropriate for cyanido or carbonyl ligands). For tetrahedral complexes, $F_4 = 5000$ (15,000) cm^{-1} refers to $\Delta(T_d) = 3700$ (11,100) cm^{-1}.

2.1. Crystal Field Terms

Figure 1 displays the relative energies of the crystal field terms (not to scale) for individual d^n configurations. These result from the GCFT calculations using the weak crystal field characterized by the crystal field poles $F_4(L) = 5000$ cm^{-1} for each ligand. For the tetragonal pyramid (C_{4v}), the angle L^a-M-L^e = 104 deg was maintained. The passage from the fully rotation group R_3 of a free atom to point group D_{3h} or C_{4v} is shown as the splitting of the atomic terms by the crystal field. The literature outlines the branching rules for such a reduction process [9].

The character tables for the point groups usually assign the dipole moment components to the IRs; these are useful in determining the selection rules for the excitation energies. For instance, within group D_{3h}, the direct product of IRs is $A'_1 \otimes A''_2 = A''_2 \in z$, meaning that the z component of the dipole moment is active in transition $A'_1 \to A''_2$, yielding the non-zero transition moment $\langle A'_1|\mu_z|A''_2\rangle \neq 0$ (orbitally allowed transition). On the contrary, $A'_1 \otimes A''_1 = A''_1 \notin x, y, z$ and thus transition $\langle A'_1|\mu_{x,y,z}|A''_1\rangle = 0$ are forbidden.

In addition to the energy levels, Figure 1 also shows the allowed/forbidden polarized electronic dipole transitions, which are displayed as solid/dashed arrows. These data can be compared with the observations of the electronic d-d spectra [10].

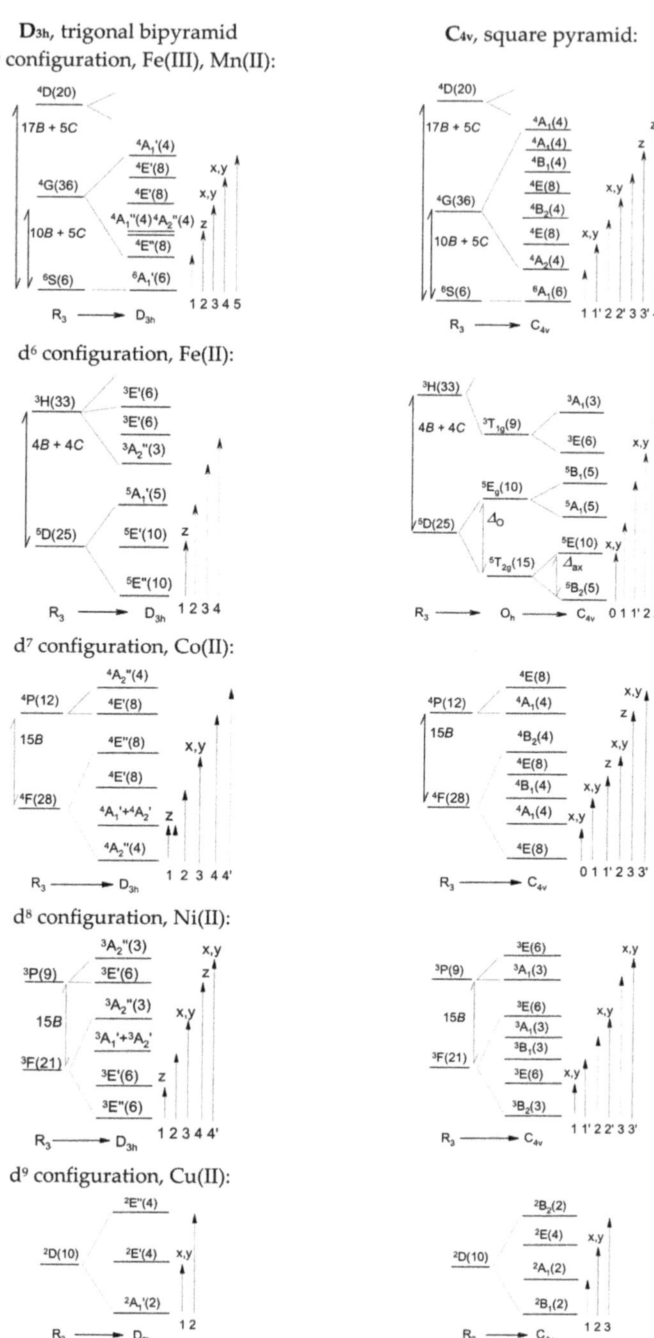

Figure 1. Crystal field terms for d^n configurations (energies not to scale). The electronic terms are labelled by exploiting the IRs of the point group, with the spin multiplicity as the superscript index and degeneracy in parentheses, e.g., $^4G(36)$. Dipole transitions: forbidden—dashed arrows, allowed—solid arrows.

2.2. Crystal Field Multiplets

The crystal field terms represent a starting point for the further precision of the energy levels: upon introduction of the spin–orbit interaction, the crystal field terms are further spit into a set of crystal field multiplets (energy levels in the zero magnetic field) [11,12]. The basis set and the resulting multiplets contain 256, 210, 120, 45 and 10 members for electron configurations d^5 through d^9, respectively. In this case, the energy levels are labelled using the Bethe notation for the IRs within double group G′. These symbols involve Γ_1 through Γ_8, and their degeneracy is shown parentheses, e.g., $\Gamma_4(2)$. (The IR tables for the double groups are useful for practical reasons).

The spin and the orbital parts of the wave function are assessed independently. For instance, in D_{3h} the level, ${}^6A_{1'}(6)$ transforms its spin according to $\{2\Gamma_4 + (\Gamma_5 + \Gamma_6)\}$. The orbital part matches $A_1 = \Gamma_1$. Finally, the spin–orbit wavefunction transforms according to the direct product $\{2\Gamma_4 + (\Gamma_5 + \Gamma_6)\}\otimes\Gamma_1$, and the result is $\{2\Gamma_4 + (\Gamma_5 + \Gamma_6)\}$. In this special case, the levels $(\Gamma_5 + \Gamma_6)$ form a complex conjugate pair that can be abbreviated as $\Gamma_{5,6}(2)$ or simply $\Gamma_5(2)$. To this end, upon passage from the D_{3h} to double group D'_3, the crystal field term ${}^6A_{1'}(6)$ is split into a set of $\{2\Gamma_4(2) + \Gamma_{5,6}(2)\}$ multiplets. However, this part of the theory says nothing about the relative energies of the final three Kramers doublets; these result from numerical calculations by GCFT.

The principal result of the CGTF calculations with spin–orbit coupling in the complete space spanned by d^n configurations is the spectrum of the crystal field multiplets. The lowest zero-field energy gaps are abbreviated as $\delta_1, \delta_2, \ldots$, provided that the energy of the ground multiplet (δ_0) is set to zero (Table 1). For the non-degenerate ground state (A or B type), the lowest multiplet gaps relate to the axial zero-field splitting parameter (D). For d^5-Fe(II) (and, analogously, d^5-Mn(III)), the sequence of the spin–orbit multiplet does not strictly follow D and $4D$ (there is a small difference (δ_a) around $4D$). For Cu(II), the ground electronic term is not split by the spin–orbit interaction; however, the spin–orbit multiplets are slightly influenced by the spin–orbit coupling. The concept of the D parameter is strictly related to the spin–Hamiltonian theory.

Table 1. Multiplet gaps (in cm−1) calculated by GCFT for pentacoordinate systems.

System	D_{3h}, Trigonal Bipyramid	C_{4v}, Square Pyramid
Fe(III), F_4 = 15,000 cm^{-1}	${}^6A_{1'}$: $\delta_1(2) = 3.68$ ($4D$), $\delta_2(2) = 5.53$ ($6D$)	${}^6A_1$: $\delta_1(2) = 1.50$ ($2D$), $\delta_2(2) = 4.51$ ($6D$)
Fe(III), F_4 = 5000 cm^{-1}	${}^6A_{1'}$: $\delta_1(2) = 0.29$ ($4D$), $\delta_2(2) = 0.44$ ($6D$)	${}^6A_1$: $\delta_1(2) = 0.10$ ($2D$), $\delta_2(2) = 0.31$ ($6D$)
Fe(II), F_4 = 5000 cm^{-1}	${}^5E''$: $\delta_1(2) = 85$, $\delta_2(2) = 180$, $\delta_{3,3'}(2) = \{270, 301\}$, $\delta_4(2) = 400$	${}^5B_2$: $\delta_1(2) = 0.31$ (D), $\delta_{2,2'}(2) = \{1.64, 1.87\}$ (~$4D$)
Co(II), F_4 = 5000 cm^{-1}	${}^4A_2''$: $\delta_1(2) = 84.7$ ($2D$)	${}^4E$: $\delta_1(2) = 220$, $\delta_2(2) = 389$, $\delta_3(2) = 697$
Ni(II), F_4 = 5000 cm^{-1}	${}^3E''$: $\delta_1(2) = 533$, $\delta_2(2) = 1191$	${}^3B_2$: $\delta_1(2) = 27.4$ (D)
Cu(II), F_4 = 5000 cm^{-1}	${}^2A_1$: $\Delta = 3020$	${}^2B_1$: $\Delta = 1691$

The effect of the spin-orbit interaction leading to the passage from the crystal-field terms to the crystal-field multiplets is depicted in Figure 2.

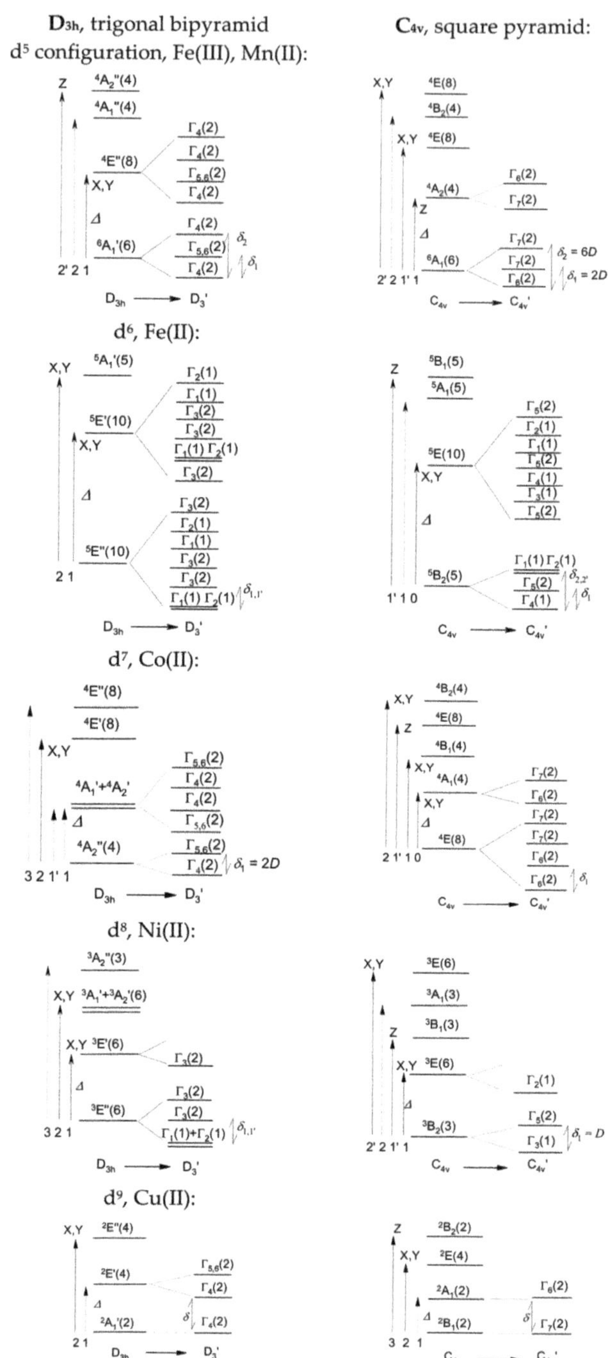

Figure 2. Crystal field multiplets for d^n configurations (energies not to scale). The crystal field multiplets are labelled by exploiting the IRs of the double group. Contributions to the Λ tensor: forbidden—dashed arrows, allowed—solid arrows.

2.3. Zero-Field Splitting

The concept of the spin Hamiltonian is a popular and very useful tool for interpretation of the spectra of electron paramagnetic resonance, as well as for analysis of DC magnetometric data. The key formulae of the spin Hamiltonian are based on consideration of only the spin kets $|S, M_S\rangle$ of non-degenerate ground term A or B. The second-order perturbation theory offers the Λ tensor in the following form:

$$\Lambda_{ab} = -\hbar^{-2} \sum_{K \neq 0} \frac{\langle 0|\hat{L}_a|K\rangle\langle K|\hat{L}_b|0\rangle}{E_K - E_0} \tag{5}$$

where K runs over all excited electronic terms, and the magnetic tensors are expressed as follows:

- the κ *tensor* (reduced, temperature−independent paramagnetic susceptibility tensor):

$$\kappa_{ab}^{\text{para}} = \mu_B^2 \Lambda_{ab} \tag{6}$$

- the g *tensor* (magnetogyric ratio tensor):

$$g_{ab} = g_e \delta_{ab} + 2\lambda\Lambda_{ab} \tag{7}$$

- the D *tensor* (spin–spin interaction tensor):

$$D_{ab} = \lambda^2 \Lambda_{ab} \tag{8}$$

This approximation fails in the case of orbital (pseudo) degeneracy. The matrix elements of the angular momentum $\langle 0|\hat{L}_a|K\rangle$ can be assessed by exploiting the symmetry of the ground and excited crystal field terms; the matrix element is non-zero only if the direct product $(\Gamma_0 \otimes \Gamma_K = \Gamma_{Lx,Ly,Lz} + \ldots)$ contains the irreducible representation of at least one component of the angular momentum. For instance, within group D_{3h}, $A_1' \otimes E'' = E'' \in L_{x,y}$ and the common character tables indicate that the result contains the irreducible representation of L_x and L_y.

The spin Hamiltonian parameters calculated via the GCFT are listed in Table 2. For Fe(III) and Mn(II), the ground electronic term ^{6}A does not allow transitions to excited terms with different spin multiplicities. Therefore, $D = 0$, $g_i = g_e$ in this approximation. In this case, the spin Hamiltonian formalism is insufficient, so $^6A_1 + {}^4T_1$ terms must be considered for the O_h symmetry [13].

Table 2. Calculated spin Hamiltonian parameters for pentacoordinate systems.

Center	D_{3h}, Trigonal Bipyramid			C_{4v}, Square Pyramid		
	D/hc/cm^{-1}	g_z, g_{xy}	$\chi_{TIP}/10^{-9}$ [SI] [1]	D/hc/cm^{-1}	g_z, g_{xy}	$\chi_{TIP}/10^{-9}$ [SI] [1]
Fe(II)	undefined	2.002, 2.096	2.05	0.50	2.111, 2.121	3.77
Co(II)	42.8	2.002, 2.500	6.34	undefined	2.218, 2.494	7.62
Ni(II)	undefined	2.002, 2.401	2.77	36.6	2.343, 2.576	5.16
Cu(II)	undefined	2.002, 2.672	1.76	undefined	2.901, 2.294	1.95

[1] SI unit for χ_{TIP} is m^3 mol^{-1}. Calculated according to the weak-field limit of F_4 = 5000 cm^{-1}.

The spin Hamiltonian is often presented in the following form:

$$\hat{H}^{\text{zfs}} = [D(\hat{S}_z^2 - \hat{S}^2/3) + E(\hat{S}_x^2 - \hat{S}_y^2)]\hbar^{-2} \tag{9}$$

where the D tensor is considered diagonal and traceless, yielding only two independent parameters: the axial zero-field splitting parameter (D) and the rhombic zero-field splitting parameter (E). This form is widely used for analysis of magnetometric and EPR data. According to convention, the rhombic part is minor: $|D| > 3E > 0$. D serves as a measure of zero-field splitting. This energy gap can also be measured also by FAR-infrared spectroscopy (FIRMS and FDMRS techniques), inelastic neutron scattering, calorimetry, etc. [14].

As mentioned above, the case of d^6-Fe(II) or d^6-Mn(III) is specific, as for C_{4v} geometry, the sequence of the spin–orbit multiplets differs depending on the exact multiplet splitting $\{0, \delta_1(2), \delta_{22'}(1 + 1)$ and the spin Hamiltonian formalism $\{0, D(2), 4D(2)\}$; the number in parentheses corresponds to the multiplicity. The ground crystal field term is $^5B_2$; the orbital and spin parts transform as $B_2 \rightarrow \Gamma_4$, $S = 2 \rightarrow \Gamma_1 + \Gamma_3 + \Gamma_4 + \Gamma_5$, and their direct product is $\Gamma_4 \otimes (\Gamma_1 + \Gamma_3 + \Gamma_4 + \Gamma_5) = \Gamma_1(1) + \Gamma_2(1) + \Gamma_4(1) + \Gamma_5(2)$. Only $\Gamma_5(2)$ is doubly degenerate, whereas the remaining multiplets are nondegenerate: $\Gamma_1(1)$, $\Gamma_2(1)$, and $\Gamma_4(1)$. The GCFT calculations for d^6-Fe(II) in the complete basis set of 210 kets obtains Γ_4 as the ground multiplet and the multiplet splitting $E(\Gamma_5) - E(\Gamma_4) = \delta_1 = 0.31$ cm^{-1}; $E(\Gamma_1) - E(\Gamma_4) = \delta_2 = 1.64$ cm^{-1}; $E(\Gamma_2) - E(\Gamma_4) = \delta_{2'} = 1.87$ cm^{-1}. This feature is reflected in the spectrum of electron paramagnetic resonance. Details about the symmetry rules are listed in Supplementary Information.

Table 3 shows a comparison of the d^n configurations from the viewpoint of the spin Hamiltonian formalism. This table is also enriched by data for d^1 to d^4 configurations, as well as data for the intermediate geometry with C_{2v} symmetry and $\tau_5 = 0.47$. Table 4 analogously summarizes data for the hexacoordinate complexes.

Table 3. Review of the SH formalism for pentacoordinate systems [1].

System	D_{3h}, Trigonal Bipyramid $\tau_5 = 1$	C_{2v}, Intermediate Geometry $\tau_5 = 0.47$	C_{4v}, Square Pyramid $\tau_5 = 0$
d^1, Ti(III)	$^2E''$, D—undefined	$^2A_2$, D—undefined	$^2B_2$, D—undefined
d^2, V(III)	$^3A_2''$, $D = 16$	$^3B_1$, $D = 19$, $E = 0.3$	^3E, D—undefined
d^3, Cr(III)	$^4E''$, D—undefined	$^4A_2$, $2D = -17$, $E = 0.6$	$^4B_2$, $2D = 6.2$
d^4, Mn(III)	$^5A_{1'}$, $D = 3.2$	$^5A_1$, $D = 3.1$, $E = 0.4$	$^5B_1$, $D = -2.9$
d^5, Fe(III)	$^6A_{1'}$, D—small	$^6A_1$, D—small	$^6A_1$, D—small
d^6, Fe(II)	$^5E''$, D—undefined	$^5A_2$, $D = -8$, $E = 1$	$^5B_2$, $D = 0.5$
d^7, Co(II)	$^4A_2''$, $2D = 85$	$^4B_1$, $2D = 102$, $E = 1$	^4E, D—undefined
d^8, Ni(II)	$^3E''$, D—undefined	$^3A_2$, $D = -105$, $E = 7$	$^3B_2$, $D = 37$
d^9, Cu(II)	$^2A_{1'}$, D—undefined	$^2A_1$, D—undefined	$^2B_1$, D—undefined

[1] The Addison structural parameter ($\tau_5 = (\beta - \alpha)/60$), where: $\beta > \alpha$ are the two greatest valence angles of the coordination center [15]. Data on D and E in cm^{-1} calculated with $F_4 = 5000$ cm^{-1}.

Table 4. Review of the SH formalism for hexacoordinate systems with tetragonal distortion [1].

System	D_{4h}, Compressed Bipyramid	O_h, Octahedron	D_{4h}, Elongated Bipyramid
d^1, Ti(III)	^2E_g, D—undefined	$^2T_{2g}$, D—undefined	$^2B_{2g}$, D—undefined
d^2, V(III)	$^3A_{2g}$, $D = 55$	$^3T_{1g}$, D—undefined	^3E_g, D—undefined
d^3, Cr(III)	$^4B_{1g}$, $2D = -0.7$	$^4A_{2g}$, $D = 0$	$^4B_{1g}$, $2D = 0.9$
d^4, Mn(III)	$^5A_{1g}$, $D = 2.5$	^5E_g(JT), D—undefined	$^5B_{1g}$, $D = -2.8$
d^5, Fe(III)	$^6A_{1g}$, D—small	$^6A_{1g}$, D—small	$^6A_{1g}$, D—small
d^6, Fe(II)	^5E_g, D—undefined	$^5T_{2g}$, D—undefined	$^5B_{2g}$, $D = 16$
d^7, Co(II)	$^4A_{2g}$, $2D = 300$	$^4T_{1g}$, D—undefined	^4E_g, D—undefined
d^8, Ni(II)	$^3B_{1g}$, $D = -4.3$	$^3A_{2g}$, $D = 0$	$^3B_{1g}$, $D = 5.3$
d^9, Cu(II)	$^2A_{1g}$, D—undefined	^2E_g(JT), D—undefined	$^2B_{1g}$, D—undefined

[1] JT points to a strong Jahn–Teller effect, owing to which a spontaneous symmetry descent proceeds. Data on D in cm^{-1} calculated with $F_4 = 5000$ cm^{-1}.

2.4. DC Magnetic Functions

The magnetic energy levels $\varepsilon_{i,a}(B_m)$ result from the diagonalization of the interaction matrix ((a) + (b) + (c) + (d) + (e)), which includes interelectronic repulsion, crystal field potential, spin–orbit coupling, and orbital and Zeeman terms in the applied magnetic field. Statistical thermodynamics offers formulae for magnetization and magnetic susceptibility when the partition function is evaluated for three reference fields: $B_m = B_0 - \delta, B_0, B_0 + \delta$ (allowing numerical derivatives):

$$Z_a(T, B_m) = \sum_i \exp[\varepsilon_{i,a}(B_m)/k_B T] \tag{10}$$

Hence, the molar magnetization is:

$$(M_{mol})_a = \frac{RT}{Z_a}\left(\frac{\partial Z_a}{\partial B_a}\right)_T \tag{11}$$

The molar magnetic susceptibility is expressed as:

$$(\chi_{mol})_{ab} = \mu_0\left(\frac{\partial (M_{mol})_a}{\partial B_b}\right)_T \tag{12}$$

where the physical constants adopt their usual meaning. The index a refers either to the Cartesian coordinates $\{x, y, z\}$ or to the grid point over a sphere along which the magnetic field is aligned, which is used to obtain the powder-sample average. Therefore, the magnetic susceptibility and magnetization are functions of discrete parameters (atomic parameters B_M, C_M, and ξ_M; ligand positions θ_L and φ_L; crystal field poles F_4(L); and eventually F_2(L)), as well as the continuous parameters, such as the reference field (B_m) and temperature (T).

The modelling of the magnetization and susceptibility for pentacoordinate d^n systems is presented in Figures 3 and 4. A counterpart of these graphs for the tetragonally distorted octahedral systems can be found elsewhere [16]. In the case of zero-field splitting with an orbitally non-degenerate ground term, the effective magnetic moment in the high-temperature limit of 300 K remains almost linear with zero slope; at low temperature, it is reduced. This is the case of d^5-D_{3h}, d^5-C_{4v}, d^6-C_{4v}, d^7-D_{3h}, and d^8-C_{4v}. For d^9-D_{3h} and d^9-C_{4v}, zero-field splitting is absent, so these systems follow the Curie law. The magnetization saturates to the value of $M_1 = M_{mol}/(N_A\mu_B) = g_{av}S$ when the zero-field splitting is small. This is the case of d^5-D_{3h}, d^5-C_{4v}, d^6-C_{4v}, d^9-D_{3h}, and d^9-C_{4v}; exceptions are d^7-D_{3h} and d^8-C_{4v}, with large zero-field splitting D parameters.

Systems with E-type orbitally doubly degenerate ground terms, such as d^6-D_{3h}, d^7-C_{4v}, and d^8-D_{3h} behave differently. The effective magnetic moment is enlarged, and it passes through a round maximum. The magnetization is also suppressed and does not reach saturation until $B = 10$ T.

A positive slope of the effective magnetic moment reflects the effect of the low-lying excited electronic terms mixed considerably with the ground term via the spin–orbit interaction. This results in temperature-independent paramagnetism, $\chi_{TIP} > 0$. This term, along with the underlying diamagnetism ($\chi_{dia} < 0$), need be subtracted from the measured temperature dependence of the magnetic susceptibility. With respect to the underlying diamagnetism, a method of additive Pascal constants is useful and frequently utilized. However, for temperature-independent paramagnetism, the amount of information is considerably limited [6,7].

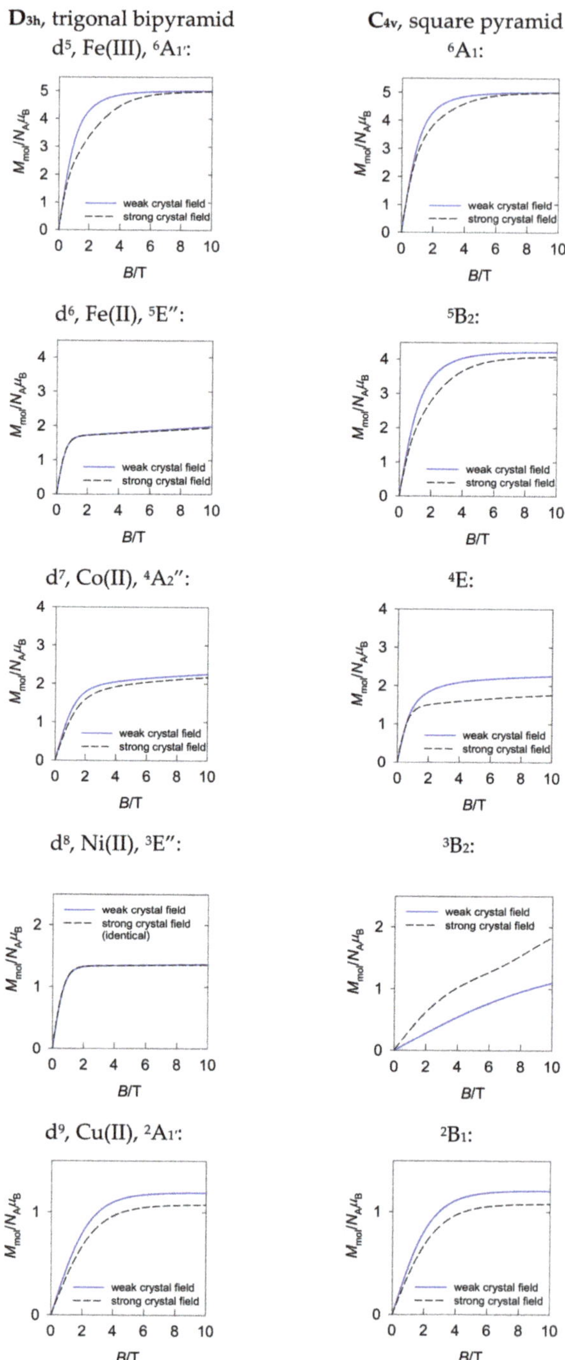

Figure 3. Magnetization functions at $T = 2.0$ K calculated by GCFT for a weak (strong) crystal field with $F_4 = 5000$ (15,000) cm^{-1}.

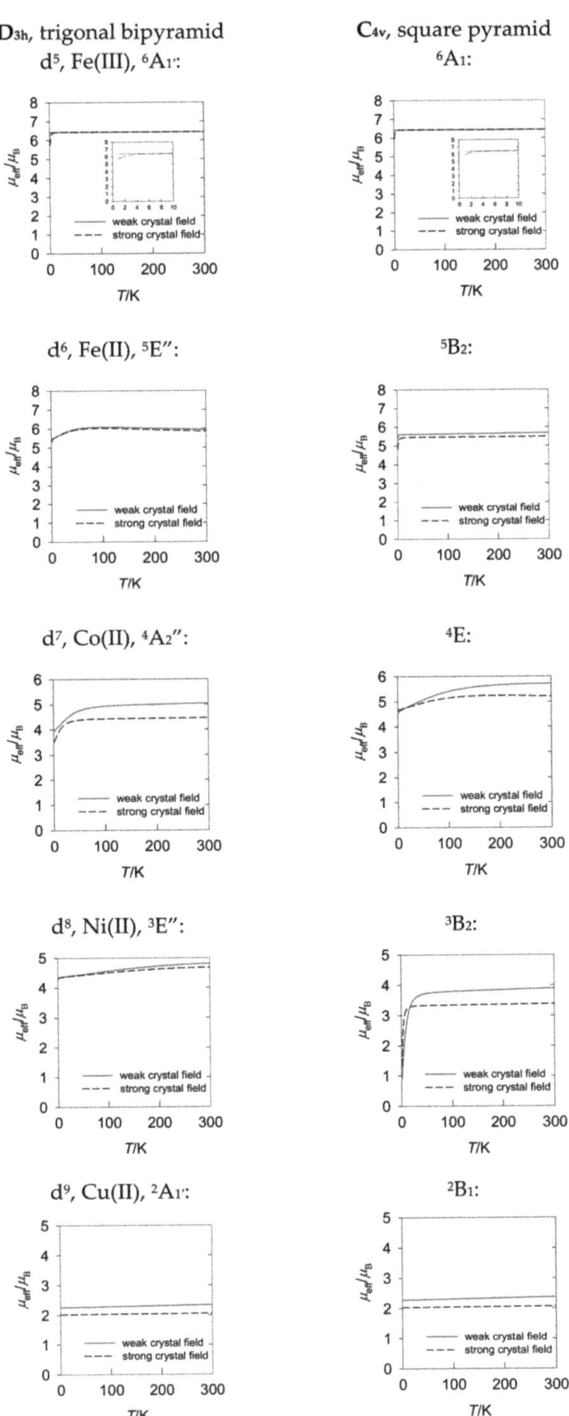

Figure 4. DC susceptibility functions at B = 0.1 T (Equation (12)) converted to the effective magnetic moments as calculated by GCFT for a weak (strong) crystal field with F_4 = 5000 (15,000) cm^{-1}.

2.5. AC Magnetic Susceptibility

In the oscillating magnetic fields (usually with a low amplitude of B_{AC} = 0.3 mT and a frequency range of $f = 10^{-2}$–10^5 Hz), the measured magnetic moment of the specimen has two components: in-phase and out-of-phase. This is easily transformed into two components of AC susceptibility: χ' (dispersion) and χ'' (absorption). The absorption component is a measure of the resistivity of the sample used to alter its magnetization; it provides information about the relaxation time, which is a function of temperature, frequency (f), and the external applied field (B_{DC}). The relaxation time can be inferred from the position of the maximum at the out-of-phase susceptibility (f''_{max}) with the following formula: $\tau = 1/(2\pi f''_{max})$. It has been reported that the sample can exhibit two or more relaxation channels and that their absorption curves can overlap or merge to form a shoulder. The whole AC susceptibility can be fitted by exploiting the generalized Debye equation [17,18]:

$$\chi(\omega) = \chi_S + \sum_k^K \frac{\chi_k - \chi_{k-1}}{1 + (i\omega\tau_k)^{1-\alpha_k}} \tag{13}$$

where K is the number of relaxation channels, χ_S is the common adiabatic susceptibility (high-frequency limit), χ_k is the thermal susceptibilities, α_k is the distribution parameters, τ_k is the relaxation times, and the circular frequency is $\omega = 2\pi f$. This complex equation can be decomposed into a real and imaginary part.

The slow magnetic relaxation includes several mechanisms that can be collected to a single equation for the reciprocal relaxation time:

$$\tau^{-1} = \tau_0^{-1}\exp(-U_{eff}/k_BT) + C_RT^n + C_{pb}T^l + AB^mT + D_1/(D_2 + B^2) \tag{14}$$

The first term describes the thermally activated Orbach process, which is associated with the height of the barrier to spin reversal (U_{eff}); the second is the Raman term, with the temperature exponent typically n = 5–9; next is the phonon bottleneck term, with $l \sim 2$; the fourth term describes the direct relaxation process, with m = 2–4; the last term refers to the quantum tunnelling of magnetization throughout the barrier to spin reversal. The reciprocating thermal behavior was recently registered with a term analogous to the phonon bottleneck but a negative temperature exponent ($l \sim -1$) [18].

The effectiveness of the slow magnetic relaxation is, as a rule, evaluated by the value of U_{eff} (when the Orbach process applies). It is assumed that it is related to the axial zero-field splitting parameter (D), which must be negative, and the molecular spin (S) [19]:

$$U_{eff} = |D|(S^2 - 1/4), \tag{15}$$

which holds true for Kramers systems with half-integral spin (e.g., $S = 3/2$ for Co^{II}); for non-Kramers systems with an integer spin, the factor $\frac{1}{4}$ is dropped (e.g., $S = 1$, for Ni^{II}). It is common practice for the U_{eff} and the pre-exponential factor (τ_0) to be subtracted using the Arrhenius-like plot ln(τ) vs. $1/T$ (Figure 5-left): a few high-temperature points are fitted by the straight line, tangential of which refers to U_{eff}. However, "high-temperature points" refer to the highest temperature among the data considered in our analysis, so there still could be points yielding a higher tangential and thus U_{eff}. A preferred approach involves plotting ln(τ) vs. ln(T), where the temperature exponent recovering the high-temperature data refers to the slope (Figure 5, right). When the temperature coefficient is $n > 9$, instead of the Raman process the Orbach process is applied.

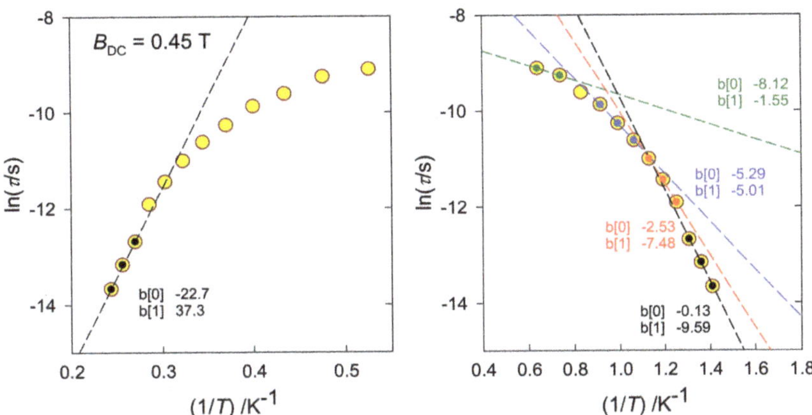

Figure 5. Contributions to the relaxation time. (**Left**): Orbach process (high-temperature, black, U_{eff}/k_B = 37 K). (**Right**): direct process (low-temperature, green, $m \sim 1$, Raman process; intermediate temperature, blue, $n > 5$; red $n < 9$). Data adapted from [20] for a mononuclear Fe^{III} complex. Straight-line formula: y = b [0] + b [1]x.

For high-spin Co(II) complexes with $S = 3/2$, eqn (15) implies a relationship of $U = 2|D|$. A collection of experimental data for a series of tetracoordinate Co^{II} complexes is shown in Figure 6 based on the analysis of higher-temperature, high-frequency relaxation data in terms of the Orbach process. Evidently, a correlation of U vs. $2|D|$ fails. D is a field-independent quantity, whereas the extracted value of U depends upon the applied magnetic field. A positive value of D contradicts the D-U paradigm; however, SIMs behavior can occur (the Raman mechanism is likely the leading term). With increased barrier to spin reversal (U), the extrapolated relaxation time (τ_0) is shortened, irrespective of the sign of the D parameter. A violation of the D-U paradigm has been discussed elsewhere with consideration of anharmonicity contributions [21].

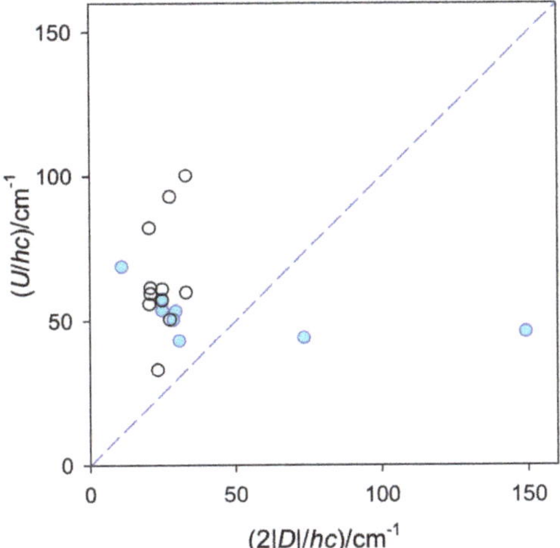

Figure 6. Collection of relaxation data for tetracoordinate Co(II) complexes, $S = 3/2$. Full points for $D < 0$, empty for $D > 0$. Dashed line—a hypothetical D-U paradigm. $(1/k_B) = 1.439\ \mathrm{K/cm^{-1}}$.

3. Discussion

The GCFT approach enables fast and "continuous" mapping of the energy levels, such as electronic terms and the spin–orbit multiplets: one is free to changing the ligand positions $\{\theta_L, \varphi_L\}$ from regular coordination polyhedra to distorted polyhedra and to alter the crystal field poles $F_4(L)$ and, eventually, $F_2(L)$. On the contrary, the modern ab initio calculations provide high-quality data on energy levels but only for the unique geometry of the complex under investigation. Therefore, it is interesting to utilize and compare both approaches.

Ab initio calculations have been performed using ORCA software [22] with respect to the experimental geometry of the complexes resulting from X-ray structural analysis (the corresponding cif files are deposited in the Cambridge Crystallographic Data Centre). The relativistic effects were included in the calculations with a second-order Douglas–Kroll–Hess (DKH) procedure. An extended basis set TZVP of Gaussian functions was used, e.g., BS1 = [17s11p7d1f] and BS2 = [17s12p7d2f1g] for Ni(II). The calculations were based on state-average complete active-space self-consistent field (SA-CASSCF) wave functions. The active space of the CASSCF calculations comprised eight electrons in five metal-based d-orbitals. The state-averaged approach was used, whereby all 10 triplet and 15 singlet states were equally weighted. The spin–orbit effects were included according to quasi-degenerate perturbation theory, whereby the spin–orbit coupling operator (SOMF) was approximated according to the Breit–Pauli form. The electronic terms were evaluated at the CASSCF + NEVPT2 level, and the multiplets by considering the spin–orbit interaction (Table 5). Effective Hamiltonian was used to evaluate the spin Hamiltonian parameters.

Table 5. Energy levels for representative Ni(II) complexes calculated by ab initio method [1].

System	Donor Set, Symmetry	SHAPE Index	Reported D/cm^{-1}	Δ/cm^{-1} NEVPT2	δ/cm^{-1} SOC	D/cm^{-1} E/D
$[Ni(Me_4cyclam)N_3]ClO_4$	NiN_4N' C_{2v}	vOC-5: 0.61 SPY-5: 0.72	+20, mag +21, EPR	a^3A: 0, b^3A: 5777	$a^3A \rightarrow$ 0, 23, 27	25 0.08
$[Ni(i\text{Prtacn})Cl_2]$	NiN_3Cl_2 $\tau_5 = 0.42$	vOC-5: 2.91 SPY-5: 3.53	+14.3, mag +15.7, EPR +15.9, FDMRS	a^3A: 0, b^3A: 5815	$a^3A \rightarrow$ 0, 17, 24	20 0.18
$[Ni(i\text{Prtacn})Br_2]$	NiN_3Br_2 $\tau_5 = 0.40$	vOC-5: 2.83 SPY-5: 3.72	+11.0, mag +13.9, EPR +13.8, FDMRS	a^3A: 0, b^3A: 6035	$a^3A \rightarrow$ 0, 13, 21	17 0.21
$[Ni(i\text{Prtacn})(NCS)_2]$	$NiN_3N'_2$ $\tau_5 = 0.44$	vOC-5: 2.56 SPY-5: 2.72	+13.8, mag +16.1, EPR +15.9, FDMRS	a^3A: 0, b^3A: 5846	$a^3A \rightarrow$ 0, 15, 27	25 0.28
$[Ni(Me_6tren)Cl]ClO_4$	$NiN_3N'Cl$ C_{3v}	TBPY-5: 0.61	−179, EPR −110, FIRMS −205, calc	a^3E: 0, 26, b^3E: 5775, 5782	$a^3E \rightarrow$ 0, 5, 512, 538, 1162, 1162	-
$[Ni(Me_6tren)Br]Br$	$NiN_3N'Br$ C_{3v}	TBPY-5: 0.92	−147, calc	a^3E: 0, 5 b^3E: 7100,7106	$a^3E \rightarrow$ 0, 3, 545, 550, 1196, 1196	-
$[Ni(MDABCO)_2Cl_3]ClO_4$	$NiCl_3N_2$, D_{3h}, $\tau_5 = 1$	TBPY-5: 0.14	−311, mag −535, EPR	a^3E: 0, 342, b^3E: 6517, 6874	$a^3E \rightarrow$ 0, 0.3, 397, 735, 1243,1252	-

[1] Abbreviations: mag—magnetometry, EPR—(high-field/high-frequency) electron paramagnetic resonance, FDMRS—frequency-domain magnetic resonance spectroscopy, FIRMS—far infrared magnetic spectroscopy; SHAPE index (consistency with the regular coordination polyhedron) [23]: TBPY—trigonal bipyramid, SPY—square pyramid, vOC—vacant octahedron; electronic terms a^3A—ground-spin triplet, b^3A—first excited spin triplet, a^3E—ground-orbital doublet, b^3E—first excited orbital doublet; ground spin-orbit multiplet at zero; δ—separation of the lowest multiplets: three from term a^3A (consistent with the spin Hamiltonian formalism), six from a^3E (beyond the spin Hamiltonian formalism). Structural and experimental data according to Refs. [24–28]. *Ab initio* calculations were carried out according to the same protocol.

The *ab initio* calculations refer to in silico state, i.e., intermolecular interactions and other solid-state effects are ignored. This is not the case for experimental magnetometric or spectroscopic data, which could be influenced by the environment. The *ab initio* data, in general, are consistent with those obtained by experimental techniques.

When comparing the CGTF calculations with *ab initio* calculations, calculated transition energies can be assessed. With a proper set of crystal field poles, the CGTF can reproduce first allowed transitions; however, the electronic spectrum, has a smaller width with respect to *ab initio* data.

An extended set of similar pentacoordinate Ni(II) complexes based on the fixed skeleton of a pentadentate Schiff base (Figure 7) was investigated by magnetometry and *ab initio* calculations with respect to the experimental geometry; these are listed in Table 6.

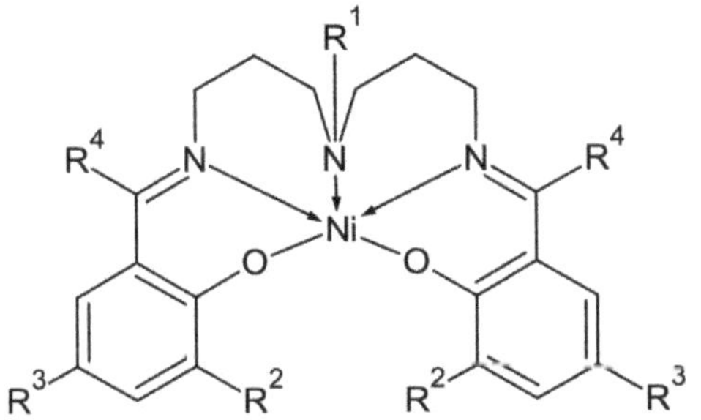

Figure 7. Schematic representations of pentacoordinate Ni(II) complexes. **1**: R^1 = R^3= $-CH_3$, R^2 = $-C(CH_3)_3$, R^4 = H; **2**: R^1 = $-CH_3$, R^2 = R^4 = H, R^3= Br; **3**: R^1 = $-CH_3$, R^2 = R^4 = H, R^3= I; **4**: R^1 = $-CH_3$, R^2 = R^3 = $-C(CH_3)_3$, R^4 = H; **5**: R^1 = $-CH_3$, R^2 = R^3 = R^4 = H; **6**: R^1 = R^3 = R^3 = H, R^4 = $-CH_3$.

Table 6. Magnetometric and *ab initio* data for a set of pentadentate Ni(II) complexes comprising Schiff base ligands [1].

System	Addison Index τ_5	SHAPE Index	D/cm^{-1} Magnetometry	δ/cm^{-1} SOC	D/cm^{-1} Calculations	E/D
1	0.52	TBPY-5: 1.583, SPY-5: 1.664	−45.1	0, 4, 51	−49.1	0.045
2	0.62	TBPY-5: 0.899, SPY-5: 2.589	−64.0	0, 5, 55	−52.6	0.044
3	0.62	TBPY-5: 0.903, SPY-5: 2.598	−60.2	0, 5, 54	−52.0	0.047
4	0.47	vOC-5: 1.881, SPY-5: 1.508	−45.1	0, 6. 44	−41.7	0.068
5	0.60	TBPY-5: 1.050, SPY-5: 2.072	−49.3	0, 5, 66	−63.2	0.037
6	0.26	vOC-5: 0.909, SPY-5: 0.913	−12.7	0, 8, 30	−25.3	0.169

[1] Data from ref. [29].

The experimentally reported and calculated D values cover a broad interval of positive and negative values over a wide range of the τ_5 parameters. These were used to plot D vs. τ_5, which can be termed the *second magnetostructural D-correlation* for Ni(II) complexes (MSDC). (The first magnetostructural D correlation for hexacoordinate Ni(II) complexes is outlined elsewhere [30].) The MSDC can be approximated by a straight line (Figure 8) when the τ_5 parameter guarantees that the ground electronic term is not orbitally quasi-degenerate (the energy gap $\Delta > 2000$ cm^{-1}). In the opposite case, the calculated D values tend to diverge. The value of the D parameter switches between positive and negative values at $\tau_5 \sim 0.2$–0.3. Furthermore, the E parameter plays a role that has not be considered so far.

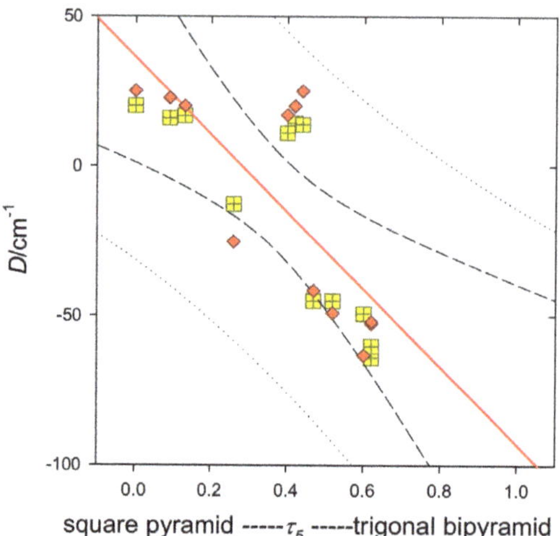

Figure 8. Dependence of the D parameter in pentacoordinate Ni(II) complexes on the distortion parameter (τ_5). Yellow squares—magnetometric data; red pica—*ab initio* calculations; solid—correlation line, dashed—confidence intervals, dotted—prediction intervals.

4. Conclusions

Experimental data on magnetic susceptibility, magnetization, and electron paramagnetic resonance require an appropriate model in order be analyzed correctly. For some shapes of coordination polyhedra, such as octahedron O_h, tetragonal bipyramid D_{4h}, trigonal antiprism D_{3d}, tetrahedron T_d, and bispehoid D_{2d}, the crystal field theory offers such a support, and the spin Hamiltonian formalism defines relationships for the set of magnetic parameters (D, E, g_x, g_y, g_z, χ_{TIP}). A dearth in the literature with respect to pentacoordinate systems, such as the trigonal bipyramid D_{3h} and tetragonal pyramid C_{4v} symmetry, is filled by this publication. The working tool is the generalized crystal field theory in the form of its fully numerical, computer-assisted tool [31]. The advantage of this approach is that the positions of the ligands can be arbitrary, making it applicable to any geometry of the chromophore and any ligands. Only the set of Racah parameters of the interelectronic repulsion (B_M and C_M), the spin–orbit coupling constant (ξ_M), polar angles (or Cartesian coordinates) of each ligand $\{\theta_L, \varphi_L\}$, the crystal field poles F_4(L) and, eventually, F_2(L) are required. This method enables evaluation of the energies of the multielectron crystal field terms, spin–orbit crystal field multiplets, and the magnetic energy levels at the applied magnetic field. Then, the magnetic susceptibility and magnetization can be evaluated as functions of the temperature field via derivatives of the partition function. The eigenvectors provide complete information about the symmetry and can be used to automatically label terms/multiplets.

Supplementary Materials: The following supporting information can be downloaded at: https://www.mdpi.com/article/10.3390/inorganics10080116/s1, Tables S1–S5: Reduction and selection rules for d^5–d^9 configurations; Table S6: Reduction of the (2S + 1) states; Table S7: Decomposition of the direct product.

Author Contributions: All three authors contributed equally to the individual parts of the manuscript. "Conceptualization, R.B.; methodology, R.B.; software, R.B.; validation, J.T and C.R.; formal analysis, J.T; investigation, J.T.; resources, C.R.; data curation, C.R.; writing—original draft preparation, R.B.; writing—review and editing, R.B.; visualization, C.R.; supervision, R.B.; project administration, J.T.; funding acquisition, C.R. All authors have read and agreed to the published version of the manuscript.

Funding: Slovak Research and Development Agency (APVV 18-0016, APVV 19-0087 and VEGA 1/0086/21, VEGA 1/0191/22) are acknowledged for their financial support.

Data Availability Statement: The experimental susceptibility and magnetization data and protocols of the CGTF and *ab initio* calculations are available from authors upon reasonable request.

Conflicts of Interest: The authors declare no conflict of interest.

References

1. Griffith, J.S. *The Theory of Transition Metal Ions*; University Press: Cambridge, UK, 1964.
2. Ballhausen, C.L. *Introduction to Ligand Field Theory*; McGraw−Hill: New York, NY, USA, 1962.
3. Figgis, B.N. *Introduction to Ligand Fields*; John Wiley & Sons Ltd.: London, UK; New York, NY, USA, 1966; p. 362, ISBN 0470258802.
4. Schäfer, H.L.; Gliemann, G. *Basic Principles of Ligand Field Theory*; John Wiley & Sons Ltd.: London, UK; New York, NY, USA, 1969; p. 550, ISBN 0471761001.
5. König, E.; Kremer, S. *Ligand Field Energy Diagrams*; Plenum Press: New York, NY, USA; London, UK, 1977; p. 454, ISBN 0306309467.
6. Boča, R. Magnetic Parameters and Magnetic Functions in Mononuclear Complexes Beyond the Spin-Hamiltonian Formalism. In *Magnetic Functions Beyond the Spin-Hamiltonian*; Structure and Bonding 117; Mingos, D.M.P., Ed.; Springer: Berlin/Heidelberg, Germany, 2006; p. 273, ISBN 3-540-26079-X.
7. Boča, R. *A Handbook of Magnetochemical Formulae*; Elsevier: Amsterdam, The Netherlands, 2012; p. 1010, ISBN 012416014X.
8. Slater, J.C. *Quantum Theory of Atomic Structure*; McGraw−Hill: New York, NY, USA, 1960; Volume 1–2.
9. Salthouse, J.A.; Ware, M.J. *Point Group Character Tables and Related Data*; University Press: Cambridge, UK, 1972; p. 93, ISBN 0521081394.
10. Lever, A.B.P. *Inorganic Electronic Spectroscopy*, 2nd ed.; Elsevier: Amsterdam, The Netherlands, 1984; p. 863, ISBN 0444423893.
11. Sugano, S.; Tanabe, Y.; Kanimura, H. *Multiplets of Transition Metal Ions in Crystals*; Academic Press: New York, NY, USA, 1970; p. 348, ISBN 0126760500.
12. Abragam, A.; Bleaney, B. *Electron Paramagnetic Resonance of Transition Ions*; Clarendon Press: Oxford, UK, 1970; p. 726, ISBN 0199651523.
13. Weissbluth, M. The Physics of Hemoglobin. In *Structure and Bonding 2*; Jorgensen, C.K., Neilands, J.B., Nyholm, R.S., Reinen, D., Williams, R.J.P., Eds.; Springer: Berlin/Heidelberg, Germany, 1967; p. 125, ISBN 978-3-540-03989-1.
14. Boča, R. Zero-field splitting in metal complexes. *Coord. Chem. Rev.* **2004**, *248*, 757–815. [CrossRef]
15. Addison, A.W.; Rao, N.T.; Reedijk, J.; van Rijn, J.; Verschoor, G.C. Synthesis, structure, and spectroscopic properties of copper(II) compounds containing nitrogen–sulphur donor ligands; the crystal and molecular structure of aqua(1,7-bis(N-methylbenzimidazol-2′-yl)-2,6-dithiaheptane)copper(II) perchlorate. *J. Chem. Soc. Dalton Trans.* **1984**, *7*, 1349–1356. [CrossRef]
16. König, E.; Kremer, S. *Magnetism Diagrams for Transition Metal Ions*; Plenum Press: New York, NY, USA, 1979; p. 555, ISBN 0306402602.
17. Boča, R.; Rajnák, C. Unexpected behavior of single ion magnets. *Coord. Chem. Rev.* **2021**, *430*, 213657. [CrossRef]
18. Rajnák, C.; Boča, R. Reciprocating thermal behavior in the family of single ion magnets. *Coord. Chem. Rev.* **2021**, *436*, 213808. [CrossRef]
19. Gatteschi, D.; Sessoli, R.; Villain, J. *Molecular Nanomagnets*; Oxford University Press: Oxford, UK, 2006; p. 408, ISBN 0199602263.
20. Rajnák, C.; Titiš, J.; Moncol', J.; Renz, F.; Boča, R. Slow Magnetic Relaxation in a High-spin Pentacoordinate Fe(III) Complex. *Chem. Commun.* **2019**, *55*, 13868–13871. [CrossRef] [PubMed]
21. Lunghi, A.; Totti, F.; Sessoli, R.; Sanvito, S. The role of anharmonic phonons in under-barrier spin relaxation of single molecule magnets. *Nat. Commun.* **2017**, *8*, 14620. [CrossRef] [PubMed]
22. Neese, F. ORCA—An Ab Initio, Density Functional and Semi-Empirical Program Package, Version 4.2.1. *WIREs Comput. Mol. Sci.* **2018**, *8*, e1327. [CrossRef]
23. Llunell, M.; Casanova, D.; Cirera, J.; Alemany, P.; Alvarez, S. *Program SHAPE*; Ver. 2.1; University of Barcelona: Barcelona, Spain, 2013.
24. Mariott, K.E.R.; Bhaskaran, L.; Wilson, C.; Medarde, M.; Ochsebbein, S.T.; Hill, S.; Murrie, M. Pushing the limits of magnetic anisotropy in trigonal bipyramidal Ni(II). *Chem. Sci.* **2015**, *6*, 6823–6828. [CrossRef] [PubMed]
25. Ruamps, R.; Maurice, R.; Batchelor, L.; Boggio-Pasqua, M.; Guillot, R.; Barra, A.L.; Liu, J.; Bendeif, E.-E.; Pillet, S.; Hill, S.; et al. Giant Ising-Type Magnetic Anisotropy in Trigonal Bipyramidal Ni(II) Complexes: Experiment and Theory. *J. Am. Chem. Soc.* **2013**, *135*, 3017–3026. [CrossRef] [PubMed]
26. Widener, C.N.; Bone, A.N.; Ozerov, M.; Richardson, R.; Lu, Z.; Thirunavukkuarasu, K.; Smirnov, D.; Chen, X.-T.; Xue, Z.-L. Direct Observation of Magnetic Transitions in a Nickel(II) Complex with Large Anisotropy. *Chin. J. Inorg. Chem.* **2020**, *35*, 1149–1156. [CrossRef]
27. Cahier, B.; Perfetti, M.; Zakhia, G.; Naoufal, D.; El Khatib, F.; Guillot, R.; Riviere, E.; Sessoli, R.; Barra, A.-L.; Guihary, N.; et al. Magnetic Anisotropy in Pentacoordinate Ni^{II} and Co^{II} Complexes: Unraveling Electronic and Geometrical Contributions. *Chem. Eur. J.* **2017**, *23*, 3648–3657. [CrossRef] [PubMed]

28. Rebilly, J.-N.; Charron, G.; Riviere, E.; Guillot, R.; Barra, A.-L.; Duran Serrano, M.; van Slageren, J.; Mallah, T. Large Magnetic Anisotropy in Pentacoordinate NiII Complexes. *Chem. Eur. J.* **2008**, *14*, 1169–1177. [CrossRef] [PubMed]
29. Nemec, I.; Herchel, R.; Svoboda, I.; Boča, R.; Trávníček, Z. Large and negative magnetic anisotropy in pentacoordinate mononuclear Ni(II) Schiff base complexes. *Dalton Trans.* **2015**, *44*, 9551–9560. [CrossRef] [PubMed]
30. Titiš, J.; Boča, R. Magnetostructural D Correlations in Hexacoordinated Cobalt(II) Complexes. *Inorg. Chem.* **2011**, *50*, 11838–11845. [CrossRef] [PubMed]
31. Boča, R. *Program MIF&FIT*; University of SS Cyril and Methodius: Trnava, Slovakia, 2022.

inorganics MDPI

Perspective

Bis(benzimidazole) Complexes, Synthesis and Their Biological Properties: A Perspective

Zdeněk Šindelář and Pavel Kopel *

Department of Inorganic Chemistry, Faculty of Science, Palacky University, 17. listopadu 12, 779 00 Olomouc, Czech Republic
* Correspondence: pavel.kopel@upol.cz; Tel.: +420-585-634-352

Abstract: Benzimidazoles are a very well-known, broad group of compounds containing nitrogen atoms in their structure that can mimic properties of DNA bases. The compounds show not only biological activities but also are used for spectral and catalytic properties. Biological activity of benzimidazoles can be tuned and accelerated in coordination compounds. This minireview is focused on preparation of bis(benzimidazoles), their complexes, and biological properties that can be found from 2015.

Keywords: bis(benzimidazole); mixed ligand complexes; antibacterial; anticancer

1. Introduction

The benzimidazole is a bicyclic molecule composed of benzene ring and imidazole ring. The compound is isostructural with naturally occurring nucleotides [1]. Its similarity to natural molecules led to the preparation of derivatives that can be utilized in medicinal chemistry. The very broad spectrum of biological activities that it treats include antimicrobial, antibiofilm, antifungal, antiviral, antioxidant, anti-inflammatory, antidiabetic, antiparasitic, anthelmintic, anticoagulant, antiallergic, antiprotozoal, anticonvulsants, anticancer and cytotoxic activities. There are drugs already used in medicine, such as Albendazole, Bendamustine, Omeprazole, Pimonbendane, Benomyl, Carbendazim, Telmisartan, Pantoprazole, Etonitazene, and Thiabendazole (some of them are depicted in Scheme 1).

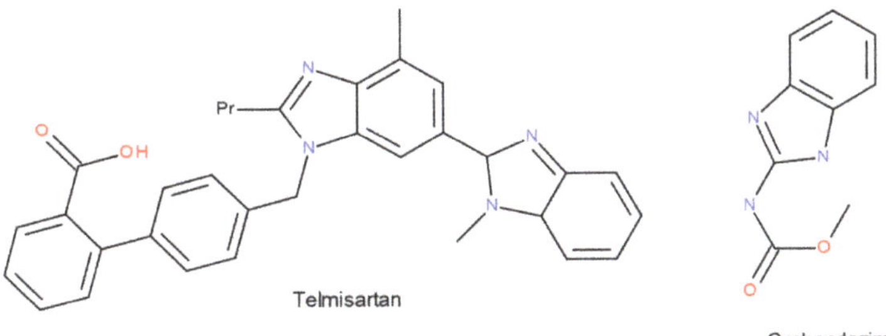

Scheme 1. Examples of commercially used drugs.

Citation: Šindelář, Z.; Kopel, P. Bis(benzimidazole) Complexes, Synthesis and Their Biological Properties: A Perspective. *Inorganics* **2023**, *11*, 113. https://doi.org/10.3390/inorganics11030113

Academic Editors: Peter Segl'a and Ján Pavlik

Received: 15 February 2023
Revised: 27 February 2023
Accepted: 7 March 2023
Published: 9 March 2023

Moreover, benzimidazoles can be utilized as optical sensors for bioimaging and in photovoltaics. There are not only many papers but also many reviews on the topic, such as the one on lanthanide complexes by Cruz-Navarro et al. [1], by Hernández-Romero et al. on first-row transition metal complexes [2], and Suarez-Moreno et al. on second- and third-row transition metal complexes [3]. The last two reviews contain information about the anticancer and antitumor activities of benzimidazole complexes.

In this review, we have focused on the preparation of bis(benzimidazoles) and their complexes that show biological activities. These compounds can be utilized as bridges among metal centers, chelating ligands with nitrogen atoms, or oxygen or sulfur coordinating atoms.

2. Bis(benzimidazole) Synthesis and Some Examples of Complex Preparation

There are many ways for benzimidazole and benzimidazole derived ligands. Some examples of the preparations are mentioned hereafter. The work of Matthews et al. described symmetric as well as asymmetric bis(benzimidazoles) [4]. The general method can be seen in Scheme 2.

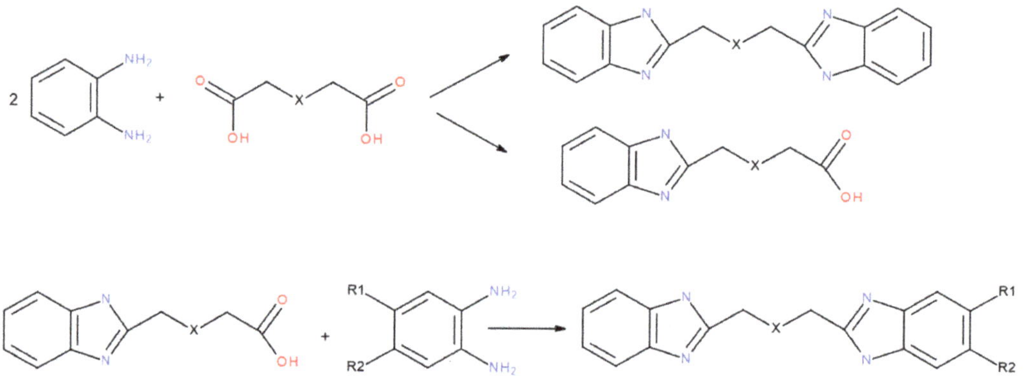

Scheme 2. General synthesis of bis(benzimidazoles). (X = O, S, NH).

For example, preparation of 4-(2-Benzimidazolyl)-3-thiabutanoic acid and 2-(1*H*-benzimidazol-2-ylmethylsulfanylmethyl)-1*H*-benzimidazole is given.

Thiodiacetic acid and o-phenylenediamine in a solution of 4 M HCl are refluxed for a total of 72 h and allowed to cool to room temperature. A green precipitate was filtered and dried. The precipitate was dissolved in distilled water. The solution was stirred and basified with ammonia solution to pH 9. A precipitate of bis(benzimidazole) was filtered and dried. The filtrate was treated with concentrated HCl until pH 7. White precipitate gave 4-(2-Benzimidazolyl)-3-thiabutanoic acid.

Similarly, 4-(2-benzimidazolyl)-3-oxabutanoic acid was prepared. These acids were used for condensation with 4-nitro-o-phenylenediamine, N-methyl-o-phenylenediamine, or 4,5-dimethyl-o-phenylenediamine to form asymmetric bis(benzimidazoles) [4]. The ligands were used for the preparation of copper(II) complexes. The complexes were prepared by equimolar addition of ligand to the copper bromide or perchlorate dissolved in methanol. The ligands were tridentate chelating.

Caymaz et al. reported synthesis of 2,2′-bis-(imidazo [1,2-*a*]pyridine-8-yl)-1*H*,1*H*′-[5,5′]-bisbenzimidazole [5]. The syntheses of compound were performed by reacting the imidazo(1,2-*a*)pyridine-8-carbaldehyde with 3,3′-diaminobenzidin (see Scheme 3).

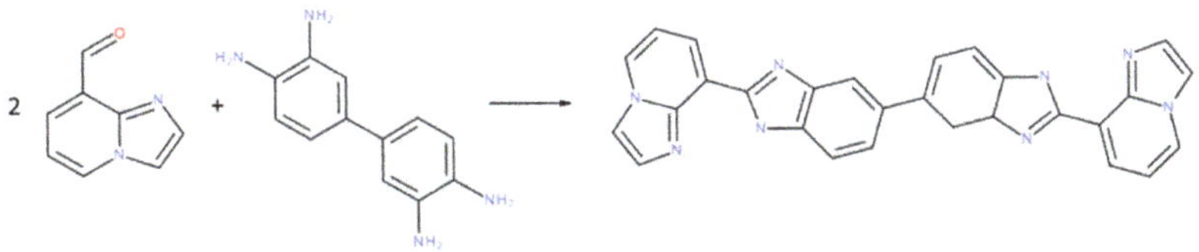

Scheme 3. 2,2′-bis-(imidazo[1,2-*a*]pyridine-8-yl)-1*H*,1*H*′-[5,5′]-bisbenzimidazole.

The compound binds to DNA grooves and has peroxide mediated DNA-cleavage properties. It was tested on cell lines HepG2, DLD-1, and MDA-MB-231, and was found to have high cytotoxic activities [5].

Synthesis of (2-((1*H*-benzo[d]imidazol-2-yl)methylthio)-1*H*-benzo[*d*]imidazol-6-yl)(phenyl)methanone [BIPM] and its Pd complex was reported by Kumar et al. [6]. In the first step, (2-mercapto-1*H*-benzo[*d*]imidazol-6-yl)(phenyl)methanone is prepared in methanol in presence of KOH by reaction of carbon disulfide with 3,4-diaminobenzophenone. Then, 2-(chloromethyl)-1*H*-benzo[*d*]imidazole was prepared by condensing chloroacetic acid and o-phenylenediamine in 4 M hydrochloric acid. Finally, (2-((1*H*-benzo[*d*]imidazol-2-yl)methylthio)-1*H*-benzo[*d*]imidazol-6-yl)(phenyl)methanone [BIPM], was obtained by a reaction of above-mentioned components in methanol (see Scheme 4).

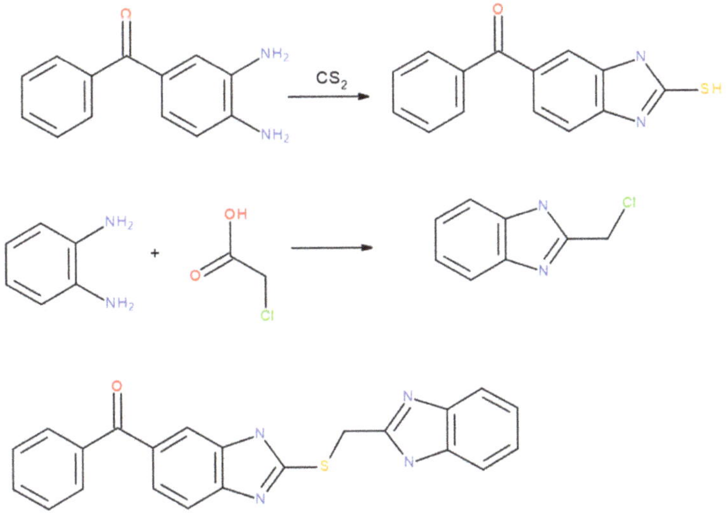

Scheme 4. Synthesis of (2-((1*H*-benzo[*d*]imidazol-2-yl)methylthio)-1*H*-benzo[*d*]imidazol-6-yl)(phenyl)-methanone (BIPM).

There are many papers on coordination compounds with 2,6-bis(1*H*-benzo[*d*]imidazol-2-yl)pyridine (BBP) and its derivatives (see Scheme 5, top left). These ligands are multidentate ligands and can be coordinated to metal atoms in different metal–ligand ratios. Substitution of the N–H bond in the benzimidazole ring by alkyl groups can lead to the formation of hydrogen bonds, and complexes are studied for interesting optical and magnetic properties [1]. BBP can be obtained by the reaction of pyridine-2,6-dicarboxylic acid and o-phenylenediamine in the presence of polyphosphoric acid (PPA). Higher yields can be obtained with phosphorus oxide or HCl and reaction in microwave oven. Instead of pyridine-2,6-dicarboxylic acid, pyridine-2,6-dicarbaldehyde can be used, but the yields are low [1].

N-substituted ligands derived from BBP can be obtained by condensing pyridine-2,6-dicarboxylic acid with N-alkyl-o-phenylenediamine derivative. The other possible way is to deprotonate N3 with a base, followed by a reaction with alkyl or aryl halides. Some examples are given below. The 2,6-bis-(6-nitrobenzimidazol-2-yl)pyridine (BNBP) (see Scheme 5, top right) was prepared by reaction of BBP with concentrated sulfuric acid and nitric acid. The complex [Ru(BBP)Cl_3] was prepared by a reaction of BBP with ruthenium(III) chloride. The nucleophilic substitution of BBP with 3,5-di-*tert*-butylbenzyl bromide or 4-*tert*-butylbenzyl chloride, in basic heated DMSO solution, led to preparation of derivatives depicted in Scheme 5 (bottom) [7].

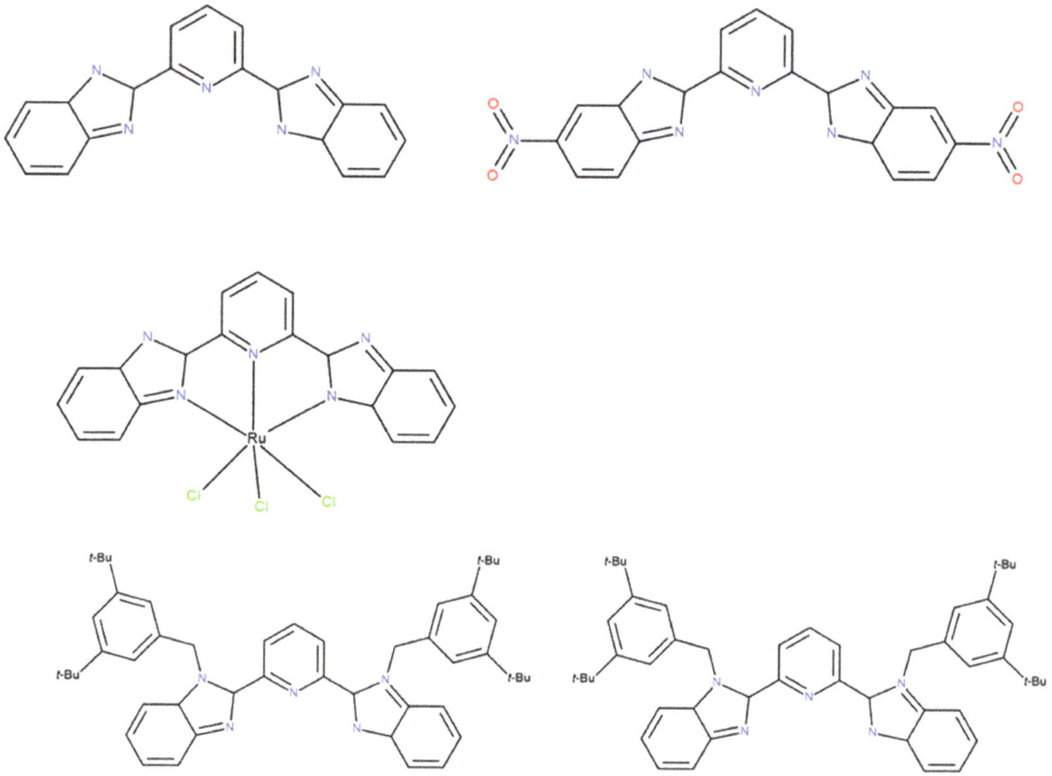

Scheme 5. 2,6-*bis*(1*H*-benzo[*d*]imidazol-2-yl)pyridine (BBP), BNBP = 2,6-bis-(6-nitrobenzimidazol-2-yl)pyridine), [Ru(BBP)Cl_3] (middle) [8]. The substituted ligands (bottom) were obtained via the nucleophilic substitution of 2,6-bis(1*H*-benzimidazole-2-yl)pyridine [7] with 3,5-di-*tert*-butylbenzyl bromide or 4-*tert*-butylbenzyl chloride in the presence of KOH in DMSO solvent at an elevated temperature.

Another example of BBP derivative on pyridine ring was recently reported by Orvos et al. The synthetic route of the ligand is outlined in Scheme 6 [9]. The 4-azidopyridine derivative was prepared from dimethyl 4-chloropyridine-2,6-dicarboxylate by the basic hydrolysis of with LiOH. Diamide was obtained with *o*-phenylenediamine using *O*-(benzotriazol-1-yl)-*N*,*N*,*N*′,*N*′-tetramethyluronium tetrafluoroborate (TBTU) in DMF. Hydrogenation on Pd/C in MeOH gave an amino derivative. NH_2-bis(benzimidazole)pyridine was obtained by heating in acetic acid. The reaction of nitrosobenzene with the amino group led to the final product.

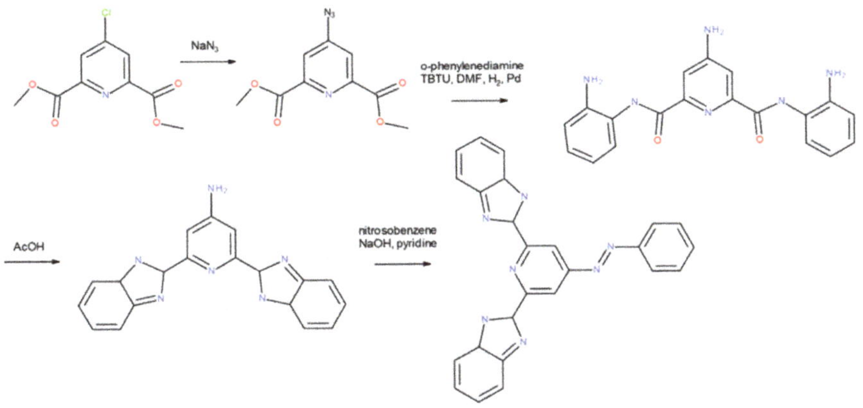

Scheme 6. The azidopyridine derivative of BBP preparation [9].

3. Biologically Active Bis(benzimidazole) Complexes

Deng et al. have prepared ruthenium complexes that have potential applications as sensitizers for use in cancer radiotherapy [8]. They prepared [Ru(BBP)Cl_3] (1), [Ru$(BBP)_2$]Cl_2 (2a), and [Ru$(BNBP)_2$]Cl_2 (2b). Complex 2b was found to be particularly effective in sensitizing human melanoma A375 cells toward radiation. Moreover, it was found that complex 2b is not toxic to normal cells. Mechanism of action is formation of intracellular reactive oxygen species (ROS) with glutathione (GSH) followed by DNA strand breaks. The subsequent DNA damage induces phosphorylation of p53 (p-p53) and upregulates the expression levels of p21, which inhibits the expression of cyclin-B, leading to G2M arrest. Moreover, p-p53 activates caspases-3 and -8, triggering cleavage of poly(ADP-ribose) polymerase (PARP), finally resulting in apoptosis [8].

BODIPY iridium(III) complexes containing 2,2′-bis(benzimidazole) show selectivity for cancerous cells over normal cells [10]. The tetranuclear (2 + 2) complexes were prepared through the self-assembly of benzimidazole and BODIPY ligands with dichloro (pentamethylcyclopentadienyl) iridium dimer. Cytotoxicity studies revealed that the complex is highly selective for cervical cancer cells (HeLa) and human glioblastoma (U87) cancer cells [10].

[MnBr$(CO)_3$L2] (3, L2 = 2,2′-bisbenzimidazole), [MnBr$(CO)_3$L3]·CH_3OH (4, L3 = BBP = 2,6-bis(benzimidazole-2′-yl)pyridine), and *fac*-[MnBr$(CO)_3$L4] (5, L4 = 2,4-bis(benzimidazole-2′-yl) pyridine) were prepared by reactions of MnBr$(CO)_5$ with appropriate ligands L2–L4, respectively, and characterized by single crystal X-ray diffraction, NMR, IR, UV-vis, and fluorescence spectroscopy [11]. The CO-release properties were investigated using the myoglobin assay and CO detection, and the results show that all of the complexes could release CO rapidly upon exposure to 365 nm UV light. The fluorescence imaging show that the Mn(I) complexes can be taken up by human liver cells (HL-7702) and liver cancer cells (SK-Hep1), and are suitable for bioimaging. A cell viability assay for SK-Hep1 shows that the anticancer activity of 3 is highest in the studied complexes [11].

Pd(II) complex with (2-((1*H*-benzo[*d*]imidazol-2-yl)methylthio)-1*H*-benzo[*d*]imidazol-5-yl)(phenyl)methanone (BIPM) was prepared by reaction of palladium acetate with BIPM in a 1:1 molar ratio [6]. BIPM is bidentate N,S chelating ligand, and the other two positions are occupied by oxygen atoms from two acetate anions. The in vitro antiproliferative effect of the BIPM and complex were tested against the MCF7, A549, Ehrlich ascites carcinoma (EAC), and Daltons lymphoma ascites (DLA) carcinoma cell lines. The mechanism is the antiangiogenic effect and promotion of apoptosis. The potential photo-induced binding mode on double-stranded calf thymus DNA and protein cleavage activity study on pBR322 DNA of the complex confirmed apoptosis. The molecular docking study proved its interaction with DNA [6].

Ruthenium mixed ligand complex with 2-(1*H*-benzimidazol-2-ylmethylsulfanylmethyl)-1*H*-benzimidazole and Schiff base (2-((*E*)-1*H*-1,2,4-triazol-5-yliminomethyl) phenol) was reported by Sur et al. [12]. The antibacterial effect of the complex was studied against *Staphylococcus aureus*, vancomycin-resistant *Staphylococcus aureus* (VRSA), methicillin-resistant *Staphylococcus aureus* (MRSA), and *Staphylococcus epidermidis*. Very high antibacterial activity was observed on growth curves and by fluorescence imaging. Moreover, in vivo tests on VRSA-infected mice proved better healing of skin wounds.

Mononuclear, binuclear, and multinuclear silver complexes of composition $[Ag_2(methacrylate)_2(Etobb)_2]\cdot CH_3CN$ **1**, [Ag(methacrylate)(Bobb)] **2**, and $[Ag_2(methacrylate)_2(Aobb)]_n$ **3**, where Etobb = 1,3-bis(1-ethylbenzimidazol-2-yl)-2-oxapropane, Bobb = 1,3-bis(1-benzyl-benzimidazol-2-yl)-2-oxapropane, Aobb = 1,3-bis(1-allylbenzimidazol-2-yl)-2-oxapropane were prepared by Zhang et al. [13]. Synthesis of the ligands is in Scheme 7. The three complexes have been prepared by reaction of silver nitrate with sodium methacrylate and corresponding ligand. Single-crystal X-ray diffraction revealed that complex **1** is binuclear, and the silver atom is coordinated by two N atoms to two Etobb ligands and oxygen of methacrylate. The complex **2** is mononuclear with a chelating ligand, but the oxygen atom of bis(benzimidazole is not involved in coordination. Complex **3** is a metal–organic compound with a diamond-like multinuclear silver center, with each silver atom bridged by two Aobb ligands and two methacrylate ions to form 1-D single-coordination polymer chain structures that extend into 2-D frameworks through π–π interactions. The binding modes of DNA were checked through absorption titration experiments of CT-DNA with complexes at 270 nm. A binding to DNA through intercalation with strong π–π stacking to the DNA base pairs was proved. Hydroxyl radical scavenging activity revealed the inhibitory effect of the complexes on $OH^{\cdot}$ radicals.

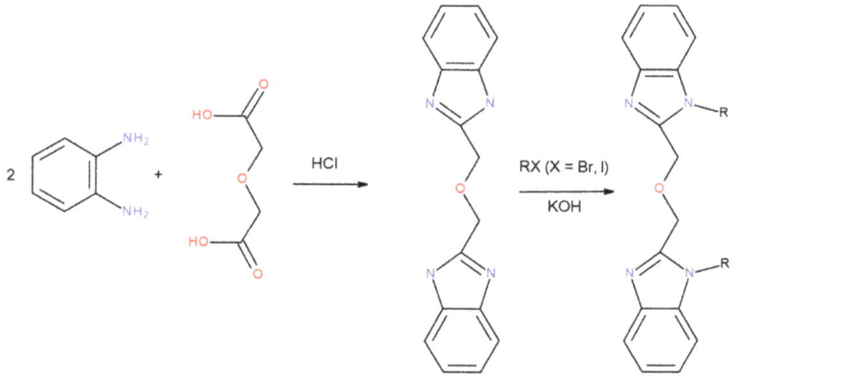

Scheme 7. Synthetic way for ligands (Etobb, Bobb, Aobb), where R = $-CH_2CH_3$ (Etobb), R = $-CH_2$-ph (Bobb), R = CH_2CHCH_2.

Rodriguez-Cordero et al. have characterized $ZnLBr_2$ complexes with bis(benzimidazole) prepared by reaction of citraconic acid with o-phenylenediamine or its derivative obtained by following reaction with benzylbromide in DMF [14]. Tetrahedral zinc coordination was proven by single crystal X-ray analysis. Zinc is coordinated by two Br atoms and a chelating N-N ligand. UV spectra of ligands and complexes showed decreases in peak intensities when increasing amounts of CT DNA. These spectral changes are consistent with intercalation or partial intercalation of the ligands and complexes into the DNA.

Pan et al. have prepared and proved structures of $[Cu(bmbp)(HCOO)(H_2O)](ClO_4)\cdot DMF$ (1), $[Co(bmbp)_2]_2(ClO_4)_4\cdot DMF\cdot H_2O$ (2) and $[Zn(bmbp)_2]_2(ClO_4)_4\cdot DMF\cdot H_2O$ (3), where bmbp is 4-butyloxy-2,6-bis(1-methyl-2-benzimidazolyl)pyridine, complexes by single crystal X-ray analysis [15]. Complex 1 has square-pyramidal geometry, and complexes 2 and 3 are distorted octahedral. The complexes and bmbp were tested on a human esophageal cancer cell line

(Eca109). Inhibition of the growth was proven, and complex 1 showed that it was the most active (IC50 = 26.09 μM).

Mixed ligand Cu(II) complexes [Cu(BBP)(L)H_2O]SO_4 (where L = 2,2′ bipyridine (bpy), and ethylene diamine (en)), have been prepared, and DNA-binding properties proved by absorption spectroscopy, fluorescence spectroscopy, viscosity measurements and thermal denaturation methods. DNA intercalation mechanism was suggested as well as the cleavage of plasmid pBR322, in the presence of H_2O_2 [16].

Very interesting ligand synthesis and copper complexes are presented by Suwalsky et al. [17]. The authors have prepared tetradentate Bis(2-methylbenzimidazolyl)(2-methylthioethyl)amine (L1) (see Figure 1A), and bis(1-methyl-2-methylbenzimidazolyl)(2-methylthioethyl)amine (L1Me). The first ligand was prepared by refluxing 1-tert-butoxycarbonyl-2-cloromethylbenzimidazole and 2-methylthioethylamine in the presence of K_2CO_3 and NaI in CH_3CN. The second ligand was prepared similarly with 1-methyl-2-chloromethylbenzimidazole. The complexes were prepared by reaction of copper perchlorates with ligands. The effect on the morphology of human erythrocytes and antiproliferative effect was tested on HeLa, REH, A546, and K-562 cells.

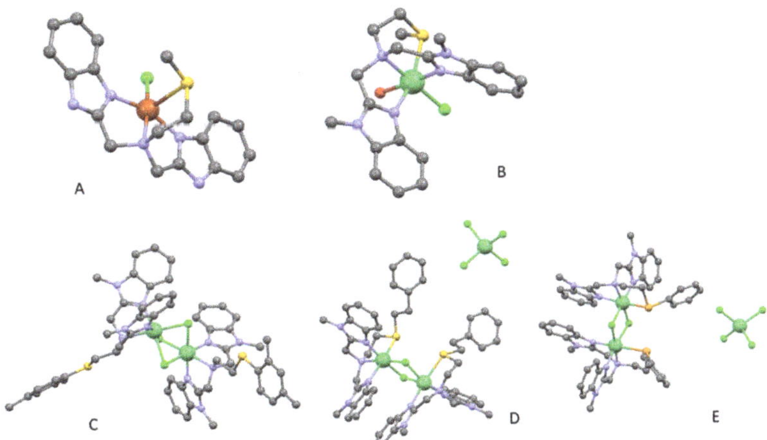

Figure 1. Molecular structures of (**A**) [L^1CuCl]; C (gray), N (blue), S (yellow), Cl (green), and Cu (red), (**B**) [Ni($L1^{Me}$)(H_2O)(Cl)]Cl, Ni (great green), O (red), (**C**) $Ni_2(L2^{Me})_2(\mu\text{-}Cl)_3$]Cl, (**D**) [$Ni_2(L3^{Me})_2(\mu\text{-}Cl)_2$][$NiCl_4$], (**E**) [$Ni_2(L4^{Me})_2(\mu\text{-}Cl)_2$][$NiCl_4$], Se (gold).

Similar Bis(benzimidazole)thio- and selenoether ligands and their nickel(II) complexes were prepared by the same group of authors [18]. This time, 1-methyl-2-(chloromethyl)benzimidazole reacted with (2-phenylethylthio)ethylamine (for L3Me) or (2-phenylseleno)ethylamine (for L4Me). Mononuclear and binuclear Ni(II) complexes were obtained by a reaction of nickel chloride with the ligands. Their structures can be seen in Figure 1.

The stability of complexes in aqueous solutions was monitored for 72 h by UV–vis spectroscopy, and these are stable in solutions. The cytotoxicity of complexes was screened against SK-LU-1 (human lung adenocarcinoma), HeLa (human cervical carcinoma), and HEK-293 (non-tumoral human embryonic kidney) cell lines using an MTT. It was found that the complexes are less cytotoxic in comparison with cisplatin, but they are selective to tumor cell lines.

1-(1*H*-benzimidazol-2-yl)-*N*-(1*H*-benzimidazol-2-ylmethyl)methanamine (abb) and 2-(1*H*-benzimidazol-2-ylmethylsulfanylmethyl)-1*H*-benzimidazole (tbb) have been prepared and characterized [19]. The trinuclear complex [$Ni_3(abb)_3(H_2O)_3(\mu\text{-ttc})$]$(ClO_4)_3$, was where $ttcH_3$ = trithiocyanuric acid was prepared and characterized by X-ray (depicted in Figure 2A1,A2). The complex and ligands were tested on bacteria strains *Staphylococcus aureus*, *Escherichia coli*, and *Saccharomyces cerevisiae*. The complex was more active than ligands.

The same complex was used together with another trinuclear complex $[Ni_3(tebb)_3(H_2O)_3(\mu\text{-}ttc)](ClO_4)_3$, tebb = 2-[2-[2-(1*H*-benzimidazol-2-yl)ethylsulfanyl]ethyl]-1*H*-benzimidazole, to study cytotoxicity on breast cell lines T-47D, MCF-7 and non-malignant HBL-100 (complex cation shown in Figure 2B) [20].

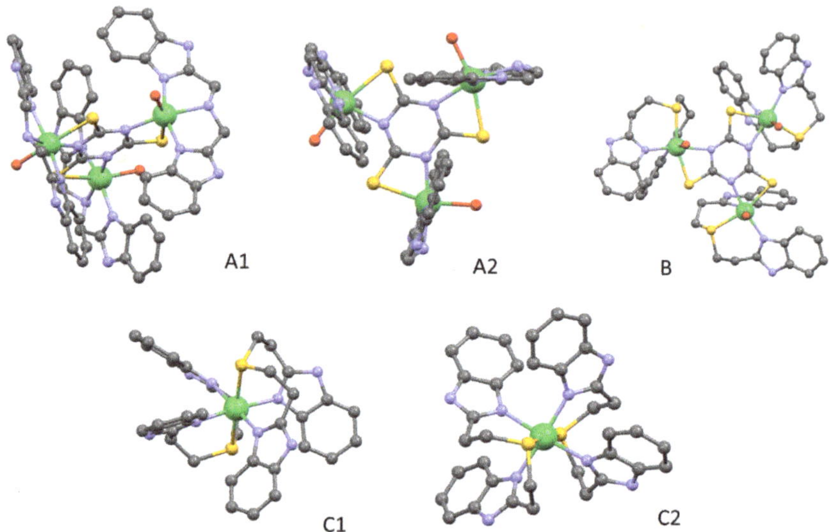

Figure 2. Molecular structures of complex cations: (**A1**,**A2**) $[Ni_3(abb)_3(H_2O)_3(\mu\text{-}ttc)]$; Ni (green), O (red), C (gray), N (blue), S (yellow), (**B**) $[Ni_3(tebb)_3(H_2O)_3(\mu\text{-}ttc)]$, (**C1**,**C2**) $[Ni(tebb)_2]$. The abb = 1-(1*H*-benzimidazol-2-yl)-*N*-(1*H*-benzimidazol-2-ylmethyl)methanamine, $ttcH_3$ = trithiocyanuric acid, tebb = 2-[2-[2-(1*H*-benzimidazol-2-yl)ethylsulfanyl]ethyl]-1*H*-benzimidazole.

It was found that complexes are very cytotoxic (24IC50 = 9.5 μM). The complex with abb was encapsulated in ferritin modified with folic acid to overcome toxicity to normal cells and enable transport to cancer cells. For a comparison of cytotoxicity of trinuclear complex with a mononuclear tebb complex, $[Ni(tebb)_2](ClO_4)_2$ has been prepared (the cation shown in Figure 2C1,C2) [21]. The complex is readily uptaken by malignant MDA-MB-231 and CACO-2 cells and is not toxic to Hs27 fibroblasts. The lowest IC50 values were found for MDA-MB-231 cells (5.2 μM). DNA cleavage, DNA fragmentation leads to the formation of reactive oxygen species [21]. Antibacterial study against *Staphylococcus aureus* and *Escherichia coli* on $[Ni_3(tebb)_3(H_2O)_3(\mu\text{-}ttc)](ClO_4)_3$ was reported by Ashrafi [22].

4. Conclusions

The goal of the mini review was to perform a literature search of the current state of the art of chemistry of bis(benzimidazole) syntheses and the preparation of complexes with the ligands. There are plenty of papers on the topic and many reviews. We have focused on complexes that were studied for their biological properties in the last 7 years.

Bis(benzimidazoles) were selected from the broad collection of papers dealing with benzimidazole complexes because bis(benzimidazoles) are structurally interesting and offer chelating and as well as bridging modes of coordination to central atoms. Plenty of these complexes have been known for ages, though these were prepared as models mimicking biological systems, so the data on biological activities was often missing. These known compounds can be further studied for biological properties and can find their potential to be used as drugs, for example, in cases where known antibiotics are insufficient to fight against resistant bacteria. From the literature, it is obvious that the complexes were mostly studied as antiproliferative, anticancer, and antitumor agents. From this point of view the studies are concerned on non-platinum metal-based drugs. Except of ruthenium, palladium and iridium complexes were found. There are some recent results on copper, cobalt, and zinc, and nickel, although nickel is mentioned in studies as a toxic metal. There are not many papers on silver complexes, and metals such as gold or iron can be other way to diversify our knowledge on biological properties of coordination compounds [2,3,23]. Very promising are combinations of ligands such as benzimidazoles with other biologically active ligands, for example Schiff bases. Some benzimidazoles contain sulfur or selenium that can increase their biological action or they can contain arms with the atoms to be more strongly coordinated to central atoms [18]. It should also be of interest to include P ligands to combine them with benzimidazoles to increase bioactivity, for example in the case of 2,6-bis(2-(diphenylphosphanyl)-1*H*-imidazol-1-yl)pyridine [24].

Fewer studies have been performed on bacteria strains. There are some papers on copper, zinc, and nickel bis(benzimidazole) complexes and their antibacterial properties, but the data are not available for all of the already known species. These can be future trends of studies on benzimidazole biological active complexes. Probable toxicity of prepared complexes can be overcome by using transporters on the base of nanoparticles that can deliver the complexes without side effects to healthy cells.

Author Contributions: Conceptualization, P.K.; resources, Z.Š. and P.K.; writing—original draft preparation, Z.Š. and P.K.; writing—review and editing, Z.Š. and P.K.; visualization, P.K.; supervision, P.K. All authors have read and agreed to the published version of the manuscript.

Funding: Authors acknowledge financial support from the institutional sources of the Department of Inorganic Chemistry, Palacky University Olomouc, Czech Republic.

Conflicts of Interest: The authors declare no conflict of interest.

Abbreviations

BIPM = 2-((1*H*-benzo[*d*]imidazol-2-yl)methylthio)-1*H*-benzo[*d*]imidazol-6-yl)(phenyl)methanone; BBP = 2,6-bis(1*H*-benzo[*d*]imidazol-2-yl)pyridine; PPA = polyphosphoric acid; BNBP = 2,6-bis-(6-nitrobenzimidazol-2-yl)pyridine; TBTU = *O*-(benzotriazol-1-yl)-*N*,*N*,*N*′,*N*′-tetramethyluronium tetrafluoroborate; Etobb = 1,3-bis(1-ethylbenzimidazol-2-yl)-2-oxapropane; Bobb = 1,3-bis(1-benzyl-benzimidazol-2-yl)-2-oxapropane; Aobb = 1,3-bis(1-allyl-benzimidazol-2-yl)-2-oxapropane; bmbp = 4-butyloxy-2,6-bis(1-methyl-2-benzimidazolyl)pyridine; abb = 1-(1*H*-benzimidazol-2-yl)-*N*-(1*H*-benzimidazol-2-ylmethyl)methanamine; tbb = 2-(1*H*-benzimidazol-2-ylmethylsulfanylmethyl)-1*H*-benzimidazole; tebb = 2-[2-[2-(1*H*-benzimidazol-2-yl)ethylsulfanyl]ethyl]-1*H*-benzimidazole; $ttcH_3$ = trithiocyanuric acid; Bcl-2 = B-cell lymphoma 2; BSA = bovine serum albumin; HBL-100 = human breast epithelial cell line; MDA-MB-231 human breast adenocarcinoma cancer cell line; MIC = minimum inhibitory concentration; MOF = metal–organic framework; ROS = reactive oxygen species; SOD—superoxide dismutase.

References

1. Cruz-Navarro, A.; Hernández-Romero, D.; Flores-Parra, A.; Rivera, J.M.; Castillo-Blum, S.E.; Colorado-Peralta, R. Structural diversity and luminescent properties of coordination complexes obtained from trivalent lanthanide ions with the ligands: Tris((1*H*-benzo[*d*]imidazol-2-yl)methyl)amine and 2,6-bis(1*H*-benzo[*d*]imidazol-2-yl)pyridine. *Coord. Chem. Rev.* **2021**, *427*, 213587. [CrossRef]
2. Hernandez-Romero, D.; Rosete-Luna, S.; Lopez-Monteon, A.; Chavez-Pina, A.; Perez-Hernandez, N.; Marroquin-Flores, J.; Cruz-Navarro, A.; Pesado-Gomez, G.; Morales-Morales, D.; Colorado-Peralta, R. First-row transition metal compounds containing benzimidazole ligands: An overview of their anticancer and antitumor activity. *Coord. Chem. Rev.* **2021**, *439*, 213930. [CrossRef]
3. Suarez-Moreno, G.V.; Hernandez-Romero, D.; Garcia-Barradas, O.; Vazquez-Vera, O.; Rosete-Luna, S.; Cruz-Cruz, C.A.; Lopez-Monteon, A.; Carrillo-Ahumada, J.; Morales-Morales, D.; Colorado-Peralta, R. Second and third-row transition metal compounds containing benzimidazole ligands: An overview of their anticancer and antitumour activity. *Coord. Chem. Rev.* **2022**, *472*, 214790. [CrossRef]
4. Matthews, C.J.; Leese, T.A.; Clegg, W.; Elsegood, M.R.J.; Horsburgh, L.; Lockhart, J.C. A route to bis(benzimidazole) ligands with built-in asymmetry: Potential models of protein binding sites having histidines of different basicity. *Inorg. Chem.* **1996**, *35*, 7563–7571. [CrossRef]

5. Caymaz, B.; Yildiz, U.; Akkoc, S.; Gercek, Z.; Sengul, A.; Coban, B. Synthesis, Characterization, and Antiproliferative Activity Studies of Novel Benzimidazole-Imidazopyridine Hybrids as DNA Groove Binders. *ChemistrySelect* **2020**, *5*, 8465–8474. [CrossRef]
6. Kumar, N.S.; Krishnamurthy, G.; Bodke, Y.D.; Malojirao, V.H.; Naik, T.R.R.; Kandagalla, S.; Prabhakar, B.T. Synthesis, characterization and tumor inhibitory activity of a novel Pd(ii) complex derived from methanethiol-bridged (2-((1*H*-benzo[*d*]imidazol-2-yl)methylthio)-1*H*-benzo[*d*]imidazol-6-yl)(phenyl)methanone. *New J. Chem.* **2019**, *43*, 790–806. [CrossRef]
7. Brachnakova, B.; Koziskova, J.A.; Kozisek, J.; Melnikova, E.; Gal, M.; Herchel, R.; Dubaj, T.; Salitros, I. Low-spin and spin-crossover iron(ii) complexes with pyridyl-benzimidazole ligands: Synthesis, and structural, magnetic and solution study. *Dalton Trans.* **2020**, *49*, 17786–17795. [CrossRef]
8. Deng, Z.Q.; Yu, L.L.; Cao, W.Q.; Zheng, W.J.; Chen, T.F. Rational Design of Ruthenium Complexes Containing 2,6-Bis(benzimidazolyl)pyridine Derivatives with Radiosensitization Activity by Enhancing p53 Activation. *ChemMedChem* **2015**, *10*, 991–998. [CrossRef]
9. Orvos, J.; Fischer, R.; Brachnakova, B.; Pavlik, J.; Moncol, J.; Sagatova, A.; Fronc, M.; Kozisek, J.; Routaboul, L.; Bousseksou, A.; et al. Pyridyl-benzimidazole derivatives decorated with phenylazo substituents and their low-spin iron(II) complexes: A study of the synthesis, structure and photoisomerization. *New J. Chem.* **2023**, *47*, 1488–1497. [CrossRef]
10. Gupta, G.; Das, A.; Lee, S.W.; Ryu, J.Y.; Lee, J.; Nagesh, N.; Mandal, N.; Lee, C.Y. BODIPY-based Ir(III) rectangles containing bis-benzimidazole ligands with highly selective toxicity obtained through self-assembly. *J. Organomet. Chem.* **2018**, *868*, 86–94. [CrossRef]
11. Hu, M.X.; Yan, Y.L.; Zhu, B.H.; Chang, F.; Yu, S.Y.; Alatan, G. A series of Mn(i) photo-activated carbon monoxide-releasing molecules with benzimidazole coligands: Synthesis, structural characterization, CO releasing properties and biological activity evaluation. *RSC Adv.* **2019**, *9*, 20505–20512. [CrossRef] [PubMed]
12. Sur, V.P.; Mazumdar, A.; Kopel, P.; Mukherjee, S.; Vitek, P.; Michalkova, H.; Vaculovicova, M.; Moulick, A. A Novel Ruthenium Based Coordination Compound Against Pathogenic Bacteria. *Int. J. Mol. Sci.* **2020**, *21*, 18. [CrossRef] [PubMed]
13. Zhang, H.; Xu, Y.L.; Wu, H.L.; Aderinto, S.O.; Fan, X.Y. Mono-, bi- and multi-nuclear silver complexes constructed from bis(benzimidazole)-2-oxapropane ligands and methacrylate: Syntheses, crystal structures, DNA-binding properties and antioxidant activities. *RSC Adv.* **2016**, *6*, 83697–83708. [CrossRef]
14. Rodriguez-Cordero, M.; Ciguela, N.; Llovera, L.; Gonzalez, T.; Briceno, A.; Landaeta, V.R.; Pastran, J. Synthesis, structural elucidation and DNA binding profile of Zn(II) bis-benzimidazole complexes. *Inorg. Chem. Commun.* **2018**, *91*, 124–128. [CrossRef]
15. Pan, R.K.; Song, J.L.; Li, G.B.; Lin, S.Q.; Liu, S.G.; Yang, G.Z. Copper(II), cobalt(II) and zinc(II) complexes based on a tridentate bis(benzimidazole)pyridine ligand: Synthesis, crystal structures, electrochemical properties and antitumour activities. *Transit. Met. Chem.* **2017**, *42*, 253–262. [CrossRef]
16. Sunita, M.; Anupama, B.; Ushaiah, B.; Kumari, C.G. Synthesis, characterization, DNA binding and cleavage studies of mixed-ligand copper (II) complexes. *Arab. J. Chem.* **2017**, *10*, S3367–S3374. [CrossRef]
17. Suwalsky, M.; Castillo, I.; Sanchez-Eguia, B.N.; Gallardo, M.J.; Dukes, N.; Santiago-Osorio, E.; Aguiniga, I.; Rivera-Martinez, A.R. In vitro effects of benzimidazole/thioether-copper complexes with antitumor activity on human erythrocytes. *J. Inorg. Biochem.* **2018**, *178*, 87–93. [CrossRef]
18. Munoz-Patino, N.; Sanchez-Eguia, B.N.; Araiza-Olivera, D.; Flores-Alamo, M.; Hernandez-Ortega, S.; Martinez-Otero, D.; Castillo, I. Synthesis, structure, and biological activity of bis(benzimidazole)amino thio- and selenoether nickel complexes. *J. Inorg. Biochem.* **2020**, *211*, 111198. [CrossRef]
19. Kopel, P.; Wawrzak, D.; Langer, V.; Cihalova, K.; Chudobova, D.; Vesely, R.; Adam, V.; Kizek, R. Biological Activity and Molecular Structures of Bis(benzimidazole) and Trithiocyanurate Complexes. *Molecules* **2015**, *20*, 10360–10376. [CrossRef]
20. Tesarova, B.; Charousova, M.; Dostalova, S.; Bienko, A.; Kopel, P.; Kruszynski, R.; Hynek, D.; Michalek, P.; Eckschlager, T.; Stiborova, M.; et al. Folic acid-mediated re-shuttling of ferritin receptor specificity towards a selective delivery of highly cytotoxic nickel(II) coordination compounds. *Int. J. Biol. Macromol.* **2019**, *126*, 1099–1111. [CrossRef]
21. Masaryk, L.; Tesarova, B.; Choquesillo-Lazarte, D.; Milosavljevic, V.; Heger, Z.; Kopel, P. Structural and biological characterization of anticancer nickel(II) bis(benzimidazole) complex. *J. Inorg. Biochem.* **2021**, *217*, 111395. [CrossRef] [PubMed]
22. Ashrafi, A.M.; Kopel, P.; Richtera, L. An Investigation on the Electrochemical Behavior and Antibacterial and Cytotoxic Activity of Nickel Trithiocyanurate Complexes. *Materials* **2020**, *13*, 1782. [CrossRef] [PubMed]
23. Ferraro, M.G.; Piccolo, M.; Misso, G.; Santamaria, R.; Irace, C. Bioactivity and Development of Small Non-Platinum Metal-Based Chemotherapeutics. *Pharmaceutics* **2022**, *14*, 954. [CrossRef] [PubMed]
24. Kumar, S.; Balakrishna, M.S. Transition metal complexes of imidazole appended pyridyline linked bisphosphine, 2,6-bis(2-(diphenylphosphanyl)-1H-imidazol-1-yl)pyridine. *Results Chem.* **2021**, *3*, 100161. [CrossRef]

MDPI

Perspective

Integrative Metallomics Studies of Toxic Metal(loid) Substances at the Blood Plasma–Red Blood Cell–Organ/Tumor Nexus

Maryam Doroudian and Jürgen Gailer *

Department of Chemistry, 2500 University Drive NW, University of Calgary, Calgary, AB T2N 1N4, Canada
* Correspondence: jgailer@ucalgary.ca

Abstract: Globally, an estimated 9 million deaths per year are caused by human exposure to environmental pollutants, including toxic metal(loid) species. Since pollution is underestimated in calculations of the global burden of disease, the actual number of pollution-related deaths per year is likely to be substantially greater. Conversely, anticancer metallodrugs are deliberately administered to cancer patients, but their often dose-limiting severe adverse side-effects necessitate the urgent development of more effective metallodrugs that offer fewer off-target effects. What these seemingly unrelated events have in common is our limited understanding of what happens when each of these toxic metal(loid) substances enter the human bloodstream. However, the bioinorganic chemistry that unfolds at the plasma/red blood cell interface is directly implicated in mediating organ/tumor damage and, therefore, is of immediate toxicological and pharmacological relevance. This perspective will provide a brief synopsis of the bioinorganic chemistry of As^{III}, Cd^{2+}, Hg^{2+}, CH_3Hg^+ and the anticancer metallodrug cisplatin in the bloodstream. Probing these processes at near-physiological conditions and integrating the results with biochemical events within organs and/or tumors has the potential to causally link chronic human exposure to toxic metal(loid) species with disease etiology *and* to translate more novel anticancer metal complexes to clinical studies, which will significantly improve human health in the 21st century.

Keywords: toxic metal(loid)s; chronic exposure; mechanism of toxicity; metallodrugs; stability in plasma; side effects; bloodstream; bioinorganic chemistry

Citation: Doroudian, M.; Gailer, J. Integrative Metallomics Studies of Toxic Metal(loid) Substances at the Blood Plasma–Red Blood Cell–Organ/Tumor Nexus. *Inorganics* **2022**, *10*, 200. https://doi.org/10.3390/inorganics10110200

Academic Editor: Navid Rabiee

Received: 9 September 2022
Accepted: 4 November 2022
Published: 7 November 2022

1. Introduction

Ever since the inception of life some 3.8 billion years ago—possibly in the vicinity of deep-sea hydrothermal vents [1]—it has been continually exposed to background concentrations of inherently toxic elements released from the earth's crust into the biosphere by natural processes, including volcanism and chemical weathering. Therefore, all organisms evolved in the presence of potentially toxic metal(loid) species and had to adapt to background concentrations over millions of years [2]. A momentous event in the 1760-1780s, however, changed the organism–earth relationship forever: the industrial revolution. This represents the onset of the production of consumer products at an ever-increasing scale, which required unprecedented amounts of energy (i.e., fossil fuels) and building blocks (i.e., chemical elements). The associated advent of the mining/metallurgy industry resulted in the increased emission of toxic metal(loid) species into the environment in many parts of the world [3,4], which has dramatically affected ecosystems [5] and, in turn, inevitably exposed certain human populations [6].

A conceptually related event which involved the interaction of an entirely different type of toxic metalloid compound with humans took place in 1911: the birth of chemotherapy by Paul Ehrlich. While his synthesis of the arsenic-containing drug Salvarsan contributed to humankind's exploration of the available 'chemical space' [7], its administration to patients who suffered from syphilis—a bacterial infection caused by *Treponema pallidum*—ushered in a fundamentally new therapeutic approach to treat human diseases by exploiting the concept of 'selective toxicity' [8,9].

What conceptually relates the exposure of humans to toxic metal(loid) species with the administration of an arsenic-containing drug to patients is the fact that both of these toxic substances infiltrate the systemic blood circulation, which was discovered by William Harvey [10]. His recognition that the bloodstream effectively represents a conveyor belt which supplies human organs with water, life-sustaining oxygen and nutrients (essential elements, vitamins, carbohydrates and lipids) to maintain human health makes this biological fluid one of the first sites where adverse interactions between toxic metal(loid) species/cytotoxic metallodrugs and constituents of the bloodstream unfold (Figure 1) [11,12]. The toxicological and pharmacological relevance of these interactions with constituents of the bloodstream is attributed to the simple fact that they collectively constitute a 'filter' which fundamentally determines which toxic-metal-containing species or metabolites thereof will impinge on organs to cause either a detrimental effect, in the case of toxic metal(loid) species, and/or a desirable pharmacological effect in the case of anticancer metallodrugs on a malignant tumor.

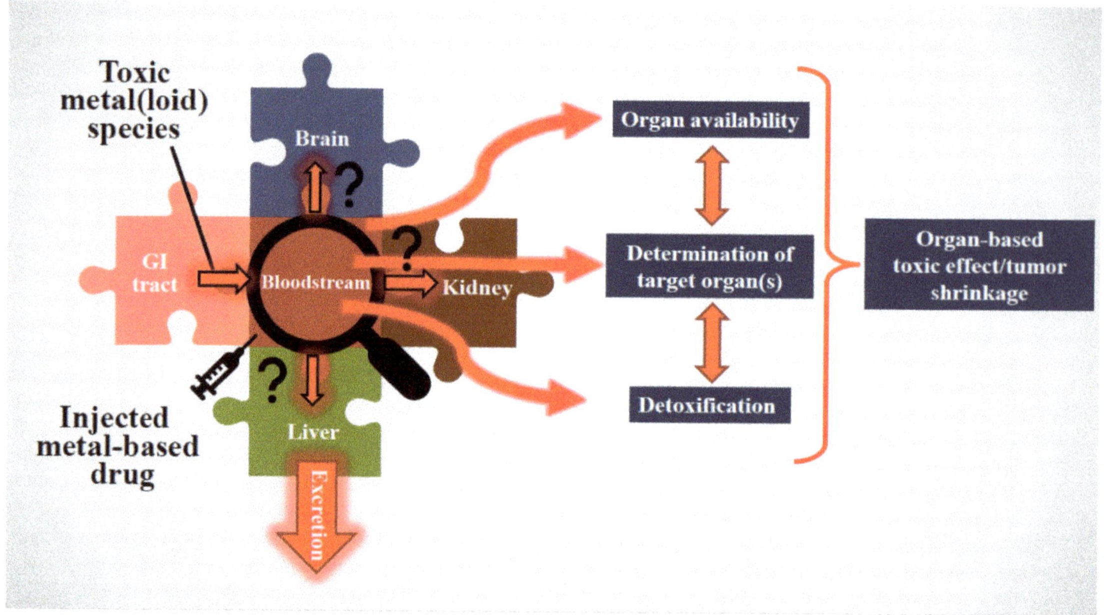

Figure 1. The central role held by the bioinorganic chemistry of toxic metal(loid) species and metallodrugs in the bloodstream in terms of linking environmental exposure to adverse health effects/organ-based diseases and in developing better drugs to treat human diseases, such as cancer (Figure modified from [13]).

Considerable progress has been made to conceptually relate toxic metal(loid) species and metallodrugs—which will from now on be referred to as toxic metal(loid) substances—in biological systems to adverse or intended human health effects. It is, therefore, timely to provide the reader with a perspective as to why the bioinorganic chemistry of toxic metal(loid) substances, specifically in the bloodstream, has become a worthwhile research avenue. Since the available experimental techniques to measure metal-containing metabolites in biological fluids are not our primary focus, the interested reader is referred to relevant reviews [14–18]. Nevertheless, a few techniques that are particularly appropriate to solve relevant problems will be outlined. Then, bioinorganic mechanisms of toxic metal(loid) species in the bloodstream will be presented, and their conceptual link to adverse health effects will be discussed. This section will be followed by an analogous brief summary of the biochemical fate of anticancer metallodrugs in the bloodstream. Finally, we underscore the importance of integrating processes involving toxic metal(loid) substances in the bloodstream with those that happen in organs in terms of understanding their impact at the whole-organism level, either negatively (toxicology) or advantageously (pharmacology).

2. Critical Importance of Studying Toxic Metal(loid) Substances in the Bloodstream (Footnote: From Now on We Use the Term 'Substances' When We Refer to Both Toxic Metal(loid) Species and Metallodrugs)

The man-made emission of toxic metal(loid) species into the environment has gradually increased since the onset of the Industrial Revolution, and perturbs the global biogeochemical element cycles of at least 11 elements today, with arsenic (As), cadmium (Cd) and mercury (Hg) species critically affecting air, water and food [19]. In fact, the input of Hg to ecosystems is estimated to have increased two- to five-fold during the industrial era [20]. These profound geochemical changes in the biosphere have prompted some scientists to refer to our age as the 'Anthropocene' [21]. Certain human populations, including children, are therefore unwittingly exposed to higher daily doses of persistent inorganic environmental pollutants through the ingestion of contaminated food [22–25], the inhalation of contaminated air [26–29] and/or dermal exposure to consumer products [30–32] than ever before. Unsurprisingly, pollution has become the world's largest environmental cause of disease and premature death, with an estimated ~9 million deaths per year [33] and is associated with staggering global healthcare costs [34]. Owing to the inherent persistence of As, Cd and Hg species and their binding to hematite and humic acids in soils [35], their concentrations in ecosystems will either remain constant—potentially for millennia [36,37]—or even gradually increase over time [38,39]. Due to the continued consumption of large quantities of Hg [40], the global-warming-induced re-mobilization of certain metal(loid) species [41] and the ongoing contamination of the food chain with these inorganic pollutants [22,25,42,43], the concomitant exposure of human populations represents a global health problem [44–46]. Owing to the severe health effects associated with human exposure to comparatively small daily doses of toxic metal(loid) species (up to 260 μg/day), the toxicology of metals has received considerable research attention [6,47], which has in turn resulted in the implementation of guidelines for maximum permissible concentrations in drinking water [48], food [22] and air [49] in many countries. As a consequence of additional human exposure to these inorganic pollutants through consumer products [30], personal care products [32], nanomaterials [50] and the global-warming-induced increase in using wastewater for food irrigation [44], an emerging challenge is the establishment of a new paradigm to assess how a lifetime of exposure affects the risk of developing chronic diseases [51]. In this context, the elucidation of the biomolecular mechanisms of toxicity associated with chronic human exposure to toxic metal(loid) species [52,53] and causally linking these mechanisms with the etiology of human illnesses that do not have a genetic origin [54] remain perhaps the two most pertinent knowledge gaps.

The development of chemotherapy to treat human diseases other than bacterial infections was—somewhat counterintuitively—accelerated during World Wars I and II, when certain warfare agents were used for the first time. In particular, the observation that soldiers' exposure to nitrogen mustards specifically targeted bone marrow cells prompted scientists to investigate if this effect may be harnessed to selectively target malignant cells. By the mid-1960s, the anticancer metallodrug cisplatin was serendipitously discovered [55] and FDA-approved in 1978. Despite cisplatin being a 'shotgun' cytotoxin (it does not differentiate between cancer and healthy cells), which is intrinsically associated with severe and dose-limiting side-effects [56], close to 50% of all cancer patients worldwide are today treated with cisplatin, carboplatin and/or oxaliplatin, either by itself or in combination with other anticancer drugs [12]. The success story of cisplatin triggered intense research efforts to develop metallodrugs for use as photochemotherapeutic drugs, antiviral drugs, antiarthritic drugs, antidiabetic drugs, drugs for the treatment of cardiovascular and gastrointestinal disorders as well as psychotropics [57], but overall metallodrugs remain a tiny minority of all medicinal drugs that are currently on the market [58,59]. The considerable burden of cancer on the world economy [60], however, necessitates the urgent development of more effective anticancer drugs that offer higher selectivity and, thus, fewer side-effects to improve the quality of life of patients during and after cancer treatment. Even though promising novel anticancer metal complexes are being developed [9,61], advancing more of these to preclinical/clinical studies remains a major problem [12]. This problem is attributed to the fact that insufficient attention is being directed to assess the effect of pharmacologically relevant doses of novel anticancer metal complexes on (a) the integrity of red blood cells (RBCs) [62], (b) their stability in blood plasma [12,63] and (c) their selectivity toward cancer cells versus healthy cells [64].

3. General Considerations Pertaining to Interactions of Toxic Metal(loid) Substances in Humans

With regard to the infiltration of the bloodstream by a toxic metal(loid) species, a reasonable question to ask is how an exceedingly small daily dose (e.g., 200 μg of inorganic As/day will eventually result in cancer [65]) can reach target organs. Interactions of toxic metal(loid) species with some bloodstream constituents are 'good', as their initial binding/sequestration (e.g., binding to human serum albumin or uptake into RBCs) will delay/preclude it from reaching a target organ. Conversely, a metallodrug that is intravenously administered is intended to reach the malignant tumor intact in order to cause maximal damage. Hence, any interaction(s) of an anticancer metallodrug with bloodstream constituents (e.g., a partial decomposition followed by plasma protein binding) is/are 'bad', as less will reach the tumor tissue.

It is important to point out that the daily human exposure to toxic metal(loid) species by inhalation and ingestion is comparatively lower (μg/day) than that of anticancer metallodrugs, which are intravenously administered (mg/day). It is, therefore, not entirely surprising that an extensive lag-time can exist between the onset of chronic human exposure to toxic metal(loid) species and the start of overt adverse health effects [52], while overt signs of toxicity manifest themselves typically within hours or days after the intravenous administration of patients with anticancer metallodrugs [56].

The challenge that pertains to both types of these toxic metal(loid) substances is the need to establish the entire sequence of bioinorganic interactions that occur at the plasma/RBC interface and to integrate them with processes that unfold in organs/tumors (Figure 2). For toxic metal(loid) species, this means revealing the biomolecular mechanisms causing organ damage; for metallodrugs, it also implies screening novel molecular structures which preferentially target the tumor cells, while leaving healthy organs unscathed. Ironically, the effect of these toxic metal(loid) substances on human health are conceptually intertwined in an interesting manner, since specific toxic metal(loid) species are established carcinogens (e.g., inorganic As and Cd [66,67]), while cancer patients are being treated with highly cytotoxic metallodrugs (e.g., cisplatin) to achieve tumor shrinkage/remission that are often associated with severe side-effects.

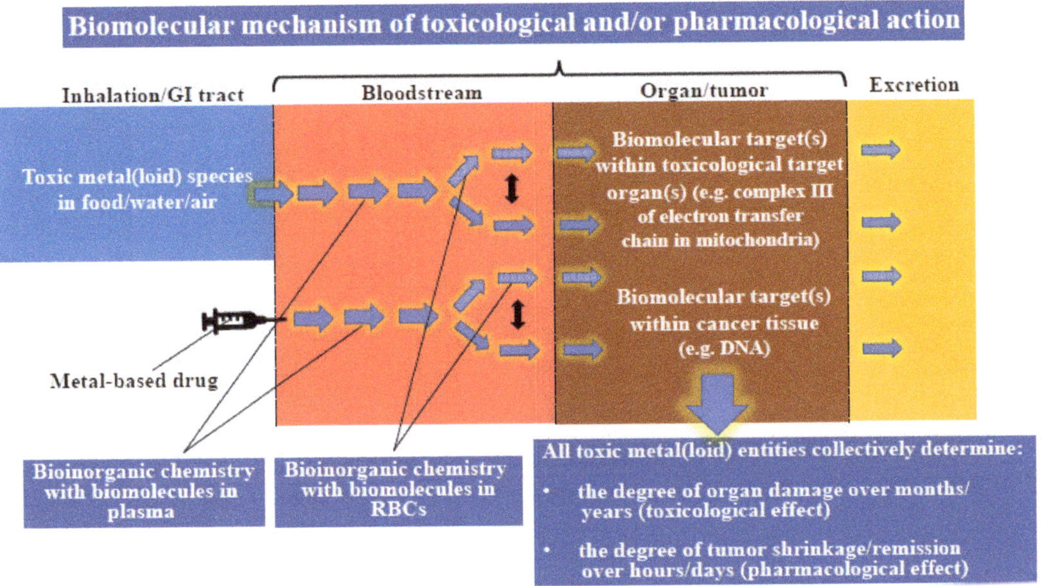

Figure 2. Conceptual relationship of bioinorganic processes in the bloodstream following (**a**) the chronic exposure of humans to a toxic metal(loid) species (toxicological effect) and (**b**) the intravenous administration by cancer patients of an anticancer metallodrug (pharmacological effect).

4. Bioinorganic Chemistry of Toxic Metal(loid) Substances in the Bloodstream In Vitro/In Vivo

One of the main difficulties of studying chemical reactions that unfold in the bloodstream is its biological complexity, which is directly related to its orchestration of the continual exchange of nutrients absorbed from the GI tract to organs and the eventual excretion of waste products via the kidneys. The bloodstream should not, however, be viewed as a seamless pipe, since it contains up to 10,000 plasma proteins [68], >400 SMW metabolites [69] and ~1600 RBC cytosolic proteins [70] which can engage in a variety of potential chemical reactions with toxic metal(loid) substances. The enormous number of potential binding sites on plasma proteins/SMW metabolites, for example, can result in their reversible vs. irreversible binding [71], while the partial or complete sequestration of toxic metal(loid) substances within RBCs would preclude their influx to target organs. Given the reducing conditions within RBCs, redox reactions of toxic metal(loid) substances within RBCs must also be considered. The intrinsic biological complexity of plasma and RBC cytosol can be overcome by using metallomics tools, which refers to hyphenated techniques that are comprised of a separation method (e.g., HPLC, 2D-PAGE) coupled to an element-specific detection technique [14–16,72]. The application of metallomics tools inherently reduces the analytical separation problem significantly, as the number of endogenous metal species within any given biological fluid (e.g., all endogenous metalloproteins) represents a sub-proteome of the proteome (i.e., all proteins within a biological fluid).

Detecting complexes containing a chemical bond between a toxic and an essential element, for example, in blood plasma, would reveal which essential trace element is targeted by a particular toxic metal(loid) species [73]. Since blood is comprised of about 50% of RBCs, it is also of importance to consider the binding of toxic metal(loid) species to the lipid bilayer membrane of intact RBCs [74] and the cytosolic proteins within RBCs [75]. In terms of anticancer metallodrugs, their metabolism in the bloodstream can lead to the formation of hydrolysis products, which can then bind to plasma proteins, such as human serum albumin (HSA). It is therefore desirable to analyze blood plasma for all parent metal-containing metabolites, which has been successfully attained for cisplatin and carboplatin [76].

In addition to analyzing plasma and RBC cytosol, one should also heed the advice from R.J.P. Williams that 'living organisms cannot be understood by studying extracted (dead) molecules. We have to study flow systems' [77]. Accordingly, conducting in vivo experiments using animal models is invaluable as it can provide an important starting point for further research. For example, it may be possible to identify a particular complex which contains a chemical bond between a toxic and an essential element [78,79] which can then initiate studies to establish the entire mechanism of chronic toxicity/pharmacological action within the bloodstream–organ system (Figure 2). To this end, our lab has demonstrated that the metabolism of As^{III} is fundamentally tied to that of the essential trace element selenium. After the intravenous injection of rabbits with As^{III} and Se^{IV}, the analysis of rabbit bile revealed an As:Se molar ratio of 1:1, which implied the in vivo formation and excretion of an As-Se compound from the liver. The latter compound was structurally characterized as the seleno-bis(S-glutathionyl) arsinium ion [$(GS)_2AsSe^-$] [78], which is formed intracellularly following Equation 1.

$$As(OH)_3 + HSeO_3^- + 8GSH \rightarrow (GS)_2AsSe^- + 3GSSG + 6H_2O \quad (1)$$

Notably, the results from this in vivo experiment were crucial to design subsequent investigations, which revealed that $(GS)_2AsSe^-$ is formed in the bloodstream [80] (Figure 3). More recent experiments revealed that this metabolite is present in the liver, the gall bladder and the GI tract [81], and experiments with rabbits revealed that this As^{III}-Se^{IV} antagonism is of direct environmental relevance [82]. Similarly, the formation and structural characterization of a Hg- and Se-containing detoxification product that is rapidly formed in rabbit blood was first observed in vivo [79]. The formation of the detoxification product $(HgSe)_{100}SelP$ (Figure 3) could not have been observed in blood plasma or RBCs alone, as its mechanism of formation involves the reduction of Se^{IV} to HSe^- within RBCs followed by its subsequent excretion into plasma where it then reacts with Hg^{2+}. These few examples underscore the vital role that in vivo studies play in the context of discovering novel metabolites of toxicological importance.

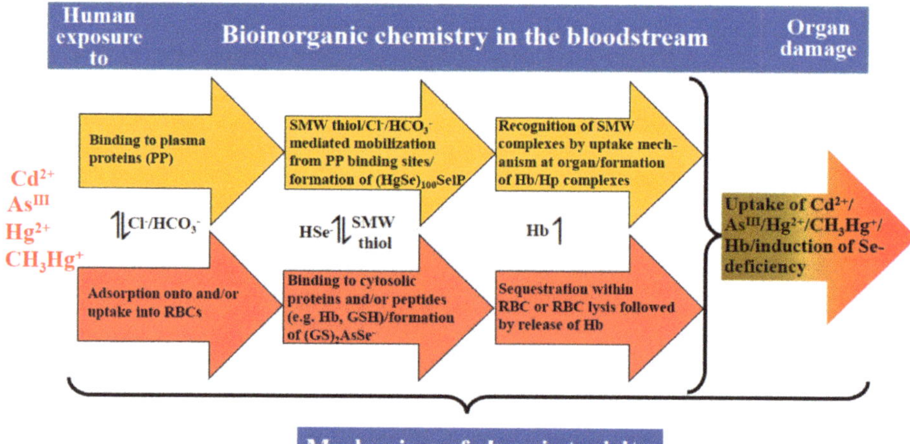

Figure 3. Bioinorganic chemistry of toxic metal(loid) species that unfold in plasma (yellow) and red blood cells (red) are critical to establish their mechanism of chronic toxicity, which determines the degree of organ damage over weeks/months/years. Abbreviations: red blood cells (RBC), small molecular weight (SMW), hemoglobin (Hb), selenoprotein P (SelP), haptoglobin (Hp), glutathione (GSH).

5. Toxic Metal(loid) Species at the Plasma–RBC–Organ Nexus

From a conceptual point of view, the 'missing link' in the toxicology of metal(loid) species ultimately lies in causally linking molecular data with human health effects [83]. At face value, this problem translates to an enormous transdisciplinary chemistry–biology problem because 'molecular data' refers to toxicologically relevant interactions of any given toxic metal(loid) species with biomolecules in the bloodstream [84] and in target organs [85] which then collectively result in a specific adverse health effect at the whole organism level (e.g., kidney damage). Understanding the associated exposure–response relationship is critically dependent on predicting how much organ damage a toxic metal(loid) species absorbed into the bloodstream (As^{III}, Cd^{2+}, Hg^{2+} and CH_3Hg^+) can ultimately inflict. This goal requires one to unravel toxicologically relevant bioinorganic events in the bloodstream, which can involve chemistry of the toxic metal(loid) species in plasma (Figure 3, yellow arrows) [84] and/or in RBCs (Figure 3, red arrows) [75]. While these events can be studied separately, one must integrate the results to establish the metabolism of the toxic metal(loid) species at the plasma–RBC interface (see the bidirectional arrows between the yellow and the red arrows in Figure 3), which ultimately determines how much of the initial dose will be able to engage in organ damage 'downstream' [86]. The dynamic bioinorganic events that unfold at the plasma–RBC–organ nexus are, in all likelihood, also responsible for the considerable lag-phase that can exist between the onset of the chronic exposure of an organism to toxic metal(loid) species and the time when adverse organ-based toxic effects manifest themselves [52,87].

From a mechanistic point of view, chronic human exposure to toxic metal(loid) species can involve three bioinorganic processes that happen in the bloodstream, which may considerably reduce the fraction of the toxic metal(loid) species that is not 'detoxified' therein, and therefore, are of toxicological relevance. One important process is the sequestration of a toxic metal(loid) species within RBCs (e.g., CH_3Hg^+ [88], Cd [89]) or the formation of non-toxic complexes with essential elements, such as selenium either in blood plasma [73] or RBCs [90] (Figure 3). Another process that needs to be considered is a toxic-metal(loid)-species-mediated lysis of RBCs, which releases Hb to plasma where it tightly binds to the plasma protein haptoglobin (Hp). Since the binding capacity of the latter is limited by its plasma concentration, however, the release of too much Hb can result in 'free' Hb in plasma which can then cause severe kidney damage [91]. Yet another process by which a toxic metal(loid) species may contribute to organ damage is the gradual induction of a trace-element deficiency based on events that unfold in the bloodstream–organ system [73,81,86].

Last but not least, one needs to consider a toxic-metal(loid)-species-induced adverse effect on the assembly of RBCs in the bone marrow, which may adversely affect the cytosolic concentrations of metalloproteins in the RBCs that are released into the blood circulation [e.g., a decreased hemoglobin (Hb) concentration therein corresponds to anemia], or may result in a decreased lifetime of RBCs in the blood circulation [92]. The exposure of mammals to a toxic metal(loid) species can also result in the hypoproduction of the hormone erythropoietin, which stimulates the production of RBCs and would, therefore, result in another form of anemia [93]. Taken together, all of these individual bioinorganic mechanisms collectively contribute to organ damage over weeks, months and possibly years depending on the dose of the toxic metal(loid) species.

From a practical view, gaining insight into the bioinorganic chemistry of toxic metal(loid)s in the bloodstream is hampered by its intrinsic biological complexity. The seemingly simple task of identifying those plasma proteins that specifically bind any given toxic metal(loid) species translates—owing to the presence of thousands of plasma proteins—into a considerable separation problem, but can be dramatically simplified by using so-called metallomics tools [13,16]. The simultaneous detection capability of some metallomics tools allows one to use the ~10 endogenous plasma/serum metalloproteins that contain transition metals (Cu, Fe and Zn) (Table 1) [94,95] and the ~4 metalloproteins that are present in RBC lysate (Table 2) as internal standards, which inherently represent molecular-weight markers and, thus, validate an obtained analytical result. Furthermore, exogenous toxic metal(loid) species, such as As^{III}, Cd^{2+}, Hg^{2+} and CH_3Hg^+ can also bind to the aforementioned metalloproteins and/or proteins in plasma and/or RBC cytosol. In blood plasma, for example, Cd^{2+} has been shown to bind to serum albumin in rats [96]. Likewise, cisplatin-derived metabolites have been demonstrated to bind to HSA in human plasma [76], but the binding appears to be irreversible. In this context, the irreversible binding of a metallodrug to HSA renders the formed metallodrug–HSA complex ineffective for uptake into tumor cells [58], unless the tumor cells can able to achieve this [97] (e.g., by macropinocytosis). Furthermore, Hg^{2+}, CH_3Hg^+ and $CH_3CH_2Hg^+$ have each been demonstrated to bind to Hb in the matrix of RBC cytosol [75]. The formation of adducts between a toxic metal species and a metalloprotein can have important toxicological ramifications. For example, the binding of $CH_3CH_2Hg^+$ to Hb is associated with its conformational change, which significantly decreases its O_2 binding capacity [98].

Table 1. Characteristics of major metalloproteins in human plasma.

Metal	Metalloprotein or Biomolecules which Contain Bound Metal(s)	Molecular Mass (kDa)	Number of Metal Atoms Bound per Protein	Reference
Fe	Ferritin	450	<4500	[94]
	Transferrin	79.9	1	[94]
	Haptoglobin–Hemoglobin complex	86–900	2	[99]
Cu	Blood coagulation factor V	330	1	[94]
	Transcuprein	270	0.5	[94]
	Ceruloplasmin	132	6	[94]
	Albumin	66	1	[94]
	Extracellular Superoxide Dismutase	165	4	[94]
	Peptides and amino acids	<5	-	[94]
Zn	α_2 macroglobulin	725	5	[94]
	Albumin	66	1	[94]
	Extracellular Superoxide Dismutase	165	4	[94]

Table 2. Characteristics of major metalloproteins in red blood cell cytosol.

Metal	Metalloprotein	Molecular Mass (kDa)	Number of Metal Atoms Bound per Protein	Reference
Fe	Hemoglobin	64.5	4	[100]
	Catalase	240	4	[101]
Zn	Carbonic anhydrase	30	1	[102]
	Superoxide Dismutase	32	1	[103]
Cu	Superoxide Dismutase	32	1	[103]

To date, the application of LC-based metallomics tools has provided important new insight into the bioinorganic chemistry of toxic metal(loid)s in the bloodstream (Table 3) by observing bi- and trimetallic complexes that contain essential and toxic metals [69], which contributed important insight into the biomolecular mechanisms that mediate organ damage (Figure 3) [86]. In terms of probing the interaction of toxic metal(loid) species with constituents of plasma, it is important to note that the latter contains about 400 small molecular weight (SMW) molecules and metabolites, including amino acids, peptides, fatty acids and nucleotides that are present at μM concentrations [104]. The variation in concentration of these SMW molecules in the bloodstream has been demonstrated to be, in part, genetically determined [104]. These SMW molecules and metabolites are likely to play an important role in the distribution of toxic metal species to target organ tissues [105] (Figure 3). Direct experimental evidence in support of this comes from a recent study in which a metallomics tool was employed to demonstrate that homocysteine (hCys) is critically involved in the delivery of CH_3Hg^+ to L-type large neutral amino acid transporters (LATs) [106], which mediate the uptake of CH_3Hg^+-Cys complexes into the brain [84]. hCys, an intermediate metabolite formed by the de-methylation of methionine [107], may therefore be also involved in the translocation of other toxic metal(loid) species to target organs. It is possible that in patients with hyperhomocysteinemia— who exhibit elevated levels of hCys in blood plasma (e.g., >15 μM) which has been linked to the development of cardiovascular disease, stroke, and Alzheimer's Disease (AD) [107]—the metabolism of toxic metal(loid)s is significantly altered and may potentially exacerbate disease progression. Other SMW species that are also likely to be implicated in the organ uptake of toxic metal(loid) species are Cl^- and HCO_3^- [108], which may therefore be indirectly implicatedin neurodegenerative diseases [109].

6. Metallodrugs/Novel Metal(loid) Complexes at the Plasma–RBC–Organ–Tumor Nexus

Metallodrugs [57], including cisplatin—the oldest and best-known metallodrug [110]—are used as therapeutics for the treatment of various diseases; most prominently, cancer. However, the severe toxic side-effects of this Pt-based anticancer metallodrug prompted the development of other Pt-based drugs, such as oxaliplatin which kills cells by inducing ribosome biogenesis stress rather than the cisplatin-mediated DNA-damage response [111]. An alternative strategy to develop better metallodrugs is to embrace different metals altogether, which has resulted in the synthesis of anticancer active Cu complexes as well as ferrocene-functionalized Ru(II) arene complexes with interesting pharmacological properties [112,113]. Despite these encouraging advances, the translation of results from in vitro studies to therapeutic success using in vivo animal models, and eventually clinical studies, remains a major challenge. Conceptually there are two reasons why metallodrugs may be inherently better-suited to target disease-relevant proteins than some one-dimensional (1D, linear) or two-dimensional (2D, planar) organic molecules. Firstly, metallodrugs allow one to build a drug molecule that provides a wider range of geometries around a metal center in three dimensions (3D conformation) compared to 75% of linear and planar organic FDA-approved drug molecules, which offers the possibility of more-selectively inhibiting the active site of a target protein [114–117]. Secondly, certain metallodrugs can induce long-lasting anticancer immune responses [118], thus offering the potential to develop 'smart' metallodrugs [119]. The latter term refers to drugs that more-selectively target cancer cells, which is in accordance with the goal of achieving 'precision medicine' [120]. Despite these potential advantages and the fact that many novel structures are reported annually, only a negligibly small number of novel metal(loid) complexes enter clinical studies [112,117,121]. This undesirable situation is attributed to a) the common misperception that, similar to cisplatin [56], all metallodrugs are associated with severe toxic side-effects and b) the challenge of translating more metal(loid) complexes through systematic in vitro and in vivo studies to successful metallodrugs. The magnitude of the latter dilemma is illustrated by the fact that certain metal complexes exhibit potent cytotoxicity toward cancer cell lines in vitro, but show no antitumoral activity in vivo and may even display nephrotoxicity [122]. These facts hint at a transdisciplinary 'chemistry–medicine' problem of equal proportions to the aforementioned 'chemistry–biology' problem (see Section 5).

The fundamental problem that is associated with the assessment of novel metal entities—at least in our opinion—is that their interactions with constituents of the bloodstream are often not sufficiently considered [123]. We would like to illustrate this important bottleneck in the assessment of novel metal entities based on what is known about the fate of cisplatin in the human bloodstream (Figure 4). While cisplatin is an ideal candidate in this context because we know more about its biochemical fate in the bloodstream than any other anticancer active metallodrug, we point out that this prototypical metallodrug lacks the desired tumor selectivity.

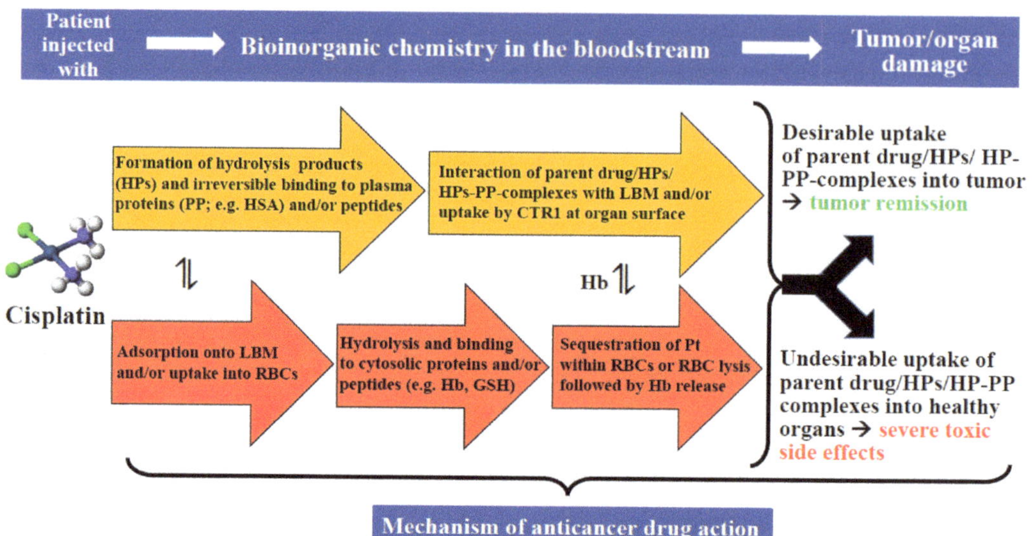

Figure 4. Bioinorganic chemistry of cytotoxic anticancer metallodrugs that unfold in plasma (yellow) and red blood cells (red) are critical to establish their mechanism of action, as they determine the degree of tumor damage/tumor shrinkage over days/weeks. Abbreviations: human serum albumin (HSA), red blood cells (RBC), hemoglobin (Hb), glutathione (GSH), lipid bilayer membrane (LBM), copper transporter 1 (CTR1).

6.1. *Stability of Metallodrugs/Novel Metal(loid) Complexes in Plasma*

A low temporal stability of a novel metal complex in plasma will dramatically reduce its probability of reaching the cancer tissue intact, and thus, increases the probability of severe toxic side-effects that are caused by the formed degradation products [120]. Compared to small organic molecules, this potential problem is much more troublesome for metallodrugs since a rupture of the bond that anchors the metal to its drug framework will compromise the integrity of the drug molecule, in turn adversely affecting not only the intended pharmacological effect and possibly also contributing to its severe adverse side-effects. Metallomics tools, which involve the hyphenation of a separation technique [e.g., capillary electrophoresis (CE), size-exclusion chromatography (SEC), high-/ultra-performance liquid chromatography (HPLC/UPLC)] with a metal-detector [e.g., inductively-coupled plasma mass spectrometry (ICP-MS), inductively coupled plasma atomic emission spectroscopy (ICP-AES)] can provide much-needed insight into the temporal stability of metallodrugs in serum [124].

The anticancer drug cisplatin, for example, undergoes several bioinorganic chemistry processes in blood plasma, including hydrolysis followed by the binding of the formed hydrolysis products to plasma proteins [76]. These dynamic bioinorganic processes gradually decrease the plasma concentration of the parent drug that is able to reach the intended tumor tissue intact, while the formed hydrolysis products are likely responsible for its severe side-effects (Figure 4) [76]. In this context, it is important to point out that the mere observation of a degradation product of a novel metal complex in blood plasma using a metallomics tool cannot provide any information about its toxicity, as this information has to be established using toxicological assays. The capability of metallomics tools to detect highly cytotoxic Pt species in plasma is reminiscent of their ability to identify potentially toxic metal species in human serum from environmentally exposed people [125] and to determine natural variations in the isotopic composition of CH_3Hg^+ and Hg^{2+} in fish tissue [126]. The application of SEC-ICP-AES has been particularly valuable to demonstrate that 70% of the bimetallic Ti- and Au-containing complex Titanocref remains intact in human plasma after incubation for 60 min at 37 °C [63], while the Ti- and Au-containing degradation products eluted bound to HSA. This comparatively novel metallomics tool has also been successfully applied to compare in vitro the metabolism of pharmacologically relevant doses of the FDA-approved platinum drugs cisplatin and carboplatin in human blood plasma [76]. The results revealed a comparatively faster hydrolysis of cisplatin followed by the binding of the formed hydrolysis products to plasma proteins, predominantly HSA, over a 24 h period compared to carboplatin. In addition, the hydrolysis products that formed from each platinum drug appeared to bind to the same plasma proteins. Importantly, these temporal changes in the metabolism of cisplatin/carboplatin provided an explanation for their considerably different toxic side-effects in patients [76] and the different biodistribution of these platinum drugs to organs after treatment.

Conceptually related to the application of SEC-ICP-AES to probe the stability of metallodrugs and novel metal complexes in plasma, this metallomics tool has also allowed researchers to visualize the deliberate modulation of the metabolism of cisplatin in human plasma by D-methionine [127], N-acetyl-L-cysteine [128] and sodium thiosulfate [129], which revealed that these SMW sulfur compounds can be used to 'neutralize' a highly toxic cisplatin hydrolysis product that is largely responsible for the severe side-effects of this Pt drug [130]. Thus, metallomics tools can be employed to obtain insight into the time-dependent in vitro metabolism of novel metal complexes in blood plasma, which is useful not only to estimate how much of the administered anticancer metal complex is likely to reach the cancer tissue intact [12,63], but also to gain insight into the formation of degradation products that may cause undesirable severe side-effects at the organ level [130,131].

6.2. Metallodrugs-/Novel-Metal(loid)-Complexes-Induced Rupture of RBCs

A novel-metal-entity-induced rupture of RBCs in the bloodstream results in the release of Hb into plasma, which can cause severe toxic side-effects, such as potentially life-threatening thrombotic effects and kidney toxicity [91]. Cisplatin, for example, is well-known to perturb the integrity of cell membranes—resulting in the damage of chicken RBCs [62]—which likely contributes to its severe side-effects in treated patients [56,123]. Based on the well-established metabolism of cisplatin in blood plasma [76], however, it is unknown if the integrity of the RBC lipid bilayer membrane is perturbed by cisplatin or its hydrolysis products. A recently developed metallomics tool can be employed to analyze blood plasma (50 μL) for a hemoglobin–haptoglobin complex to estimate the RBC rupture that is induced by a pharmacologically relevant dose of a metallodrug when added to whole blood [132].

6.3. Assessment of the Selectivity of Metallodrugs/Novel Metal(loid) Complexes to Cancer Cells

The selectivity of a novel metal(loid) complex toward the intended cancer can be assessed in cell-culture studies by measuring the IC_{50} value in cancer cells and healthy cells (e.g., fibroblasts). The advent of single cell (sc) ICP-MS over the last couple of years has demonstrated that it can—at least in principle—be employed to observe the uptake of cisplatin into sensitive and resistant cancer cells [133]. Remarkably, this inherent capability offers the ability to probe the selectivity of novel metal(loid) complexes under near-physiological conditions in vitro, with the caveat that, in the presence of the bloodstream, the same selectivity may not be obtainable in vivo.

Faced with the urgent need to accelerate more novel metal(loid) complexes that exhibit the desired biological activity to clinical studies, the application of certain metallomics tools (SEC-ICP-AES) can provide useful information about potentially adverse effects of a metallodrug in the bloodstream (i.e., RBC rupture; degradation in plasma), while other metallomics tools (scICP-MS) can allow for probing the selectivity of the metallodrug to preferentially 'hit' cancer cells compared to healthy cells in vitro [133]. The analysis of a set of novel metal(loid) complexes with different metallomics tools, therefore, offers the prospect of identifying those that are stable in plasma, do not compromise the integrity of RBCs (in blood) and exhibit selectivity toward cancer cells (Table 3). The complexes that fulfill these criteria can then be forwarded to preclinical and clinical studies. While knowledge about the mechanism of action of an anticancer drug at the biomolecular level is unquestionably useful [134], it is not necessarily a priority when screening novel metal complexes. Instead, their selective delivery to the tumor tissue should be the primary focus [63,135].

7. Integrative Metallomics Studies

In order to make progress in terms of linking 'molecular data' of potentially toxic metal(loid) species and cytotoxic metallodrugs with undesirable or desirable human health effects, it is necessary to establish the complete sequence of *bioinorganic chemical reactions* in various *biological compartments* of the bloodstream–organ system (i.e., plasma, RBCs and organs) (Figures 3 and 4) (Table 3). With regard to the exposure of an organism to a toxic metal(loid) species, one potential challenge is to measure an outcome in a target organ (e.g., nutrient deficiency) [69,86,108], while it is comparatively easy to observe the shrinkage/remission of a tumor after treatment with a metallodrug [136]. Our limited understanding of the 'molecular toxicology' of toxic metal(loid) species and metallodrugs in the bloodstream must be attributed to the fact that most studies are conducted using either blood plasma or RBC lysate for practical reasons, since the shelf life of whole blood is rather limited. This approach to addressing the bioinorganic chemistry in individual biological compartments can intrinsically not answer the most crucial question of which specific metal species will impinge on a toxicological target organ and/or tumor in the whole organism. In a sense, this undesirable situation is somewhat reminiscent of the problem that proteomics researchers faced who chose to study purified proteins in isolation for many years without realizing that addressing the biological complexity requires one to study protein–protein interactions to describe the events that unfold inside any given cell [137]. Since it is ultimately the flux of toxic metal(loid) substances that impinge on toxicological target organs/tumors and determine the damage [detrimental in the context of toxic metal(loid) species, but desirable in the context of metallodrugs], a better understanding of the corresponding bioinorganic chemistry in the bloodstream will contribute to advancing toxicology [108], ecotoxicology [138] and pharmacology [97]. Integrating the bioinorganic mechanisms that unfold in the bloodstream with biomolecular uptake mechanisms that are located at the organ surface [139] will inevitably advance human health, as these processes play a crucial role in terms of a) causally linking human exposure to toxic metal(loids) to disease and b) ultimately achieving 'precision oncology' [120]. Some metallomics tools, such as X-ray-based spectroscopy, even offer the capability to probe deeper inside the organ/tumor tissue to obtain information on the sub-cellular compartment that a specific toxic metal(loid) or metallodrug targets in a healthy cell or cancer cell [140]. The innovative applications of metallomics tools thus represent an essential first step in terms of unravelling the potential pharmacology and the mechanism of action of novel metal(loid) complexes to further advance the important role that integrated metallomics are destined to play in health and disease [15].

Table 3. Overview of studies which exemplify the usefulness of different metallomics tools to probe the interaction of toxic metal(loid) substances with different biological fluids and/or cells.

Toxic Metal(loid) Species in Biological Fluid	Metallomics Tool(s)	Investigated Species	Obtained Information	Reference
Blood plasma (in vitro)	SEC-ICP-AES	Cd^{2+}	Cd^{2+}-driven displacement of Zn^{2+} from a Zn metalloprotein	[16]
	SEC-ICP-AES	CH_3Hg^+	formation of hCys-CH_3Hg^+ complexes	[84]
Red blood cell cytosol (in vitro)	SEC-ICP-AES and XAS	Hg^{2+}, CH_3Hg^+, $CH_3CH_2Hg^+$, Cd^{2+}	formation of stable complexes of Hg^{2+}, CH_3Hg^+ and $CH_3CH_2Hg^+$ with hemoglobin	[75]
	SEC-ICP-AES and XAS	CH_3Hg^+ and $(GS)_2AsSe^-$	formation of $(GS)_2AsSe$-$HgCH_3$	[90]
Bile (in vivo)	XAS	injection of hamsters with As^{III} and Se^{IV}	detection of $(GS)_2AsSe^-$	[81]
Metallodrug in Biological Fluid/Compartment	**Metallomics Tool**	**Investigated Metallodrug**	**Obtained Information**	**Reference**
Blood plasma (in vitro)	SEC-ICP-AES	cisplatin and carboplatin, Titanocref	stability/degree of hydrolysis/degradation	[76] [63]
	SEC-ICP-AES	cisplatin	formation of complexes between a cisplatin-derived hydrolysis product and thiosulfate	[129]
	SEC-ICP-AES	cisplatin	formation of complex between a cisplatin-derived hydrolysis product and N-acetyl-L-cysteine	[128]
	SEC-ICP-AES	cisplatin	formation of complexes between a cisplatin-derived hydrolysis product and D-methionine	[127]
	SEC-ICP-AES	cisplatin	modulation of the metabolism of cisplatin in serum of cancer patients with human serum albumin (HSA)	[141]
	SEC-ICP-AES	(2,2′:6′2″-terpyridine) platinum (II) complexes	binding to rabbit serum albumin	[131]
	SEC-ICP-AES	cisplatin and NAMI-A	comparative metabolism	[142]
Whole blood (in vitro)	SEC-ICP-AES SEC-GFAAS	no metallodrugs investigated	dose-dependent effect of metallodrug on RBC lysis	[99] [132]
Healthy and cancer cells (cell culture)	scICP-MS	cisplatin	selectivity of metallodrug	[133]

8. Conclusions

Throughout life, the dynamic flow of toxic metal(loid) species inevitably connects every mammalian organism to the surface geochemistry of our home planet. Because of the paucity of information about possible mechanistic links that functionally connect environmental exposure to toxic metal(loid)s with adverse pregnancy outcomes [143], neurodevelopment in children [144], adverse effects on organs [145] and human diseases of unknown etiology [54,108], gaining insight into the underlying bioinorganic chemistry in the environment–bloodstream–organ system [146] is critical, especially given the rapidly growing problem of dealing with electronic waste—of which we generate 53.5 million metric t per year [147]. Elucidating this bioinorganic chemistry not only holds the prospect of causally linking human environmental exposure to toxic metal(loid) species with the etiology of neurodegenerative diseases including Alzheimer's Disease [148] and to curb emissions into the environment, but also to develop affordable, practical solutions [149]. Relatedly, the development of better anticancer metal-base drugs hinges on improving our screening process of novel metal complexes to include events that unfold in the bloodstream and, in turn, advance more drug candidates to preclinical studies [12]. The integration of bioinorganic chemistry events in the blood plasma–RBC–organ system is therefore critical to causally link human exposure to toxic metal(loid) species with diseases [150] and to help discover the next generation of metallodrugs.

Funding: Funding for some of the studies cited in the article was provided by the Natural Sciences and Engineering Research Council (NSERC) of Canada.

Institutional Review Board Statement: Studies that were conducted at the University of Calgary were conducted in accordance with the Declaration of Helsinki, and approved by the Life & Environmental Sciences Animal Care Committee (LESACC) and the Conjoint Health Research Ethics Board (CHREB).

Informed Consent Statement: Informed consent was obtained from all subjects that were involved in studies conducted at the University of Calgary.

Data Availability Statement: The data presented in this study that were obtained at the University of Calgary are available on request from the corresponding author.

Acknowledgments: Maryam Doroudian is supported by a grant from the Natural Sciences and Engineering Research Council (NSERC) of Canada. We greatly appreciate one particular reviewer's feedback which helped to improve the overall quality and clarity of our manuscript.

Conflicts of Interest: The authors declare no conflict of interest.

References

1. Sousa, F.L.; Neunkirchen, S.; Allen, J.F.; Lane, N.; Martin, W.F. Lokiarcheon is hydrogen dependent. *Nat. Microbiol.* **2016**, *16034*, 1–3.
2. Li, J.J.; Pawitwar, S.S.; Rosen, B.P. The organoarsenical biocycle and the primordial antibiotic methylarsenite. *Metallomics* **2016**, *8*, 1047–1055. [CrossRef]
3. Vandal, G.M.; Fitzgerald, W.F.; Boutron, C.F.; Candelone, J.-P. Variations in mercury deposition to Antarctica over the past 34,000 years. *Nature* **1993**, *362*, 621–6623. [CrossRef]
4. McConnell, J.R.; Edwards, R. Coal burning leaves toxic heavy metal legacy in the Arctic. *Proc. Natl. Acad. Sci. USA* **2008**, *105*, 12140–12144. [CrossRef] [PubMed]
5. Izatt, R.M.; Izatt, S.R.; Bruening, R.L.; Izatt, N.E.; Moyer, B.A. Challenges to achievement of metal sustainability in our high-tech society. *Chem. Soc. Rev.* **2014**, *43*, 2451–2475. [CrossRef]
6. Campbell, P.G.C.; Gailer, J. Effects of Non-essential Metal Releases on the Environment and Human Health. In *Metal Sustainability: Global Challenges, Consequences and Prospects*; Izatt, R.M., Ed.; John Wiley & Sions Ltd.: Chichester, UK, 2016; pp. 221–252.
7. Llanos, E.J.; Leal, W.; Luu, D.H.; Jost, J.; Stadler, P.F.; Restrepo, G. Exploration of the chemical space and its three historical regimes. *Proc. Natl. Acad. Sci. USA* **2019**, *116*, 12660–12665. [CrossRef]
8. Stern, F. Paul Ehrlich: The Founder of Chemotherapy. *Angew. Chem. Int. Ed.* **2004**, *43*, 4254–4261. [CrossRef]
9. Marzo, T.; La Mendola, D. Strike a balance: Between metals and non-metals, metalloids as a source of anit-infective agents. *Inorganics* **2021**, *9*, 46. [CrossRef]
10. Bolli, R. William Harvey and the discovery of the circulation of the blood. *Circul. Res.* **2019**, *124*, 1300–1302.
11. Gailer, J. Reactive selenium metabolites as targets of toxic metals/metalloids in mammals: A molecular toxicological perspective. *Appl. Organometal. Chem.* **2002**, *16*, 701–707. [CrossRef]
12. Sarpong-Kumankomah, S.; Gailer, J. Application of a novel metallomics tool to probe the fate of metal-based anticancer drugs in blood plasma: Potential, challenges and prospects. *Curr. Top. Med. Chem.* **2021**, *21*, 48–58. [CrossRef]
13. Doroudian, M.; Gailer, J. Toxic metal(loid) species at the blood-organ interface. In *Environmental and Biochemical Toxicology*; Turner, R.J., Gailer, J., Eds.; DeGruyter: Munich, Germany, 2022; Volume 1, p. 335.
14. Zheng, X.; Cheng, Y.; Wang, C. Global mapping of metalloproteomes. *Biochemistry* **2021**, *60*, 3507–3514. [CrossRef]

15. Zhou, Y.; Li, H.; Sun, H. Metalloproteomics for biomedical research: Methodology and applications. *Annu. Rev. Biochem.* **2022**, *91*, 449–473. [CrossRef] [PubMed]
16. Gomez-Ariza, J.L.; Jahromi, E.Z.; Gonzalez-Fernandez, M.; Garcia-Barrera, T.; Gailer, J. Liquid chromatography-inductively coupled plasma-based metallomic approaches to probe health-relevant interactions between xenobiotics and mammalian organisms. *Metallomics* **2011**, *3*, 566–577. [CrossRef] [PubMed]
17. Theiner, S.; Schoeberl, A.; Schweikert, A.; Keppler, B.K.; Koellensperger, G. Mass spectrometry techniques for imaging and detection of metallodrugs. *Curr. Opin. Chem. Biol.* **2021**, *61*, 123–134. [CrossRef]
18. Arruda, M.A.Z.; Jesus, J.R.D.; Blindauer, C.A.; Stewart, A.J. Speciomics as a concept involving chemical speciation and omics. *J. Proteom.* **2022**, *263*, 104615. [CrossRef] [PubMed]
19. Pacyna, J.M. Monitoring and assessment of metal contaminants in the air. In *Toxicology of Metals*; Chang, L.W., Ed.; CRC Press: Boca Raton, Florida, FL, USA, 1996; pp. 9–28.
20. Jonsson, S.; Andersson, A.; Nilsson, M.B.; Skyllberg, U.L.; Lundberg, E.; Schaefer, J.K.; Akerblom, S.; Bjoern, E. Terrestrial dischanrges mediate trophic shifts and enhance methylmercury accumulation in estuarine biota. *Sci. Adv.* **2017**, *3*, e1601239. [CrossRef] [PubMed]
21. Steffen, W.; Crutzen, P.J.; McNeill, J.R. The anthropocene: Are humans now overwhelming the great forces of nature? *Ambio* **2007**, *36*, 614–621. [CrossRef]
22. Kasozi, K.I.; Otim, E.O.; Ninsiima, H.I.; Zirintunda, G.; Tamale, A.; Ekou, J.; Musoke, G.H.; Muyinda, R.; Matama, K.; Mujinga, R.; et al. An analysis of heavy metals contamination and estimating the daily intakes of vegetables in Uganda. *Toxicol. Res. Appl.* **2021**, *5*, 1–15. [CrossRef]
23. Borchers, A.; Teuber, S.S.; Keen, C.L.; Gershwin, M.E. Food safety. *Clin. Rev. Allergy Immunol.* **2010**, *39*, 95–141. [CrossRef]
24. Schartup, A.T.; Thackray, C.P.; Qureshi, A.; Dassuncao, C.; Gillespie, K.; Hanke, A.; Sunderland, E.M. Climate change and overfishing increase neurotoxicant in marine predators. *Nature* **2019**, *572*, 648–650. [CrossRef] [PubMed]
25. Dufault, R.; LeBlanc, B.; Schnoll, R.; Cornett, C.; Schweitzer, L.; Wallinga, D.; Hightower, J.; Patrick, L.; Lukiw, W.J. Mercury from chlor-alkali plants: Measured concentrations in food product sugar. *Environ. Health* **2009**, *8*, 1–6. [CrossRef] [PubMed]
26. McFarland, M.J.; Hauer, M.E.; Reuben, A. Half of US population exposed to adverse lead levels in early childhood. *Proc. Natl. Acad. Sci. USA* **2022**, *119*, e2118631119. [CrossRef] [PubMed]
27. Guan, W.J.; Zheng, X.Y.; Chung, K.F.; Zhong, N.S. Impact of air pollution on the burden of chronic respiratory diseases in China: Time for urgent action. *Lancet* **2016**, *388*, 1939–1951. [CrossRef]
28. Mills, N.L.; Donaldson, K.; Hadoke, P.W.; Boon, N.A.; MacNee, W.; Cassee, F.R.; Sandstroem, T.; Blomberg, A.; Newby, D.E. Adverse cardionvascular effects of air pollution. *Nat. Clin. Pract. Card.* **2009**, *6*, 36–44. [CrossRef]
29. Steckling, N.; Tobollik, M.; Plass, D.; Hornberg, C.; Ericson, B.; Fuller, R.; Bose-O'Reilly, S. Global burden of disease of mercury used in artisanal small-scale gold mining. *Ann. Glob. Health* **2017**, *83*, 234–247. [CrossRef]
30. Guney, M.; Zagury, G.J. Contamination by ten harmful elements in toys and children's jewelry bought on the North American market. *Environ. Sci. Technol.* **2013**, *47*, 5921–5930. [CrossRef]
31. Travasso, C. Skin whitening cream may contain mercury, and lipstick may contain chromium and nickel, Indian study shows. *Brit. Med. J.* **2014**, *348*, g1330. [CrossRef]
32. Hamann, C.R.; Boonchai, W.; Wen, L.; Sakanashi, E.; Chu, C.-Y.; Hamann, K.; Hamann, C.P.; Sinniah, K.; Hamann, D.; Linda, L. Spectrometric analysis of mercury content in 549 skin-lightening products: Is mercury toxicity a hidden global health hazard? *J. Am. Acad. Dermayol.* **2013**, *70*, 281–287. [CrossRef]
33. Landrigan, P.J.; Fuller, R.; Acosta, N.J.R.; Adeyi, O.; Arnold, R.; Basu, N.; Balde, A.B.; Bertollini, R.; Bose-O'Reilly, S.; Boufford, J.I.; et al. The Lancet commission on pollution and health. *Lancet* **2018**, *391*, 462–512. [CrossRef]
34. Trasande, L.; Liu, Y. Reducing the staggering costs of environmental disease in children, estimated at $76.6 billion in 2008. *Health Affair.* **2011**, *30*, 863–870. [CrossRef] [PubMed]
35. North, A.E.; Sarpong-Kumankomah, S.; Bellavie, A.R.; White, W.M.; Gailer, J. Environmentally relevant concentrations of aminopolycarboxylate chelating agents mobilize Cd from humic acid. *J. Environ. Sci.* **2017**, *57*, 249–257. [CrossRef] [PubMed]
36. Coulthard, T.J.; Macklin, M.G. Modeling long-term contamination in river systems from historical metal mining. *Geology* **2003**, *31*, 451–454. [CrossRef]
37. Pruvot, C.; Douay, F.; Herve, F.; Waterlot, C. Heavy metals in soil, crops and grass as a source of human exposure in the former mining areas. *J. Soils Sediments* **2006**, *6*, 215–220. [CrossRef]
38. Lamborg, C.H.; Hammerschmidt, C.R.; Bowman, K.L.; Swarr, G.J.; Munson, K.M.; Ohnemus, D.C.; Lam, P.J.; Heimbuerger, L.-E.; Rijkenberg, M.J.A.; Saito, M.A. A global ocean inventory of anthropogenic mercury based on water column measurements. *Nature* **2014**, *512*, 65–68. [CrossRef] [PubMed]
39. Streets, D.G.; Horowitz, H.M.; Jacob, D.J.; Lu, Z.; Levin, L.; ter Schure, A.F.H.; Sunderland, E.M. Total mercury released to the environment by human activities. *Environ. Sci. Technol.* **2017**, *51*, 5969–5977. [CrossRef]
40. Ogunseitan, O.A. Mercury safety reform in the 21st centrury: Advancing the new framework for toxic substance control. *Environ. Sci. Policy Sustain. Dev.* **2017**, *59*, 4–13. [CrossRef]
41. Hawkings, J.R.; Linhoff, B.S.; Wadham, J.L.; Stibal, M.; Lamborg, C.H.; Carling, G.T.; Lamarche-Gagnon, G.; Kohler, T.J.; Ward, R.; Hendry, K.R.; et al. Large subglacial source of mercury from the southwestern margin of the Greenland Ice Sheet. *Nat. Geosci.* **2021**, *14*, 496–502. [CrossRef]

42. Schaefer, H.R.; Dennis, S.; Fitzpatrick, S. Cadmium. Minimization strategies to reduce dietary exposure. *J. Food Sci.* **2020**, *85*, 260–267. [CrossRef]
43. Sebastian, A.; Prasad, M.N.V. Cadmium minimization in rice: A review. *Agron. Sustain. Dev.* **2014**, *34*, 155–173. [CrossRef]
44. Khan, S.; Cao, Q.; Zheng, Y.M.; Huang, Y.Z.; Zhu, Y.G. Health risks of heavy metals in contaminated soils and food crops irrigated with wastewater in Beijing, China. *Environ. Pollut.* **2008**, *152*, 686–692. [CrossRef] [PubMed]
45. Cristol, D.A.; Brasso, R.L.; Condon, A.M.; Fovarque, R.E.; Friedman, S.L.; Hallinger, K.K.; Monroe, A.P.; White, A.E. The movement of aquatic mercury through terrestrial food webs. *Science* **2008**, *320*, 335. [CrossRef] [PubMed]
46. Obrist, D.; Agnan, Y.; Jiskra, M.; Olson, C.L.; Colegrove, D.P.; Hueber, J.; Morre, C.W.; Sonke, J.E.; Helmig, D. Tundra uptake of atmospheric elemental mercury drives Arctic mercury pollution. *Nature* **2017**, *547*, 201–204. [CrossRef] [PubMed]
47. Nyanza, E.C.; Bernier, F.P.; Martin, J.W.; Manyama, M.; Hatfield, J.; Dewey, D. Effetcs of prenatal expsoure and co-exposure to metallic or metalloid elements on early infant neurodevelopment outcomes in areas with small-scale gold mining activities in Northern Tanzania. *Environ. Int.* **2021**, *149*, 106104. [CrossRef] [PubMed]
48. Mitchell, E.; Frisbie, S.; Sarkar, B. Exposure to multiple metals from groundwater-a global crisis: Geology, climate change, health effects, testing, and mitigation. *Metallomics* **2011**, *3*, 874–908. [CrossRef]
49. Sanchez-Rodas, D.; Sanchez de la Campa, A.; Martinez, M.M. The role of metalloids (As, Sb) in airborne particulate matter related to air pollution. In *Environmental and Biochemical Toxicology*; Turner, R.J., Gailer, J., Eds.; DeGruyetr: Munich, Germany, 2022; pp. 169–189.
50. Turner, R.J. Toxicity of nanomaterials. In *Environmental and Biochemical Toxicology*; DeGruyter: Munich, Germany, 2022; pp. 221–247.
51. Rappaport, S.M. Discovering environmental causes of disease. *J. Epidemiol. Community Health* **2012**, *66*, 99–102. [CrossRef]
52. Weiss, B.; Clarkson, T.W.; Simon, W. Silent latency period in methylmercury poisoning and in neurodegenerative disease. *Environ. Health Perspect.* **2002**, *110*, 851–854. [CrossRef]
53. Sturla, S.J.; Boobis, A.R.; Fitzgerald, R.E.; Hoeng, J.; Kavlock, R.J.; Schirmer, K.; Whelan, M.; Wilks, M.F.; Peitsch, M.C. Systems toxicology: From basic research to risk assessment. *Chem. Res. Toxicol.* **2014**, *27*, 314–329. [CrossRef]
54. Marth, J.D. A unified vision of the building blocks of life. *Nature Cell. Biol.* **2008**, *10*, 1015–1016. [CrossRef]
55. Rosenberg, B.; Van Camp, L.; Krigas, T. Inhibition of cell division in *Escherichia coli* by electrolysis products from a platinum electrode. *Nature* **1965**, *205*, 698–699. [CrossRef]
56. Oun, R.; Moussa, Y.E.; Wheate, N.J. The side effetcs of platinum-based chemotherapy drugs: A review for chemists. *Dalton Trans.* **2018**, *47*, 6645–6653. [CrossRef] [PubMed]
57. Mjos, K.D.; Orvig, C. Metallodrugs in medicinal inorganic chemistry. *Chem. Rev.* **2014**, *114*, 4540–4563. [CrossRef] [PubMed]
58. Sooriyaarachchi, M.; Morris, T.T.; Gailer, J. Advanced LC-analysis of human plasma for metallodrug metabolites. *Drug Discov. Today Technol.* **2015**, *16*, e24–e30. [CrossRef] [PubMed]
59. Hambley, T.W. Metal-based therapeutics. *Science* **2007**, *318*, 1392–1393. [CrossRef] [PubMed]
60. Poirier, A.E.; Ruan, Y.; Walter, S.D.; Franco, E.L.; Villeneuve, P.J.; King, W.D.; Volesky, K.D.; O'Sullivan, D.E.; Friedenreich, C.M.; Brenner, D.R. The future burden of cancer in Canada: Long-term cancer incidence projections. *Cancer Epidemiol.* **2019**, *59*, 199–207. [CrossRef]
61. Casini, A.V.A.; Meier-Menches, S.M. *Metal-Based Anticancer Agents*; Royal Society of Chemistry: London, UK, 2019.
62. Kutwin, M.; Sawosz, E.; Jaworski, S.; Kurantowicz, N.; Strojny, B.; Chwalibog, A. Structural damage of chicken red blood cells exposed to platinum nanoparticles and cisplatin. *Nanoscale Res. Lett.* **2014**, *9*, 257. [CrossRef]
63. Sarpong-Kumankomah, S.; Contel, M.; Gailer, J. SEC hyphenated to a multielement-specific detector unravels the degradation pathway of a bimetallic anticancer complex in human plasma. *J. Chromatogr. B* **2020**, *1145*, 122093. [CrossRef]
64. Gandin, V.; Ceresca, C.; Esposito, G.; Indraccolo, S.; Porchia, M.; Tisano, F.; Santini, C.; Pellei, M.; Marzano, C. Therapeutic potential of the phosphino Cu(I) complex (HydroCuP) in the treatment of solid tumors. *Sci. Rep.* **2017**, *7*, 13936. [CrossRef]
65. Marcus, W.L.; Rispin, A.S. Threshold carcinogenicity using arsenic as an example. In *Advances in Modern Environmental Toxicology*; Cothern, C.R., Mehlmans, M.A., Marcus, W.L., Eds.; Princeton Scientific Publishing Company, Inc: Princeton, NJ, USA, 1988; Volume 15, pp. 133–158.
66. Kitchin, K.T. The role of protein binding of trivalent arsenicals in arsenic carcinogenesis and toxicity. *J. Inorg. Biochem.* **2008**, *102*, 532–539. [CrossRef]
67. Waisberg, M.; Joseph, P.; Hale, B.; Beyersmann, D. Molecular and cellular mechanisms of cadmium carcinogenesis. *Toxicology* **2003**, *192*, 95–117. [CrossRef]
68. Anderson, N.L.; Anderson, N.G. The human plasma proteome: History, character, and diagnostic prospects. *Mol. Cell. Proteom.* **2002**, *1*, 845–867. [CrossRef] [PubMed]
69. Bridle, T.G.; Kumarathasan, P.; Gailer, J. Toxic metal species and 'endogenous' metalloproteins at the blood-organ interface: Analytical and bioinorganic aspects. *Molecules* **2021**, *26*, 3408. [CrossRef] [PubMed]
70. Righetti, P.G.; Boschetti, E. The Proteominer and the fortyniners: Searching for gold nuggest in the proteomic arena. *Mass Spec. Rev.* **2008**, *27*, 596–608. [CrossRef] [PubMed]
71. Gailer, J. Metal species in biology: Bottom-up and top-down LC approaches in applied toxicological research. *ISRN Chromatography* **2013**, *2013*, 21. [CrossRef]

72. Sarpong-Kumankomah, S.; Miller, K.; Gailer, J. Biological chemistry of toxic metals and metalloids, such as arsenic, cadmium and mercury. In *Encyclopedia of Analytical Chemistry*; John Wiley & Sons, Ltd.: Hoboken, NJ, USA, 2020; pp. 2006–2020.
73. Gailer, J. Arsenic-selenium and mercury-selenium bonds in biology. *Coord. Chem. Rev.* **2007**, *251*, 234–254. [CrossRef]
74. Kerek, E.; Hassanin, M.; Prenner, E.J. Inorganic mercury and cadmium induce rigidity in eukaryotic lipid extracts while mercury also ruptures red blood cells. *Biochim. Biophys. Acta* **2018**, *1860*, 710–717. [CrossRef]
75. Gibson, M.A.; Sarpong-Kumankomah, S.; Nehzati, S.; George, G.N.; Gailer, J. Remarkable differences in the biochemical fate of Cd^{2+}, Hg^{2+}, CH_3Hg^+ and thimerosal in red blood cell lysate. *Metallomics* **2017**, *9*, 1060–1072. [CrossRef]
76. Sooriyaarachchi, M.; Narendran, A.; Gailer, J. Comparative hydrolysis and plasma binding of cis-platin and carboplatin in human plasma in vitro. *Metallomics* **2011**, *3*, 49–55. [CrossRef]
77. Manley, S.A.; Gailer, J. Analysis of the plasma metalloproteome by SEC-ICP-AES: Bridging proteomics and metabolomics. *Expert. Rev. Proteom.* **2009**, *6*, 251–265. [CrossRef]
78. Gailer, J.; George, G.N.; Pickering, I.J.; Prince, R.C.; Ringwald, S.C.; Pemberton, J.E.; Glass, R.S.; Younis, H.S.; DeYoung, D.W.; Aposhian, H.V. A metabolic link between arsenite and selenite: The Seleno-bis(S-glutathionyl) Arsinium Ion. *J. Am. Chem. Soc.* **2000**, *122*, 4637–4639. [CrossRef]
79. Gailer, J.; George, G.N.; Pickering, I.J.; Madden, S.; Prince, R.C.; Yu, E.Y.; Denton, M.B.; Younis, H.S.; Aposhian, H.V. Structural basis of the antagonism between inorganic mercury and selenium in mammals. *Chem. Res. Toxicol.* **2000**, *13*, 1135–1142. [CrossRef] [PubMed]
80. Manley, S.A.; George, G.N.; Pickering, I.J.; Glass, R.S.; Prenner, E.J.; Yamdagni, R.; Wu, Q.; Gailer, J. The seleno bis (*S*-glutathionyl) arsinium ion is assembled in erythrocyte lysate. *Chem. Res. Toxicol.* **2006**, *19*, 601–607. [CrossRef] [PubMed]
81. Ponomarenko, O.; LaPorte, P.F.; Singh, S.P.; Langan, G.; Fleming, D.E.B.; Spallholz, J.; Alauddin, M.; Ahsan, H.; Ahmed, S.; Gailer, J.; et al. Selenium-mediated arsenic excretion in mammals: A synchrotron-based study of whole-body distribution and tissue-specific chemistry. *Metallomics* **2017**, *9*, 1585–1595. [CrossRef] [PubMed]
82. Gailer, J.; Ruprecht, L.; Reitmeir, P.; Benker, B.; Schramel, P. Mobilization of exogenous and endogenous selenium to bile after the intravenous administration of environmentally relevant doses of arsenite to rabbits. *Appl. Organometal. Chem.* **2004**, *18*, 670–675. [CrossRef]
83. Maret, W.; Moulis, J.M. The bioinorganic chemistry of cadmium in the context of its toxicity. In *Cadmium: From Toxicity to Essentiality, Metal Ions in Life Sciences*; Sigel, A.S.H., Sigel, R.K.O., Eds.; Springer Science+Business Media Dordrecht: Basel, Switzerland, 2013; Volume 25, pp. 1–29.
84. Bridle, T.G.; Doroudian, M.; White, W.; Gailer, J. Physiologically relevant hCys concentrations mobilize MeHg from rabbit serum albumin to form MeHg-hCys complexes. *Metallomics* **2022**, *14*, mfac010. [CrossRef] [PubMed]
85. Wang, Y.; Fang, J.; Leonard, S.S.; Rao, K.M.K. Cadmium inhibits the electron transfer chain and induces reactive oxygen species. *Free Rad. Biol. Med.* **2004**, *36*, 1434–1443. [CrossRef]
86. Sarpong-Kumankomah, S.; Gibson, M.A.; Gailer, J. Organ damage by toxic metals is critically determined by the bloodstream. *Coord. Chem. Rev.* **2018**, *374*, 376–386. [CrossRef]
87. Thijssen, S.; Maringwa, J.; Faes, C.; Lambrichts, I.; Van Kerkhove, E. Chronic exposure of mice to environmentally relevant, low doses of cadmium leads to early renal damage, not predicted by blood or urine cadmium levels. *Toxicology* **2007**, *229*, 145–156. [CrossRef]
88. Bakir, F.; Damluji, S.F.; Amin-Zaki, L.; Murtafdha, M.; Khalidi, A.; Al-Rawi, N.Y.; Tikriti, S.; Dhahir, H.I.; Clarkson, T.W.; Smith, J.C.; et al. Methylmercury poisoning in Iraq. *Science* **1973**, *181*, 230–241. [CrossRef]
89. Nordberg, G.F. Historical perspectives on cadmium toxicity. *Toxicol. Appl. Pharmacol.* **2009**, *238*, 192–200. [CrossRef]
90. Korbas, M.; Percy, A.J.; Gailer, J.; George, G.N. A possible molecular link between the toxicological effects of arsenic, selenium and methylmercury: Methylmercury(II) seleno bis (*S*-glutathionyl) arsenic(III). *J. Biol. Inorg. Chem.* **2008**, *13*, 461–470. [CrossRef] [PubMed]
91. Rother, R.P.; Bell, L.; Hillmen, P.; Gladwin, M.T. The clinical sequelae of intravascular hemolysis and extracellular plasma hemoglobin. *J. Am. Med. Assoc.* **2005**, *293*, 1653–1662. [CrossRef] [PubMed]
92. Kostic, M.M.; Ognjanovic, B.; Dimitrijevic, S.; Zikic, R.V.; Stajn, A.; Rosic, G.L.; Zivkovic, R.V. Cadmium-induced changes of antioxidant and metabolic status in red blood cells of rats: In vivo effects. *Eur. J. Haematol.* **1993**, *51*, 86–92. [CrossRef] [PubMed]
93. Horiguchi, H.; Oguma, E.; Kayama, F. Cadmium induces anaemia through interdependent progress of hemolysis, body iron accumulation, and insufficient erythropoietin production in rats. *Toxicol. Sci.* **2011**, *122*, 198–210. [CrossRef]
94. Manley, S.A.; Byrns, S.; Lyon, A.W.; Brown, P.; Gailer, J. Simultaneous Cu-, Fe-, and Zn-specific detection of metalloproteins contained in rabbit plasma by size-exclusion chromatography-inductively coupled plasma atomic emission spectroscopy. *J. Biol. Inorg. Chem.* **2009**, *14*, 61–74. [CrossRef] [PubMed]
95. Sarpong-Kumankomah, S.; Knox, K.B.; Kelly, M.E.; Hunter, G.; Popescu, B.; Nichol, H.; Kopciuk, K.; Ntanda, H.; Gailer, J. Quantification of human plasma metalloproteins in multiple sclerosis, ischemic stroke and health controls reveals an association of haptoglobin-hemoglobin complexes with age. *PLoS ONE* **2022**, *17*, e0262160. [CrossRef]
96. Suzuki, K.T.; Sunaga, H.; Kobayashi, E.; Shimojo, N. Mercaptoalbumin as a selective cadmium-binding protein in rat serum. *Toxicol. Appl. Pharmacol.* **1986**, *86*, 466–473. [CrossRef]
97. Massai, L.; Pratesi, A.; Gailer, J.; Marzo, T.; Messori, L. The cisplatin/serum albumin systen: A reappraisal. *Inorg. Chim. Acta* **2019**, *495*, 118983. [CrossRef]

98. de Magalhanes Silva, M.; de Araujo Dantas, M.D.; de Sila Filho, R.C.; dos Santos Sales, M.V.; de Almeida Xavier, J.; Leite, A.C.R.; Goulart, M.O.F.; Grillo, L.A.M.; de Barros, W.A.; de Fatima, A.; et al. Toxicity of thimerosal in biological systems: Conformational changes in human hemoglobin, decrease of oxygen binding, increase of protein glycation and amyloid formation. *Intern. J. Biol. Macromol.* **2020**, *154*, 661–671. [CrossRef]
99. Sarpong-Kumankomah, S.; Gailer, J. Identification of a haptoglobin-hemoglobin complex in human blood plasma. *J. Inorg. Biochem.* **2019**, *201*, 110802. [CrossRef]
100. Muckenthaler, M.U.; Rivella, S.; Hentze, M.W.; Galy, B. A red carpet for iron metabolism. *Cell* **2017**, *168*, 344–361. [CrossRef] [PubMed]
101. Gebicka, L.; Krych-Madej, J. The role of catalases in the prevention/promotion of oxidative stress. *J. Inorg. Biochem.* **2019**, *197*, 110699. [CrossRef] [PubMed]
102. Supuran, C.T. Structure and function of carbonic anhydrases. *Biochem. J.* **2016**, *473*, 2023–2032. [CrossRef] [PubMed]
103. Ordonez, Y.N.; Montes-Bayon, M.; Blanco-Gonzalez, E.; Sanz-Medel, A. Quantitative analysis and simultaneous activity measurements of Cu, Zn-superoxide dismutase in red blood cells by HPLC-ICPMS. *Anal. Chem.* **2010**, *82*, 2387–2394. [CrossRef] [PubMed]
104. Shin, S.-Y.; Fauman, E.B.; Petersen, A.-K.; Krumsiek, J.; Santos, R.; Huang, J.; Arnold, M.; Erte, I.; Forgetta, V.; Yang, T.-P.; et al. An atlas of genetic influences on human blood metabolites. *Nat. Genet.* **2014**, *46*, 543–550. [CrossRef]
105. Thomas, D.J.; Smith, J.C. Effects of coadministered low-molecular weight thiol compounds on short-term distribution of methylmercury in the rat. *Toxicol. Appl. Pharmacol.* **1982**, *62*, 104–110. [CrossRef]
106. Simmons-Wills, T.A.; Koh, A.S.; Clarkson, T.W. Transport of a neurotoxicant by molecualr mimicry: The methylmercury-L-cysteine complex is a substrate for human L-type large neutral amino acid transporter (LAT) 1 and LAT 2. *Biochem. J.* **2002**, *367*, 239–246. [CrossRef]
107. Jakubowski, H. Homocysteine modification in protein structure/function and human disease. *Physiol. Rev.* **2019**, *99*, 555–604. [CrossRef]
108. Hill, A.; Gailer, J. Linking molecular targets of Cd in the bloodstream to organ-based adverse health effects. *J. Inorg, Biochem.* **2021**, *216*, 111279. [CrossRef]
109. Costa, L.G.; Aschner, M.; Vitalone, A.; Syversen, T.; Soldin, O.P. Developmental neuropathology of environmental agents. *Annu. Rev. Pharmacol. Toxicol.* **2004**, *44*, 87–110. [CrossRef]
110. Kelland, L. The resurgence of platinum-based cancer chemotherapy. *Nat. Rev. Cancer* **2007**, *7*, 573–584. [CrossRef] [PubMed]
111. Bruno, P.M.; Liu, Y.; Park, G.Y.; Murai, J.; Koch, C.E.; Eisen, T.J.; Pritchard, J.R.; Pommier, Y.; Lippard, S.J.; Hemann, M.T. A subset of platinum-containing chemotherapeutic agents kill cells by inducing ribosome biogenesis stress rather that by engaging a DNA damage response. *Nat. Med.* **2017**, *23*, 461–471. [CrossRef] [PubMed]
112. Santini, C.; Pellei, M.; Gandin, V.; Porchia, M.; Tisato, F.; Marzano, C. Advances in copper complexes as anticancer agents. *Chem. Rev.* **2014**, *114*, 815–862. [CrossRef] [PubMed]
113. Mu, C.; Prosser, K.E.; Harrypersad, S.; MacNeil, G.A.; Panchmatia, R.; Thompson, J.R.; Sinha, S.; Warren, J.J.; Walsby, C.J. Activation by oxidation: Ferrocene-functionalized Ru(II)-arene complexes with anticancer, antibacterial, and antioxidant properties. *Inorg. Chem.* **2018**, *57*, 15247–15261. [CrossRef]
114. Morrison, C.N.; Prosser, K.E.; Stokes, R.W.; Cordes, A.; Metzler-Nolte, N.; Cohen, S.M. Expanding medicinal chemistry into 3D space: Metallofragments as 3D scaffolds for fragment-based drug discovery. *Chem. Sci.* **2020**, *11*, 1216–1225. [CrossRef]
115. Prosser, K.E.; Stokes, R.W.; Cohen, S.M. Evaluation of 3-dimensionality in approved and experimental drug space. *ACS Med. Chem. Lett.* **2020**, *11*, 1292–1298. [CrossRef]
116. Frei, A.; Zuegg, J.; Elliott, A.G.; Baker, M.; Braese, S.; Brown, C.; Chen, F.; Dowson, C.G.; Dujardin, G.; Jung, N.; et al. Metal complexes as a promising source for new antibiotics. *Chem. Sci.* **2020**, *11*, 2627–2639. [CrossRef]
117. Kenny, R.G.; Marmion, C.J. Toward multi-targeted platinum and ruthenium drugs-a new paradigm in cancer drug treatment regimens? *Chem. Rev.* **2019**, *119*, 1058–1137. [CrossRef]
118. Fronik, P.; Poetsch, I.; Kastner, A.; Mendrina, T.; Hager, S.; Hohenwallner, K.; Schueffl, H.; Herndler-Brandstetter, D.; Koellensperger, G.; Rampler, E.; et al. Structure-activity relationships of triple-action platinum(IV) prodrugs with albumin-binding properties and immunomodulating ligands. *J. Med. Chem.* **2021**, *64*, 12132–12151. [CrossRef]
119. Englinger, B.; Pirker, C.; Heffeter, P.; Terenzi, A.; Kowol, C.R.; Keppler, B.K.; Berger, W. Metal drugs and the anticancer immune response. *Chem. Rev.* **2019**, *119*, 1519–1624. [CrossRef]
120. Hambley, T.W. Transporter and protease mediated delivery of platinum complexes for precision oncology. *J. Biol. Inorg. Chem.* **2019**, *24*, 457–466. [CrossRef] [PubMed]
121. Bruijnincx, P.C.A.; Sadler, P.J. New trends for metal complexes with anticancer activity. *Curr. Opin. Chem. Biol.* **2008**, *12*, 197–206. [CrossRef] [PubMed]
122. Moretto, J.; Chauffert, B.; Ghiringhelli, F.; Aldrich-Wright, J.R.; Bouyer, F. Discrepancy between in vitro and in vivo antitumor effect of a new platinum(II) metallointercalator. *Invest. New Drugs* **2011**, *29*, 1164–1176. [CrossRef] [PubMed]
123. Mandal, R.; Kalke, R.; Li, X.-F. Interaction of oxaliplatin, cisplatin, and carboplatin with hemoglobin and the resulting release of a heme group. *Chem. Res. Toxicol.* **2004**, *17*, 1391–1397. [CrossRef] [PubMed]
124. Timerbaev, A.R.; Hartinger, C.G.; Aleksenko, S.S.; Keppler, B.K. Interactions of antitumor metallodrugs with serum proteins: Advances in characterization using modern analytical methodology. *Chem. Rev.* **2006**, *106*, 2224–2248. [CrossRef]

125. Levine, K.E.; Young, D.J.; Afton, S.E.; Harrington, J.M.; Essader, A.S.; Weber, F.X.; Fernando, R.A.; Thayer, K.; Hatch, E.E.; Robinson, V.C.; et al. Development, validation, and application of an ultra-performance liquid chromatography-sector field inductively coupled plasma mass spectrometry method for simultaneous determination of six organotin compounds in human serum. *Talanta* **2015**, *140*, 115–121. [CrossRef]
126. Entwisle, J.; Malinovsky, D.; Dunn, P.J.H.; Goenaga-Infante, H. Hg isotope ratio measurements of methylmercury in fish tissues using HPLC with off line cold vapour generation MC-ICPMS. *J. Anal. At. Spetcrom.* **2018**, *33*, 1645–1654. [CrossRef]
127. Sooriyaarachchi, M.; Narendran, A.; White, W.H.; Gailer, J. Chemoprotection by D-methionine against cis-platin-induced side-effects: Insight from in vitro studies using human plasma. *Metallomics* **2014**, *6*, 532–541. [CrossRef]
128. Sooriyaarachchi, M.; Narendran, A.; Gailer, J. N-acetyl-L-cysteine modulates the metabolism of cis-platin in human plasma in vitro. *Metallomics* **2013**, *5*, 197–207. [CrossRef]
129. Sooriyaarachchi, M.; Narendran, A.; Gailer, J. The effect of sodium thiosulfate on the metabolism of *cis*-platin in human plasma in vitro. *Metallomics* **2012**, *4*, 960–967. [CrossRef]
130. Sooriyaarachchi, M.; George, G.N.; Pickering, I.J.; Narendran, A.; Gailer, J. Tuning the metabolism of the anticancer drug cisplatin with chemoprotective agents to improve its safety and efficacy. *Metallomics* **2016**, *8*, 1170–1176. [CrossRef] [PubMed]
131. Harper, B.W.J.; Morris, T.T.; Gailer, J.; Aldrich-Wright, J.R. Probing the interaction of bisintercalating (2,2′:6′2″)platinum(II) complexes with glutathione and rabbit plasma. *J. Inorg. Biochem.* **2016**, *163*, 95–102. [CrossRef] [PubMed]
132. Miller, K.; Sarpong-Kumankomah, S.; Egorov, A.; Gailer, J. Sample preparation of blood plasma enables baseline separation of iron metalloproteins by SEC-GFAAS. *J. Chromatogr. B* **2020**, *1147*, 122147. [CrossRef] [PubMed]
133. Corte Rodriguez, M.; Alvarez-Fernandez Garcia, R.; Blanco, E.; Bettmer, J.; Montes-Bayon, M. Quantitative evaluation of cisplatin uptake in sensitive and resistant individual cells by single-cell ICP-MS (sc-ICP-MS). *Anal. Chem.* **2017**, *89*, 11491–11497. [CrossRef]
134. Temaj, G.; Chichiarelli, S.; Eufemi, M.; Altieri, F.; Hadziselimovic, R.; Farooki, A.A.; Yaylim, I.; Saso, L. Ribosome-directed therapies in cancer. *Biomedicines* **2022**, *10*, 2088. [CrossRef]
135. He, L.; Chen, T.; You, Y.; Hu, H.; Zheng, W.; Kwong, W.-L.; Zou, T.; Che, C.-M. A cabcer-targeted nanosystem for delivery of Gold(III) complexes: Enhanced selectivity and apoptosis inducing efficacy of a gold (III) porphyrin complex. *Angew. Chem. Int. Ed.* **2014**, *53*, 12532–12536.
136. Soignet, S.L.; Maslak, P.; Wang, Z.-G.; Jhanwar, S.; Calleja, E.; Dardashti, L.J.; Corso, D.; DeBlasio, A.; Gabrilove, J.; Scheinberg, D.A.; et al. Complete remission after treatment of acute promyelocytic leukemia with arsenic trioxide. *N. Engl. J. Med.* **1998**, *339*, 1341–1348. [CrossRef]
137. Gierasch, L.M.; Gershenson, A. Post-reductionist protein science, or putting Humpty Dumpty back together again. *Nat. Chem. Biol.* **2009**, *5*, 774–777. [CrossRef]
138. Mani, U.; Prasad, A.K.; Kumar, V.S.; Lal, K.; Kanojia, R.; Chaudhari, B.P.; Murthy, R.C. Effect of fly ash on biochemical and histomorphological changes in rat liver. *Ecotoxicol. Environ. Saf.* **2007**, *68*, 126–133. [CrossRef]
139. Puris, E.; Gynther, M.; Auriola, S.; Huttunen, K.M. L-type amino acid transporter I as a target for drug delivery. *Pharm. Res.* **2020**, *37*, 88. [CrossRef]
140. Xiong, X.; Liu, L.-Y.; Mao, Z.-W.; Zou, T. Approaches towards understanding the mechanism-of-action of metallodrugs. *Coord. Chem. Rev.* **2022**, *453*, 214311. [CrossRef]
141. Morris, T.T.; Ruan, Y.; Lewis, V.A.; Narendran, A.; Gailer, J. Fortification of blood plasma from cancer patients with human serum albumin decreases the concentration of cis-platin derived toxic hydrolysis products in vitro. *Metallomics* **2014**, *6*, 2034–2041. [CrossRef] [PubMed]
142. Sooriyaarachchi, M.; Wedding, J.A.; Harris, H.H.; Gailer, J. Simultaneous observation of the metabolism of cisplatin and NAMI-A in human plasma in vitro by SEC-ICP-AES. *Metallomics* **2014**, *19*, 1049–1053. [CrossRef] [PubMed]
143. Au, F.; Bielecki, A.; Blais, E.; Fisher, M.; Cakmak, S.; Basak, A.; Gomes, J.; Arbuckle, T.E.; Fraser, W.D.; Vincent, R.; et al. Blood metal levels and third trimester maternal plasma matrix metalloproteinases (MMPs). *Chemosphere* **2016**, *159*, 506–515. [CrossRef] [PubMed]
144. Grandjean, P.; Landrigan, P.J. Developmental neurotoxicity of industrial chemicals. *Lancet* **2006**, *368*, 2167–2178. [CrossRef]
145. Orr, S.E.; Bridges, C.C. Chronic kideny disease and exposure to nephrotoxic metals. *Int. J. Mol. Sci.* **2017**, *18*, 1039. [CrossRef] [PubMed]
146. Nogara, P.A.; Oliveira, C.S.; Schmitz, G.L.; Piquini, P.C.; Farina, M.; Aschner, M.; Rocha, J.B.T. Methylmercury's chemistry: From the environment to the mammalian brain. *Biochim. Biophys. Acta Gen. Subj.* **2019**, *1863*, 129284. [CrossRef]
147. Saha, A.L.; Kumar, V.; Tiwari, J.; Rawat, S.; Singh, J.; Bauddh, K. Electronic waste and their leachates impact on human health and environment: Global ecological threat and management. *Environ. Technol. Innov.* **2021**, *24*, 102049.
148. Bjorklund, G.; Tinkov, A.A.; Dadar, M.; Rahman, M.M.; Chirumbolo, S.; Skalny, A.V.; Skalnaya, M.G.; Haley, B.E.; Ajsuvakova, O.P.; Aaseth, J. Insights into the potential role of mercury in Alzheimer's disease. *J. Mol. Neurosci.* **2019**, *67*, 511–533. [CrossRef]
149. Chapman, L.; Chan, H.M. The influence of nutrition on methyl mercury intoxication. *Environ. Health Perspect.* **2000**, *108*, 29–56.
150. Rossignol, D.A.; Genius, S.J.; Frye, R.E. Environmental toxicants and autism spetrum disorders: A systematic review. *Transl. Psychiatry* **2014**, *4*, e360. [CrossRef] [PubMed]

www.ingramcontent.com/pod-product-compliance
Lightning Source LLC
LaVergne TN
LVHW071000170726
843515LV00006B/1037

* 9 7 8 3 0 3 6 5 8 0 1 5 9 *

MDPI
St. Alban-Anlage 66
4052 Basel
Switzerland
Tel. +41 61 683 77 34
Fax +41 61 302 89 18
www.mdpi.com

Inorganics Editorial Office
E-mail: inorganics@mdpi.com
www.mdpi.com/journal/inorganics